Sacred Cows in Science
No Objectivity Allowed

Edited By

E. Norbert Smith, Ph.D.

Tonya Holmes Shook: Publishing Coordinator

Printed in the United States of America

Smith, E. Norbert   1941

      ISPN   1456585169

      EAN 13   978-1456585167

Dedication

    This book is dedicated to the memory of Alfred Lothar Wegener (1880-1930). He was a notable scientist, explorer, geophysicist, meteorologist and record holding balloonist. He earned his Ph.D. in astronomy in 1905 from what is today Humboldt University. His lectures on the thermodynamics of the atmosphere became standard in meteorology textbooks. He made several expeditions to Greenland to study the polar air circulation long before the jet stream was generally accepted. In 1930 he led his last expedition to Greenland for a yearlong study to monitor arctic weather. He felt responsible for the success of the expedition and contributed 1.5 million dollars (by today's standard) of his own money. His last message was that his party had run out of fuel which led to his untimely death.

    He is best known for introducing the highly controversial theory of continental drift in 1912. In doing so he was attacking a huge sacred cow of science that assumed fixity of the continental positions. Because he lacked demonstrable scientific evidence for his theory, it was rejected until the 1950's when numerous discoveries finally provided convincing scientific support of continental drift. Unfortunately this did not occur until 20 years after his death.

The reason for his continental drift theory is easy to comprehend. When many of us first saw a large world map in school, it was obvious Africa looked like it was at one time joined to South America and had somehow separated. I commented about this to my grade school teacher and was ridiculed. It sounded like a preposterous idea. What could move an entire continent? This is an excellent example of a sacred cow in science and one that died a slow and painful death. Anyone going against scientific tradition is often ridiculed. This was true in Wegener's day and remains so today. In spite of what we are taught in school, science is NOT objective and new ideas often come with a huge cost.

Wegener's work, published in 1925, was received with disdain and caused such strong objection that the American Association of Petroleum Geologists organized a symposium specifically to oppose the continental drift hypothesis. As a result, Wegener's hypothesis fell out of favor for decades. He was openly ridiculed, even though his idea of continental drift was eventually accepted by the scientific community. George G. Simpson was one of his most outspoken and influential opponents. Others jumped on the bandwagon.

In the 1950's the new science of paleomagnetism provided data showing India and the southern hemisphere had been connected as predicted long before by Wegener. The scientific view changed abruptly in 1964 when the Royal Society held a symposium on the topic. In the 1960's evidence of seafloor spreading and the theory of plate tectonics finally gave Wegener recognition as the founding father of one of the major scientific revolutions of the 20th century. Another sacred cow had fallen to objective scientific evidence. Many more exist and need to be slain.

References

Spaulding, Nancy E.; Namowitz, Samuel N. (2005). *Earth Science*. Boston: McDougal Littell.

McIntyre, Michael; Eilers, H. Peter; Mairs, John (1991). *Physical geography*. New York: Wiley. pp. 273.

Dansgaard W (2004). *Frozen Annals Greenland Ice Sheet Research*. Odder, Denmark: Narayana Press. pp. 124.

Jacoby, W. R. (January 1981). Modern concepts of earth dynamics anticipated by Alfred Wegener in 1912. *Geology* 9: 25-27.

Simpson, G.G. (1943). Mammals and the Nature of Continents. *American Journal of Science* 241: 1-31.

du Toit, A. (1944). Tertiary Mammals and Continental Drift. *American Journal of Science* 242: 145-63.

Frankel, H. (1987). The Continental Drift Debate in H.T. Engelhardt Jr and A.L. Caplan. *Scientific Controversies: Case Solutions in the resolution and closure of disputes in science and technology*. Cambridge University Press.

In a time of universal deceit, telling the truth is a revolutionary act.
George Orwell

Table of Contents

List of illustrations

Foreword
Fritz Ward, Ph.D.

Science was at one time defined by its method. Carefully controlled experiments, provisional conclusions, and considered debate once defined the field. But those days have passed. Today, science is defined by public policy statements, consensus, and a set of metaphysical assumptions that cannot be directly tested. Students are told that science is in conflict with "faith" or, worse yet, that faith operates in a different "magisterial" with no real application to the world we inhabit. They are told that "science" is a set of conclusions held by the "consensus" of scientists and that these conclusions, confirmed by "peer review" should direct public policy. In short, students are now told, and many in the general population believe, that science is a set of beliefs held by people with special access to knowledge unavailable to ordinary humans.

How did this turn of events come about? To some extent it began with increased specialization in the twentieth century. Generalists were disparaged and specialists became the gatekeepers of scientific knowledge. State and private universities encouraged this specialization by demanding "peer reviewed" publications that were increasingly read only by a small number of other specialists. This trend is widely acknowledged, but there is a more fundamental reason for the changing perception of science. As Tom Bethell pointed out, in his popular *The Politically Incorrect Guide to Science*, it is direct government funding of scientific research that has fundamentally changed how science is done. Whereas science used to be a subject of debate (and anyone who attends some science conventions will find that it still is) today a few scientists have the ability to limit the research of those who would dissent from them. Nowhere is this more evident than in the "science" of global warming where recently released emails show climate researchers actively manipulated peer review and conspired to hide raw data from their critics. So much for the vision of science in which the scientific method could, at least in theory, offer replicable results regardless of the individual who performed the experiment.

Despite the dominance of paradigms supported by government funding, a growing number of scientists are critical of the new practices of science. Unable to publish in traditional settings, they still have found other avenues to disseminate their work, most notably internet blog sites. Their dissent is more than simply a challenge to the monopoly accorded to particular viewpoints by government bureaucrats. In making science available to the masses, they are challenging the specialization of science itself and many of its underlying assumptions. When all is said and done, science is a human enterprise, seeking the answers to "how" questions ("why" questions are more the domain of religion, though the two fields overlap more than is commonly supposed). These dissenters are taking back science for humanity as a whole.

This book is part of the new dissenting tradition in science. Some of its papers, most notably those by Stan Robertson, Jerry Bergman and Norbert Smith are moderately technical, but all dissent from prevailing notions in academe. Other papers challenge today's science on more fundamental grounds. The essays by Charles Imes and Fritz Ward both reveal the extent to which the practices of science, widely thought of as "objective", are in fact tied to religious and political norms of the present era. Both essays to some extent condemn the influence non-scientific reasoning has on today's scientific process. Other essays in this volume, however, suggest science should reconsider its rejection of traditional religious thought, most notably Christianity. The arguments of Steve Kern and Jay Hall fall into this category, while still other papers in this book challenge some of the boundaries of science. Does science speak to social issues like homosexuality and pornography? Many in academia would argue not, but when viewed as a method, rather than a set of politically correct conclusions, science may indeed have much to offer in these areas as well.

Suffice to add, not all the authors are in agreement with each other. Readers of this volume will note some essays which take a radically different perspective than others. Some might even be surprised by this. But dissent in science does not imply agreement. Indeed, in the brave new world of modern science, acknowledging genuine dissent is in and of itself a radical concept. But all the essays in this volume should be read closely. They offer well reasoned perspectives not found in text books. The arguments presented herein are at once thought provoking and illuminating. Even if the reader should disagree with one or more of the authors, the fact remains that, as the dedication to this book reveals, holding a minority position in science is hardly proof one is wrong. But refusing to consider dissenting perspectives almost certainly insures errors and mistakes. It also undermines the important human element in our search for understanding.

Acknowledgements

This book was inspired by Ben Stein's groundbreaking documentary, "Expelled, no Intelligence Allowed." Unfortunately the movie was not widely seen and received sharp criticism from the media as well the scientific community. It fell short of the anticipated impact hoped for by many Christians. Still, it revealed the prejudice that abounds in scientific circles for anyone doubting evolution. The movie was yet another example of what happens when sacred cows of science are attacked.

Many of us have known for years that science is NOT objective in spite of what we are taught in public schools and university science classes. To go against accepted scientific traditions, especially evolution can be dangerous and has ended the promising careers of many educators. According to Jerry Bergman's excellent book, *Slaughter of the Dissidents* over 3,000 university professors have been denied tenure or outright fired for rejecting evolution dogma. I know this from personal experience for I was one of them. With my teaching and scientific research career ended, I became a truck driver until my retirement five years ago. The same thing has happened to some public school history and science teachers as well.

The purpose of this book is to shed light on a few of the many sacred cows that remain in science. I could not have done this project alone, but had the help of many professional friends. This is the first book for which I am only the editor; it was my friends who did the difficult work. My only criterion was that each author selected a topic about which he or she was passionate. The result is much better, more diverse and deeper than I could have done alone. As the book began to take shape other topics were suggested and some of the authors invited their friends to contribute. I sincerely thank each of the contributing authors for you time, encouragement and in some cases your excitement for this project. Kudos to all contributors or to use a quaint Oklahoma idiom, "Ya'll done good"

Overview

The term "sacred cow" is fitting to use in the title of this book and comes from the veneration of cows by the Hindus of India. As early as 1910 the term was used as a metaphor to describe a person, organization or institution that is unreasonably immune from criticism. I have long used the term when describing certain aspects of macroevolution, but it also applies equally well today for many other areas of science. We are taught science is above all objective. It is not. Scientists are human and we all have preconceived notions and a worldview that colors how we interpret the things around us. A scientist's worldview can influence the experiments they perform and how data are analyzed and interpreted. It is also a factor in which papers that scientific journal editors select for publication. Perhaps because scientific journals are readily available around the world and especially now on the Internet many people believe they are widely read. This is simply not true. Scientific articles are written for and understood by select few specialists. As a doctoral student, I was able to get several papers published about my alligator research. My major professor rightly deflated my ego by reminding me that such scientific papers are typically read and understood by fewer than a dozen people. While this may have been an under exaggeration, the point is valid. Very few people have the interest or background to read and understand most technical papers. Such papers are written for a very small number of specialists. It is hoped this book will reach a much wider audience and is an attempt to expose some of the sacred cows in science. It is aimed primarily to graduate students and professors interested in truth in scientific discovery.

Seventeen professional friends contributed nineteen essays that address a variety of topics. No doubt some readers will disagree with some of the points presented. Such disagreement may lead to debate which can often be helpful. We often learn the most from people with different backgrounds and, yes from those with differing worldviews. Also included are brief bios about each author along with their email address. Feel free to comment directly to me or to any of the authors I also include a brief *Editors note* at the beginning of each essay to provide a bit of context. Again any comments or suggestions are welcome and contact information is provided.

The book is divided into three sections: life science, physical science and behavioral science with a brief introduction at the beginning of each section. Within each section are essays covering a variety of topics. I asked contributors to address areas about which they feel science is wrong and about topics for which they were passionate. I also suggested references be provided not only to support comments in the essay, but also so readers can dig deeper in the various topics discussed. In some ways this project was more difficult than other books I have written, in part because some of the authors were over extended and delays were unavoidable. Still, the diversity viewpoints exceed anything I could have accomplished alone. This effort has been both frustrating and deeply rewarding. To each member of the sacred cow herd let me extend a heartfelt THANK YOU. I also sincerely appreciate your patience as I slowly fit all the parts together and attempted to standardize the format while leaving the individual writing style largely intact. Any remaining errors are mine and I hope readers will contact me with corrections and suggestions for improvement in the event there is a revised edition in the future.

Section One

Life Science

Charles Robert Darwin (1809-1882)

Life Science Introduction

Ideas have consequences. This is certainly true regarding the scientific and societal impact of Charles Darwin's theory of evolution by natural selection (Darwin, 1859). The best known example is Hitler's murder of 5 million Jews in an attempt to "improve the human species" Today abortion is the most tragic and far-reaching application of Darwin's concept of evolution. Worldwide approximately 46 million abortions are performed each year. The connection to evolution is obvious. People are taught man is but an animal and has no intrinsic value. The unfit, even the unwanted should be eliminated, without consequent or guilt. Evolution also holds there can be no life after death. Women are told the fetus is not a child, but merely a blob and should be eliminated if having it would be an inconvenience. Abortion "clinics" and doctors have made millions of dollars by eliminating unwanted children. Abortion is increasingly seen as nothing more than a contraceptive. It seems many in today's society have resurrected the mantra of the 1960's, "If it feels good, do it"

Darwinian evolution is the unequivocal leader of the sacred cow herd. There is not even a close second. In spite of the claim that science is objective, there is an unspoken, but universal rule to the contrary in academia. Stated in today's jargon that rule is, "Don't dis Darwin" The public is largely unaware that over 3,000 university professors and many high school teachers have been denied tenure, promotion or have been outright fired for doubting Saint Darwin. Tens of thousands more have rejected evolution dogma for scientific reasons, but have wisely remained silent. Some were fired merely for assigning readings in scientific journals by authors opposing evolution. I know of what I speak for I was one of those denied tenure, ending my teaching career. With the academic door slammed rudely in my face, I became a truck driver until my retirement five years ago.

Even more tragic is over 200 promising young graduate students have been denied access to graduate school for rejecting evolution dogma. For them, the ivory towers of academia will forever be off limits. It seems the media purposely ignores these tragic events and the general public is unaware that such religious persecution is rampant today in America. We are taught that America is the land of free speech. It is not. Ben Stein's excellent movie, "Expelled: No Intelligence Allowed" documented this modern form of religious persecution, but seems to have had little effect and was either not seen by enough people. Some wrongly dismissed it as fiction. Jerry Bergman describes hundreds of cases of modern religious persecution in his excellent and well documented book, *Slaughter of the Dissidents*. This is the first in a six volume series about this important and growing problem in academia. The second volume, *Silencing the Dissidents* is finished and will soon be available from Amazon.com or your local bookstore. I strongly recommend both of these excellent books as well as the rest of the series. As educated caring adults we must be informed and inform others. It is our responsibility since the mass media has dropped the ball.

Contrary to what many believe, evolution is taught as absolute fact throughout the life sciences. One of my professors in graduate school said there is more evidence supporting evolution than there is for gravity. I secretly wished that particular professor would have tested that thesis…by jumping off a cliff, but my wish remained unfulfilled. Unfortunately, evolution dogma has reached far outside the biology classroom and has influenced medicine, psychology, education, history and even the interpretation of current events. Let's consider some examples.

Evolution has influenced medicine and several ways and some of them have been counterproductive and even lethal. Perhaps the most telling is the evolutionary concept of vestigial organs and its influence on medical practice. This has long been my favorite of the so called "evidences for evolution," for it is easily shown to lack real scientific support. It also illustrates the deception common in textbooks and biology classrooms. This argument is easy to understand. If animals actually evolved over time from one kind of animal to another, we would expect to see some organs no longer needed as well as nascent organs or organs in the process of becoming useful. No nascent organs have been described, but the argument of vestigial organs is still widely used to prop up evolution dogma.

Robert Wiedersheim (1848-1923) was not a good student, barely passing his final exams and his academic advancement was slow. In 1876, he became an anatomist at the University of Freiburg and soon became a comparative anatomy expert publishing several textbooks. He became widely known and respected in 1893 for publishing a list of eighty-six useless or vestigial organs (see Wiedersheim, 1893) found in the human body. In his own words, he said each of those human organs had, "lost their original physiological significance" He theorized they were vestiges of past human evolution and called them "vestigial" Over the next few decades, he and others added additional organs and the list grew to one-hundred and then to one hundred and eighty useless organs by the time of the infamous Scopes Monkey Trial. I searched diligently for the actual list of one hundred and eighty organs mentioned in many textbooks, but was unsuccessful in finding such a list. This seems to be yet another false myth widely disseminated by evolutionists to support failing evolution dogma.

Wiedersheim claimed the human body was a veritable walking museum of evolutionary history. He picked up and expanded on Darwin's concept of rudimentary organs as mentioned in *The Descent of Man* (See Darwin, 1871). Original organs included on Wiedersheim's list as useless included the human appendix, adenoids, tonsils, parathyroid gland, pineal gland, pituitary gland, thymus, valves in veins and many other important organs and structures. As knowledge replaced ignorance, uses were found for these organs. Today, no one claims any useless organs exist in the human body. Yes, I can live without my appendix. I can also live without my right arm, but this does not mean it is without purpose. For decades, medical doctors actually accepted the evolutionary myth that the appendix was useless and during other surgical procedures, even healthy appendixes were removed, increasing surgical complications and prolonging recovery. Also, the doctors obviously profited monetarily from the needless surgery.

What is astounding is many of these structures are still mentioned in university textbooks, in fact of ample scientific evidence of their utility. Perhaps the best know of these so called "useless" organs is the human appendix. It still occurs in some textbooks and is often mentioned by university professors who are without excuse for perpetuating such false teaching.

The human appendix is a finger sized hollow tube near the end of the small intestine and entrance of the colon, and has long been taught in biology classrooms to be a relic of our vegetarian ancestors and no longer serves a function. We now know it serves many functions. For example, over fifty years ago we find these words from the prestigious *Quarterly Review of Biology*; "there is no longer any justification for regarding the vermiform appendix as a vestigial structure" (Straus, 1947). As a blind reservoir, it continually repopulates the colon with important bacteria lost in excrement. The walls of the appendix contain lymph tissue, known to be important in the immune response to disease. Recent evidence indicates it also lubricates oil needed for lubrication of the large intestine.

So it is with each of the so-called vestigial organs on the original list published by Wiedersheim. Yet, this argument is still sometimes heard in classrooms and is in some textbooks. Again, evolutionists must cling to this myth because the actual evidence supporting macroevolution is non-existent. Those pesky facts keep getting in the way of good dogma. Let us look briefly at some of the other organs originally listed as useless.

Psychology is another discipline that has been heavily influenced by evolution. Perhaps this can be best illustrated by two of the recognized leaders of psychology. In 1906 Ivan Pavlov (1849-1936) cut holes in dog's cheeks and inserted tubes to measure salivation. He conditioned the dogs to come to the sound of a metronome. Eventually he found salivation would increase with the sound of the metronome alone and food was not necessary. Throughout his life, Pavlov studied animals and in his view, man was nothing more than an animal and could be manipulated in much the same way as the animals in his studies.

The best known modern American behavioral psychologist is Burrhus Frederic Skinner (1904-1990). Better known as B. F. Skinner, he is best known for the invention of the Skinner box and work with laboratory rats. Without exception he was quick to transfer his findings with rats to humans. Even if evolution were true, it is a huge leap from a rat to a human. Although not widely known, he and his wife forced their own daughter to live much of her early life in a Skinner box. In his classic book, *Walden Two*, first published in 1948 he describes a utopian world based on the application of his theories. He felt ALL societal problems could be solved by the application of scientific findings of human behavior. Many of his unsupported views have been adopted as truth by those in education, psychology and the current environmental movement today.

Education has long been influenced by erroneous evolution dogma. Perhaps this explains why education is always changing and often seems faddish. Educational psychologists perform a few simple experiments in Skinner boxes with rats and are quick to apply their findings on school children. Is it any wonder the education system of the United States ranks lowest among all the industrial nations of the world? When I was teaching at a small community college all, incoming high school graduates were given exams to evaluate their reading skills and general knowledge. A full 85% required 1-2 years of remedial course work before they could enroll in our watered down "college level" courses. The future of our nation will depend more and more on Christian schools and home schooling.

Evolution's Influence on Students

As a life-long teacher, I have a heart for students of all ages. Removal of the Bible, prayer and the Ten Commandments from our public schools has had tragic and far-reaching effects. Today, children are taught man is the product of mindless evolution and is but an animal. Life after death is impossible and we are no longer responsible for our actions. Eternal judgment, heaven and hell are considered myths of the unlearned and superstitious. Life is nothing more than "matter in motion" Morality and ethics are no longer taught in our public schools. Instead, students are taught there is no such thing as right and wrong, for everything is relative. It all depends on the person and situation. With this sort of philosophy being proclaimed to our children, we should not be surprised that school and university shootings are becoming pandemic. Metal detectors and bomb sniffing dogs are now commonplace in our government schools. How much are we willing to tolerate? Where will it end? More importantly, what is the root cause for these terrible changes?

Consider the tragic Columbine high school massacre that occurred in Colorado on April 20, 1999. One of the shooters, Eric Harris, admitted being motivated by the teachings of Charles Darwin. From his website, these insightful words were posted: *You know what I love? Natural selection! It's the best thing that ever happened on the earth, getting rid of all the stupid and weak.* He and Dylan Klebold had planned that fateful event for over a year. They knew that day would be their last. Eric Harris wore a T-shirt displaying the most famous words of Charles Darwin. Emblazoned in large letters on the shirt was *Natural Selection.* Consider also the date of the massacre. It was not chosen at random. It coincided with the birthday of another infamous Darwin follower and Eric Harris' personal hero, Adolf Hitler. Ideas can have profound consequences for good or for evil.

School shootings are not limited to the United States. On November 7, 2007, an eighteen year-old student in Finland shot and killed seven students and the school principal before turning the gun on himself. The shooter, Pekka Eric Auvinen, died shortly after the shooting in a nearby hospital of self-inflicted wounds. As with the Columbine shooting, this event had been planned for months. Also like the Columbine shooters, Auvinen was an avid follower of Darwin.

On Sunday December 9, 2007, another copycat killer, 24-year old Matthew Murray, killed four people in Colorado. Two were killed at a mission training camp near Denver. Later the same day, two more people were killed at New Life Church in Colorado Springs before he was shot by an armed female church security guard. On an Internet posting, Murray said he intended to kill as many Christians as he could and admitted following the example of the Columbine shooters. Once again, an attempt was made to "assist" Darwin's natural selection with devastating results. As with Hitler and others, each of these murderers intended to hasten the natural selection process by removing those considered the unfit by the shooters. Ideas have consequences. Removing God, prayer, and the Ten Commandments from our public schools has had terrible consequences. Perhaps it is time to revisit those terrible decisions.

Life Science Chapters

Following are a series of essays dealing with some of the sacred cows in the life sciences. The first essay is by Ed Blick and is humorous look at Charles Darwin. I hope you enjoy it as much as I did. It made me laugh out loud. Thanks for the smiles Ed.

The second essay is serious and deals with Haeckel's biogenic law and the whole evolutionary argument of vestigial organs. It is an excellent read and is well documented for those wanting to dig deeper. For over one hundred years vestigial organs were thought to be one of the major evidences supporting evolution, but as Dr. Bergman shows the argument has no solid scientific support. Indeed, if macroevolution were true, vestigial organs MUST exist, but alas, there are none. Another sacred cow is in serious trouble.

The next essay is by Charles Jackson and clearly argues that micro-evolution cannot lead to major evolutionary changes. The distinction between microevolution or genetic drift and selection for genetic differences has long been thought to be the driving force for macro evolutionary changes. If this is not the case, then evolution is lacks a rational mechanism.

Mutations have long been thought to provide the raw material that enables evolution to bring about new kinds of plants and animals. There has been an explosion in our understanding of genetics in recent years, including of course the human genome project. We now understand the interaction between genetics and cell biology as never before. Jonathan Bartlett shows such mutations are not random, again causing macroevolution to seen as lacking a mechanism. Once again as knowledge is replacing ignorance the evidence for evolution is falling by the wayside. The more we learn about the complexity of all living things the more ridiculous is the idea that life could evolve from non-living material. Even more telling is evolution has no rational explanation for the extreme complexity seen is all living things. The more we understand the more impossible evolution has become, which is why tens of thousands of former evolutionists no longer accept evolution. They have abandoned it like rats from a sinking ship and with recent discoveries of the complexity of all living cells, this departure if accelerating.

My long time personal friend George Howe provides an in depth look at homology. He was my Sunday school teacher when I was a lowly doctoral student at the University of California at Los Angeles and we have stayed in touch over the decades. He is a well published and respected botanist and once again finds the arguments from homology used to support evolution are without support. His essay is well documented for those wanting to learn more on this important topic. Dr. Howe has also been one of the leaders in the current Creation science arena. He served as president of the respected Creation Research Society and served many years as science editor for the *Creation Research Society Quarterly*.

The next essay is perhaps one of the more unusual and is by my good friend Pastor Steve Kern. I have long thought dinosaurs and man lived together. Of course the major reason is the Bible clearly teaches ALL animals and man were created during Creation week. There is also an excellent description of both a tyrannosaurus and the even larger Brachiosaurus found in Job 40 and 41. Even as a young boy I knew the behemoth and leviathan were not merely a hippopotamus and crocodile as some preachers and Bible scholars suggested. Pastor Steve agrees and has written an excellent essay.

Let me add he has also written a most unusual series of children's adventure stories set before the Genesis Flood when men used dinosaurs for riding and even to pull plows and wagons as oxen were used after the flood. *Eden's Veil* is the first in the series. All are excellent for kids AND adults. He has also written an excellent book for "big people" called *No other Gods*. It is an excellent read and provides insight into our increasingly anti-Christian culture today. See his Bio for additional details.

I have included an essay dispelling the lingering belief that frightened animals must flee or fight and display an increased heart rate. I was the first to discover the passive fear response where animals often hide and show a marked reduction in heart rate. Death feigning is the extreme response displayed by the opossum during which the heart rate slows as much as 98%. This sacred cow has long passed. For more details see my book on the topic: *Passive Fear: Alternative to Flight or Fight*.

Another myth exposed by my research is that the heart always slows during diving. This response had even been dubbed, "diving bradycardia" but is a misnomer. It is yet another example of the passive fear response. I discovered it with alligators that normally dove underwater without reducing their heart rate. Quite by accident a fellow graduate student studying the hatching success of nesting shore birds approached where my alligator was swimming. Upon seeing the canoe the telemetered alligator dove with a marked reduction in heart rate. I later demonstrated the same thing with swamp rabbits. When forced underwater, their heart does indeed slow, but when they voluntarily dive there is no slowing of the heart. The same has been demonstrated in diving seals and many other animals. Another sacred cow falls into oblivion. This is also discussed in greater depth in *Passive Fear: Alternative to Flight or Fight*.

The last two life science essays will to some be considered controversial and are written my good friend and medical doctor, J. Y. Jones. He is an ophthalmologist and has made countless trips to developing countries restoring sight to the blind. He is also an avid large game hunter and hunting guide and has written extensively about hunting. He provides an unusual insight into animal rights and the issue of endangered species. I strongly recommend his most recent book, *Worship Not the Creature: Animal Rights and the Bible*. Even if you disagree with some of his conclusions, it will provide fresh insight into what should be done to protect the many endangered species and how the present government attempt has failed miserably.

Perhaps his work is dear to my heart because the American alligator, which I spent most of my life studying, was once an endangered species and had made a remarkable comeback. Today they are again hunted legally and yes, I have alligator meat in my freezer. Doesn't everyone? I even served alligator at my 50th high school reunion … and my classmates loved it and wanted more. What can I say? It tastes better than steak!

References

Darwin, Charles (1959) *The Origin of Species by Means of Natural Selection, or the Preservation of Favoured Races in the Struggle for Life*.

Darwin, Charles (1871) *The Descent of Man, and Selection in Relation to Sex*.

Straus, William L, Jr. (1947) *The Quarterly Review of Biology* June 1947, Vol. 22, No. 2: pp. 148

Wiedersheim, R. (1893) *The Structure of Man: An Index to His Past History.* Second Edition. Translated by H. and M. Bernard. London: Macmillan and Co. 1895.

Laughing at Darwin
Edward F. Blick, Ph.D.

Editor's note

Darwin Day 2009 celebrated the 150th anniversary of the publication of the *Origin of Species* and 200th year since Darwin's birth. Darwin fawning droned on for weeks in the media and many universities continued praising Saint Darwin throughout the year. Professor Ed Blick wrote the following reality check to set the record straight using a humorous, but fact filled and insightful approach. It helps set the mood for this unconventional look at some of the issues in modern science and dispels many of the myths that seem to persist in evolution dogma. His use of humor is truly remarkable. Thanks Ed … you're the man.

Introduction

We all love to laugh, its good medicine. We laughed at the Queen in Lewis Carroll's *Alice in Wonderland*, who said "I sometimes believe in six impossible things before breakfast" The Darwinists are even more hilarious, they not only believe, but also teach more than six impossible fairy tales in their biology classes. The history of their pathetic attempt to pump life into the Lenin-like corpse of evolution is full of laughs. Let's peel back the skin of this rotten baloney and laugh at how this sausage was made.

Charles Darwin was born into wealth, spent two years in medical school, dropping out after spending too much time in bars. He had some divinity training but failed to make it as an Anglican minister. He was never a scientist but took a position as a naturalist on a ship and later wrote his racist books, *The Origin of the Species* and *The Preservation of Favored Races* and *Descent of Man*. He was ignorant of genetics. He married his first cousin. All seven of his children either died young or had mental or physical disorders.

Without any facts, he conjured up his Pangenesis theory. He assumed that species changed to other species because all cells produced Gemmules. Gemmules supposedly arose by some kind of reaction to the environment. Each of these gemmules entered the sex cells of the sperm or egg (it must have been crowded in there), which later were transmitted to the offspring. Big problem! No one could find Darwin's imaginary Gemmules and Pangenesis died shortly after birth!

Reality Begins

In 1870, Adam Sedgewick, leading geologist of England, wrote Darwin: "I read your book with more pain than pleasure. Parts I laughed at till my sides were sore; others I read with absolute sorrow, because I think them utterly false … you deserted the true method of induction" Induction is reasoning from facts to theory, but Darwin reasoned from theory to facts, but neither he nor anyone else could find the facts! His writings were conjecture piled upon conjecture. "Maybe" and "Perhaps" form the basis of his books!

Darwin's writings were not science but philosophical musings. But something had to be done to keep the world believing Darwinism. In 1874, future Nazi-like preacher Ernest Haeckel tried by faking drawings of embryos (which he claimed repeated "fish to reptile to mammal" evolution). Later that year felloe embryologist Wilhem His, Sr., in much detailed, exposed the hoax in detail in his book *Unsere Korperform*. His's scholarly books are considered the foundation of modern embryology. For the past century other embryologists have denounced Haeckel's drawings as utter foolishness. "The theory of capitulation … should be defunct today", Stephen J. Gould, (1980). Believe it or not, but Haeckel's "most famous fakes in biology" are used as proofs of evolution in biology books today. Some abortion doctors even use "recapitulation: to convince would-be mothers to abort their baby …" Why, that's not a human in your womb, you're in the first trimester, that is only a fish, or reptile in the second trimester! Don't school boards ever read these biology books? Haeckel's forgeries are like gonorrhea, a gift that keeps on giving!

The next attempt to resurrect Darwinism came in 1872, when the British ship HMS Challenger dredged ocean sediments for four years looking for half-formed fossils. None were found, and since none had ever been found on land, the evolutionary fairy tale of the gradual production of billions of fossils in sedimentary strata was quietly set aside. The Challenger did provide a momentary hope. It dredged up some blob from the ocean floor and Darwinists leaped for joy. It was a live microbe, some kind of a missing link! They named it *Bathybuis Haeckel* after the old king of biological fakery, Ernest Haeckel. However in 1875 a chemist discovered it was not any form of life, but a chemical precipitate of sulphate of lime (gypsum). So, true to form, the discovery was carefully swept under the rug and hidden from the public.

In the meantime Darwin had returned to Jean-Baptiste Lamarck (1744-1829) who thought that giraffes developed long necks by stretching to reach those leaves on the top of trees. This theory died again in 1883, when German biologist Leopold Weisman cut off the tails of white mice in nineteen successive generations and the tails always reappeared. Similarly, after four thousand years of circumcision, Jewish men still had foreskins. More bad news for poor old Saint Darwin!

Darwin Resurrection

Who can rescue Darwinism? Quick, before the "unwashed" discover the emperor has no clothes. Finally in 1930, Austin H. Clark tried to plug the gap with a new theory, "Zoogenesis." Clark was a well-respected Darwinist at the Smithsonian Institute. He had written books and six-hundred articles in five languages. However to his dismay, he could never find any evidence of macroevolution in animals or plants. In his 1930 book, *The New Evolution: Zoogenesis* he cited fact after fact proving macroevolution could not have occurred. He concluded therefore plants and animals must have sprung fully formed from dirt and water! The evolutionary world was stunned into silence. Clark was the Carl Sagan (or Ophra) of his day. He supposedly knew all the answers. Quickly they buried Clark's theory.

The next batter was world famous geneticist Richard Goldschmidt, who attempted to come to the rescue of the embarrassed Darwinians by attempting to prove macroevolution was caused by mutations. For twenty-five years he was the godfather to millions of generations of gypsy moths. He zapped them with X-rays and chemicals. He found mutations produced nothing but deformities. No new species! He concluded rats were still rats and rabbits were still rabbits. In his 1940 book, *The Material Basis for Evolution*, Goldschmidt exploded the ammunition box of evolutionary theory. He literally tore the theory to pieces. No one knew how to answer him and they cannot answer him today. He was an honest atheist who faced the facts. But not wanting to acknowledge God, he proposed a new mechanism of evolution called *The Hopeful Monster Mechanism*. One day an alligator laid an egg and a turkey hatched out! You've got to remember boys and girls this is science!

Continued Deceit

For the next thirty years evolutionists were dazed and in turmoil because they had, 1. no proof that evolution had ever occurred, 2. no reasonable mechanism to explain evolution, and 3. zillions of missing links! They had bitter arguments among themselves about possible theories. The embarrassment of Goldschmidt's crude *Hopeful Monster Mechanism* caused Harvard's Stephen Gould in 1972 and a little later, Steven Stanley, of John's Hopkins University, to "smarten up" Goldschmidt's "ugly theory" by giving it a new name, "Punctuated

Equilibrium" (Gould) and the even better "high-fallutin" scientific name, "Quantum Speciation" (Stanley). But it was still a monster by any name.

The discovery in the 1950's of the DNA by Francis Crick and James Watson crushed the hopes of biological evolutionists. It provided clear evidence that every species is locked into its own coding pattern. Only variation within a kind (microevolution) can occur. Mathematicians showed the odds against forming DNA by chance were "quad-zillions and quad-zillions to one." Evolution by chance was impossible! But atheist Crick was not ready to believe in God. He dreamed up a new theory ... are you ready for this? Some unknown "space alien" sprinkled sperm in our solar system and eventually creatures evolved on some planet (Krypton?). Then these "evolved space creatures" built a "Noah's Ark" rocket ship and after a long journey, zoomed down to the earth, to unloaded their zoo. Crick named his new theory "Panspermia." This, boys and girls, is called science or ... maybe a fairy tale! Now NASA's "Life in Space Program" believes this baloney and is spending billions of our tax dollars shooting up probes in our solar system looking for this "sperm donor"!

There you have it, the skeletons in Evolution's closet. The kooky theories of "Pangenesis," "Gemmules," "Lamarkism," "Zoogenesis," "Hopeful Monster Mechanism," "Punctuated Equilibrium," "Quantum Speciation" and "Panspermia" are all just guesses. None were proven. They make good fodder for fairy tale writers. They are a barrel of laughs!

How can supposedly reasonable men believe this weird stuff and then try to pass it off as science, when it is really a cult religion? They've emptied out the stables and dumped it on the gullible public. Most Americans believe people with Ph.D.'s in science are unbiased, honest and seek the truth. But they are fallen creatures like the rest of humanity. They can have biases, be dishonest and seek only to further their own goals, honorable or dishonorable.

The Darwinists have a well-oiled propaganda machine to keep their true goals hidden from the taxpayers, two-thirds of whom believe in creation, and pay their salaries. Darwinists have web sites set up to deflect criticism of evolution and to further their legislative and judicial goals, which are to kill God and elevate humanism to His throne.

Darwinists try to hide their atheist religion from the majority of Americans, who believe in God. One of the Darwinist web sites has enlisted Jimmy Carter, our worst ever ex-president, to proselyte Christians and baptize them into "The Church of Darwin" (in the name of the unholy trinity, Darwin, Haeckel and Nietzsche?). These new converts are called theistic evolutionists. At the 1959 "Darwinian Centennial Celebration," Julian Huxley's keynote address focused on the total repudiation of God. Huxley was asked why the world, a hundred years ago, leaped at Darwin's book *The Origin of the Species*. He answered it freed us from God's sexual mores! Evolution is a religion of no God!

Darwinists have given up public debates because they've lost hundreds of them in the 1970s and '80s. Why did they lose? As a participant in two of them I will tell you. They lost because they had no proof of macroevolution. Amazing! No Proof! They usually tried old debate tricks of personal attacks on their opponents, i.e. "You can't be a scientist because you believe the Bible", etc. But they lost because audiences were shocked. Shocked that the Darwinists had no proof! And they have none today!

In editorials and letters to the editor, the Darwinists produce no proofs. So they commonly try to bluff us Okie rubes with pompous statements like, "evolution has been proved as much as 'gravity' and it is believed by all scientists." Get real ... sure, and the moon is made of green cheese! It's all bluff, designed to shut up critics and convert us to their atheistic religion. Hitler and his propaganda chief Joseph Goebbels, would have been proud. You tell a lie long enough and loud enough and people will believe it! Unfortunately, a lot of Americans have swallowed the lie, including about half of our college graduates. Our courts and media are full of Darwinists. Their bulldog, the ACLU, is working overtime to wipe God from all of public life. Humanism over all is their goal!

Tragically the Darwinists have made great strides in wrecking western civilization. In the first half of the twentieth century, Darwinism hijacked the militant policies of evolutionism ever seen. Germany. The religion of Darwin, Nietzsche and Haeckel became the religion of Hitler and his Nazi gang. The result was in the murder of millions in their attempt to produce the Aryan super race and a victorious Germany. World War II was the most violent form of evolutionism ever seen.

In the last half century, evolution hijacked America and its schools and inflicted a great defeat on American culture. Crime has skyrocketed, homosexuality and gay marriage have been mainstreamed, and our morals have submerged into a cesspool. Why? Kids brainwashed with this kooky nonsense are taught that they evolved from apes, so they act like apes. If it feels good, do it.

Not only are the Darwinians scrambling to deflect attacks from creationists, but also they are also arguing with each other over their different theories. "So heated is the debate that one Darwinian says there are times when he thinks about going into a field with more intellectual honesty, the *used car business*" (*Newsweek*, April 8, 1985, p. 80).

"I suppose that nobody will deny that it is a great misfortune if **an entire branch of science becomes addicted to a false theory**. But this is what has happened in biology ... I believe that one day the Darwinian **myth will be**

29

ranked as the greatest deceit in the history of science" (Soren Lovtrup, *The Refutation of a Myth*, 1987).

Author Bio

Edward Blick has three degrees in engineering from the University of Oklahoma. He served four years in the U.S. Air Force as a Weatherman. His industrial experience includes working at McDonnell Aircraft as an aerodynamicist on the Mercury capsule, our first space craft. He also was a space engineer for Lockheed Missile & Space Co. working on the "Polaris Reentry" missile. He taught and performed research at the University of Oklahoma from 1958-2007 in four areas, School of Aerospace, Nuclear & Mechanical & Nuclear Engineering, School of Meteorology, School of Medicine, and School of Petroleum and Geological Engineering. He wrote and published 135 technical papers, one book on the Bible and Science, and two engineering textbooks. He also served as the Associate Dean of Engineering. For the past 45 years he has enjoyed presenting slide shows on the "Bible vs. Evolution" and during the last four years, "The Global Warming Lie." You may e-mail him at: edblick@cox.net.

The Rise and Fall of Haeckel's Biogenetic Law

Jerry Bergman, Ph.D.

Editor's note

This is an excellent way to begin exposing some of the sacred cows in science. I must confess the first time I saw Haeckel's fraudulent drawings in my biology textbook, it gave me cold chills. I KNEW God did not resort to trickery, but the evidence seemed unequivocal. I have known Dr. Bergman since a few days after his own tenure denial over twenty years ago. Unlike many of us, he was undaunted, persevered and stayed in academia. He has published nearly one thousand scientific papers and books and has become one of the leaders of the modern Creation science movement. He currently serves as the Biology Editor for the prestigious *Creation Research Society Quarterly*. As you can see in this chapter, he documents his writings in great detail. Perhaps he is best known for his powerful recent book, *Slaughter of the Dissidents*. A sequel, *Silencing the Dissidents* is finished and will be available soon. Several more in the series are planned. I strongly recommend reading and sharing several copies. I keep ten copies loaned out most of the time and am anxious to get the second volume.

Regarding Haeckel's fraudulent drawings a pressing question remains. Since the drawings have been known to be fraudulent since 1869, why were they presented as factual in biology textbooks for so many decades? The answer is obvious. Fraudulent support of evolution is the only kind available to brainwash biology students. Evolution must be accepted by faith, because the actual facts of biology keep getting in the way of evolution dogma.

Abstract

Darwinists once commonly believed that all basic life forms that existed in our past multi-millions of years of evolution were repeated in the first few months between conception and birth. Called the biogenetic law, this belief concluded that embryos always rapidly pass through their evolutionary history, starting with the one cell stage, then, in the case of humans, developing into the fish stage, the reptile stage, the mammal stage, ape stage and, finally, into a human-child stage. This theory, commonly called the "ontogeny recapitulates phylogeny" law, was cited as a major proof of evolutionism for over a century. Recent discoveries in the field of embryology and a reevaluation of the evidence for the theory, has shown it is without foundation and now has largely been discarded by embryologists. Many of the biogenetic law claims commonly used in pre 1960's textbooks, including the gill slits, tail, and yolk sac, were also found to be invalid.

Introduction

Related to the vestigial organ argument is the belief that some structures in the developing fetus and embryo are useless. Many Darwinists once believed that, as the human embryo developed, it passed through most of its major past evolutionary stages from which it was believed to have evolved. Called *The Recapitulation Law*, the now discredited theory taught that human life begins as a single cell as did the first life forms, then develops into a fish stage, next a reptile stage, then a mammal stage, an ape stage and, before birth, the embryo is at the highest life form that evolution has so far achieved the human stage. Darwin did much to popularize the theory and included a detailed discussion of it in his *Decent of Man* (1871).

Although by the end of the 1915 the theory was disproved by Professor Rütimey, Professor William His and others, it is still occasionally mentioned as valid proof for evolution. Some Darwinists claimed as recently as 1987 that many of the "higher" evolved animal embryos passed through "identical" ancient evolutionary stages before acquiring their modem features (Kent, 1987). The recapitulation law teaches that each successive stage in the development of an individual represented one of the adult forms that appeared in its evolutionary history. Gill depressions in the neck of the human embryo were, for example, believed to prove that we had a fish-like ancestor.

Foremost it was Professor Ernst Haeckel who gave to the world the generalization that: *ontogeny (individual development) recapitulates (repeats) phylogeny (evolutionary descent)*, which means "the development of the individual repeats the evolution of the race" (Moore, 1963, p. 608). This notion later became known simply as *recapitulation* or *The Biogenetic Law* (Hickman et al., 1996, p. 161). A review of older biology textbooks reveals that the biogenetic law was once considered "one of the most important sources of biological evidence for evolution" and for this reason until recently was almost always discussed in textbooks that covered biological evolution (Smallwood, 1930, p. 392). In the Words of Princeton's Professor Conklin: "Ontogeny, or the origin of individuals, and phylogeny, or the origin of races, are two aspects of one and the same thing, namely, organic development. There is a remarkable parallelism between the two, and in particular the factors or causes of development are essentially the same in both" (1928, p. 64). In other words, the embryonic stages of an animal reveal the past evolution stages that the animal has passed though "in the course of its evolution. Embryonic development is a brief and condensed repetition of a series of ancestral stages through which the race has passed. Or, as often stated, ontogeny (the development of the individual) recapitulates phylogeny (the development of the race)" (Haupt, 1940, p. 345).

Some authorities have tried to claim that even plant embryo development documents recapitulation. An example is: "Germinating moss and fern spores produce a short filament of green cells which resembles a filamentous green algae. Soon the moss protonema develops into the male and female leafy shoots, while the filament of fern cells develops into the mature prothallus. For a brief period, though, mosses and ferns both pass through a stage reminiscent of the algae from which we think they evolved" (Kimball, 1965, p. 546).

Many modern biology texts still try to argue that the resemblances embryos display to their putative ancestors as they develop is critical evidence for evolution. The biogenetic law was even widely taught in popular lay books, such as Dr. Spock's *Baby and Child Care* which sold over forty million copies in thirty-nine languages. Under the subheading "They're repeating the whole history of the human race," Spock wrote that watching a baby grow is "full of meaning" because the development of each child retraces the entire history of the human race, physically and spiritually, step by step. Babies start off in the womb as a single tiny cell, just the way the first living thing appeared in the ocean. Weeks later, as they lie in the amniotic fluid in the womb, *they have gills like fish* and tails like amphibians. Toward the end of the first year of life, when they learn to clamber to their feet, they're celebrating that period millions of years ago when our ancestors got up off all fours and learned to use their fingers with skill and delicacy" (Spock and Rothenberg, 1998, p. 18, emphasis mine).

This passage implies that the embryo breathes by using its gills to extract oxygen from the amniotic fluid! As will be documented, this once common conclusion has now known to be incorrect (McNamara, 1999). Nonetheless, the influence of the biogenetic law in convincing the public of the validity of macroevolution has been enormous and, even though it has been refuted, is still commonly mentioned in science textbooks: So great was the desire on the part of some to strengthen this [biogenetic] idea, that a classic series of drawings showing embryonic similarities was produced in which the resemblances of the embryos of fish and man were remarkable. They were so remarkable, in fact, that further investigation showed that overzealous artistry had indicated a few resemblances that did not quite exist! (Moore, 1963, p. 608)

Basis for the Theory

Extensive comparisons of most animals has confirmed that a great deal of similarity exists in both the structure and function of body morphology including skeletons, muscles, nerves, body organs, and cell ultra-structure. For example, Haupt argues for evolution by claiming that fish gill slits exist during human embryo development stage when the human heart is two chambered and the circulatory system distinctly fish-like. The heart then passes through a three-chambered stage, characteristic of amphibians and reptiles, and finally becomes four chambered, as in birds and in other mammals. Similarly the human brain, in its embryonic development, passes through a series of stages corresponding to adult conditions in the lower vertebrate groups (1940, p. 347). Embryos of different species share similar structures, and they often appear physically similar, at least superficially, especially in their earlier stages of development. Although many animal kinds look superficially very similar in the zygote, cleavage, blastula, gastrula and other stages of development, profound differences exist.

All vertebrates, from amphioxus to humans, share fundamental resemblances in fetal development but, as biological research has progressed, more and more differences between the life forms the animal supposedly evolved from were discovered, eventually disproving the biogenetic law. One major difference is the organism's DNA, which can differ by many hundreds of thousands or millions of base pairs. This contrast eventually resulted in the more obvious morphological differences that result from the divergence that occurs in the animal's later developmental stages (Richardson et al., 1997).

Even vertebrate eggs of many animals display profound differences. Most obvious is they vary greatly in size, ranging from the microscopic eggs of mammals to the enormous eggs of birds. Eggs also differ in the conditions required for them to develop: some eggs begin as naked cells independent of their parents; others are enclosed in both protective membranes and shells and are incubated by their parents. Still other eggs, such as those of most mammals, develop within the body of the mother. Associated with these varying developmental conditions are many differences in egg size, rate of development, and methods of nutrition.

History of Biological Recapitulation

The theory that the "ontogeny" of the fetus "recapitulates" or duplicates the evolution of the organism was probably first expounded in modern times by Kielmeyer in 1793 partly from the observation that a frog tadpole resembles a fish (Rusch, 1969). The theory was further developed by embryologist Karl Von Baer (1792-1876) and Müller in 1864. In chapter 14 of *The Origin of Species* Darwin further developed the idea that the embryo's evolutionary history was written in its developmental stages (Darwin, 1859). It was later elaborated and popularized by German professor of comparative anatomy Ernst Haeckel (1834-1919) spelled Häckel in German (Rusch, 1969). Haeckel, an accomplished artist, produced a series of illustrations that have been widely reproduced in textbooks throughout the world. Haeckel's drawings were published in his own books, then in the 1901 book *Darwin, and After Darwin* by George Romanes. From this source they were widely reproduced in English language biology texts for the next century, no doubt because his pictures appeared to provide clear evidence for macroevolution.

The biogenetic law has proved critically important in converting people to Darwinism, and has been cited as a major proof of evolution in science textbooks from high school to graduate school for over a century (Taylor, 1984). One reason recapitulation was so popular with text book authors was because it was a simple, easily grasped concept that could be effectively illustrated by diagrams that superficially appeared to prove Darwinism. Darwin and Huxley were both very impressed by Haeckel's illustrations, although as research revealed more and more flaws in his idea, many scientists realized Haeckel's biogenetic law went "far beyond anything resembling science," and later became "an embarrassment to Darwin himself" (Milner, 1990, p. 205).

In a review of the history of the biogenetic law and why it was important, Conklin claimed the "law" that taught every animal climbs up its own ancestral tree when developing from egg to adult was a "god-send" for evolutionists. As a result, the study of embryology was pursued with "feverish zeal" because the method "promised to reveal more important secrets of the past than would the unearthing of all the buried monuments of antiquity—in fact nothing less than a complete genealogical tree of all the diversified forms of life which inhabit the earth. It promised to reveal not only the animal ancestry of man and the line of his descent but also the ... origin of his mental, social, and ethical faculties" (Conklin, 1928, p. 70).

The biogenetic law influenced the direction of embryology research and, importantly, it also distorted conclusions about past research findings (De Beer, 1958). As Rusch (1969, p. 27) noted, "in most cases, recapitulation was considered to be sufficient cause for the various stages in embryological development" and as a result discouraging research into the true causes. Nonetheless, doubts about the theory began to emerge very early in its history:

For a time embryology was studied chiefly to learn the course of past evolution, but owing to the highly speculative character of such studies and to the differences of opinion as to what were original (palingenetic) and what were acquired (coenogenetic) characters, there gradually arose a widespread skepticism concerning the value of embryology for this purpose (Conklin, 1928, pp. 70-71). In 1889, Gegenbaur voiced the "growing opinion among zoologists" that if coenogenetic (recently evolved traits) are "intermingled with palingenetic, then we cannot regard ontogeny as a pure source of evidence regarding phyletic relationships. Ontogeny accordingly becomes a field in which an active imagination may have full scope for its dangerous play, but in which positive results are by no means everywhere to be attained. To attain such results the palingenetic and the coenogenetic phenomena must be sifted apart, an operation that requires more than one critical *granum salis*" (quoted in Conklin, 1928, pp. 70-71).

Since then serious problems with the "law" accumulated, and more and more scientists have discounted the theory, some even declaring that no evidence exists that ontogeny ever recapitulates phylogeny and that Haeckel's "biogenetic law" has no foundation in fact (Rusch, 1969, p. 28). Many of the major difficulties in the theory were well known as early as 1928: "Inasmuch as many phenomena of development are mere adaptations to the conditions of embryonic or larva life and could never have been present in adult animals, Haeckel separated such characters, which he called "coenogenetic," from the truly ancestral ones, which he called "palingenetic," Unfortunately there was no certain method of always distinguishing these *two* types of embryonic characters, but in spite of this difficulty embryology was supposed to afford a short and easy method of determining the ancestral history of every group" (Conklin, 1928, p. 70).

Other problems with the theory include the fact that no certain criterion existed by which the various ancestral features existing in development could be distinguished from the "recently acquired ones, and what one embryologist regarded as ancestral another might consider a recent addition. Furthermore, when there were no living or fossil animals resembling certain embryological forms the fancy was given free rein to invent hypothetical ancestors corresponding to such forms. As a result of such speculations multitudes of phylogenetic trees sprang up in the thin soil of embryological fact and developed a capacity of branching and producing hypothetical ancestors which was in inverse proportion to their hold on solid ground" (Conklin, 1928, p. 70). Unfortunately, in his enthusiasm to prove the law and, thereby, vindicate evolution, the biogenetic law's major popularizer, resorted to outright fraud.

Fraud Proven

Many of Haeckel's drawings that he used to support his biogenetic law now have been proven to be grossly fraudulent. Richardson, an embryologist at St. George's Hospital Medical School in London, has concluded that generations of biology students were "misled by a famous set of drawings of embryos published one hundred-twenty-three years ago by the German biologist Ernst Haeckel. They show vertebrate embryos of different animals passing *through* identical stages of development. But the impression they give, that the embryos are exactly alike, is wrong … [Richardson] hopes once and for all to discredit Haeckel's work, first found to be flawed more than a century ago" (Pennisi, 1997, p. 1435).

The fraud was evidently actually first exposed in 1868 by University of Basel comparative anatomy Professor L. Rtitimyer, and then again in 1874 by the leading embryologist then, Wilhelm His Sr. (1831-1904). Dr. His, a comparative

36

embryologist and professor of anatomy at the University of Leipzig, concluded that both Haeckel's drawings and his conclusions were a gross distortion of the facts. In a review of His' work, Taylor argued that he proved Haeckel had engaged in blatant fraud and, therefore, Haeckel had eliminated himself from the ranks of scientific research workers of any stature (His 1874, 163). His, whose work still stands as the foundation of our knowledge of embryological development, was not the first to point out the deficiencies of Haeckel's work, nor indeed was he the last, yet Haeckel's fraudulent drawings have continued to the present day to be reproduced throughout the biological literature (Taylor, 1984, pp. 276-277). Cambridge University biologist Michael Pitman even claimed that, after Haeckel formulated his "fundamental biogenetic law" in 1868, he claimed that the entire animal kingdom was descended from an organism resembling the gastrula—an early stage in the embryonic development of most animals. To support his case he began to fake evidence. Charged with fraud by five professors and convicted by a university court at Jena, he agreed that a small percentage of his embryonic drawings were forgeries; he was merely filling in and reconstructing the missing links when the evidence was thin, and he claimed unblushingly that "hundreds of the best observers and biologists lie under the same charge" (Pitman, 1984, p. 120).

Assmuth and Hull (1915) even wrote a whole book on Haeckel's "many frauds and forgeries," concluding that "Haeckel knowingly and deliberately falsified documents in an effort to convince readers of the validity of evolution. Nonetheless, for scientific reasons alone, the early promising start of the new theory soon fell on hard times even though many biologists contemporary to Haeckel thought that embryology: would be a golden key to problems of phylogeny. Yet there was much unsound biology associated with the Biogenetic Law, and *few aspects of evolutionary science have been so heavily attacked in recent years*" (Dodson, 1960, p. 51, emphasis mine).

The major reason for the attack was that as "biological knowledge increased … the biogenetic law has been subjected to considerable criticism" (Carlson, 1996, p. 39). It soon became clear that the law is lethally flawed even though some Darwinists still clung to remnants of it today. Its flaws were openly discussed in mainline textbooks as early as 1963. "The similarities of embryological development among multicellular animals were intensively studied during the latter half of the nineteenth century. These studies led to the conclusion that the embryonic development of the individual repeated the evolutionary history of the race. Thus, it was thought to be possible to trace the evolutionary history of a species by a study of its embryonic development. This idea was so attractive as to gain the status of a biological principle … Today the idea of embryonic resemblances is viewed with caution. We can see and demonstrate similarities between embryos of related groups ... However ... the old idea that a human passes through fish, amphibian, and reptile stages during early development is not correct" (Moore, 1963, p. 608).

Thanks to the work of Richardson et al. the many fatal flaws in Haeckel's work have again resurfaced. Richardson concluded from his extensive study of Haeckel's work, that it may be "one of the most famous fakes in biology" (1997, p. 91). Haeckel even once admitted that he "used artistic license in preparing his drawings," but Haeckel's confession was either forgotten or ignored by those who wanted to use his biogenetic law to support evolution (quoted in Pennisi, 1997, p. 1435).

Examples of the distortion include Haeckel's embryo drawings showed the "tailbud" stage was close to identical in all of the different species that he drew. Richardson's team found that tailbud stage embryos, which were thought to correspond to a conserved stage of evolution, actually showed many major variations in form due to allometry, heterochromy, and differences in body plan and somatic number. These variations foreshowed critical differences in the adult body form (Richardson et al., 1997, p. 91). Richardson concluded that studying the many *differences* in embryos may prove to be far more fruitful than focusing on their similarities.

Many types of embryos share certain features at the early stages of development, including what appears to be a tail-like structure on their posterior and certain identifiable body segments. Embryos of different animal types possess many major differences which also negate the biogenetic theory. Evidence in favor of the biogenetic law was exploited by Haeckel, and the wealth of examples against the law was ignored. For example, by the time human embryos have developed to the extent of having the number of body segments shown in Haeckel's drawing's, they possess prominent protrusions called *limb buds* which later develop into limbs. These structures are absent in Haeckel's drawings.

Haeckel not only left out limb buds in his drawings, but even added structures to make the embryos of different animals appear more similar than they actually were. For example, he added a curl to the bird embryo "tail" so it would more closely resemble a human "tail." Haeckel even added certain features to the select few examples he used to prove his law, all of which were chosen because they seemed to prove his recapitulation theory. Furthermore, in the examples Haeckel used, he fudged the scale by as much as tenfold in order to exaggerate similarities among species. A comparison of Haeckel's drawings with accurate drawings or photographs show how enormously distorted, actually outright fraudulent, his drawings actually were (Pennisi, 1997, p. 1435).

In most cases Haeckel also neglected to name the species he drew to illustrate his theory, falsely implying that the one example he chose was representative of the entire group. Even closely related embryos, such as those of different types of fish, can vary greatly in appearance both at the embryo stage and in their developmental pathway (Carlson, 1996).

We now know that far more variation exists in vertebrate embryo development than was once assumed. For this reason, by focusing on these variations, Richardson's work is "a great service to developmental biology" to help us better understand development (Gilbert, quoted in Pennisi, 1997, p. 1435). As a result of Richardson's work and that of others, Haeckel's 1874 phylogenic tree based in part on the biogenetic law and strongly influenced by "Darwin's theory of common descent … including the unilateral progression of evolution toward humans … have since been refuted" (Hickman et al., 1996, p. 15). Dobson demonstrated that the biogenetic law failed when applied to echinoderms, an organism that was important in establishing the biogenetic law in the first place. He wrote, "the recent comprehensive study of echinoderm embryology by Fell reveals extensive differences among various groups of echinoderms, and these differences are referable to embryonic adaptations. Fell even casts doubt on the echinoderm-chordate relationship, for the hemichordate larva does not fit into the scheme of larval relationships which he has worked out" (Dobson, 1960, p. 52).

The current most optimistic status of the "ontogeny recapitulates phylogeny" law was summarized by Trefil as follows. "Nineteenth-century biologists noted that, as an embryo of an advanced organism grows, it passes through stages that look very much like the adult phase of less advanced organisms … In the nineteenth century, this so-called biogenetic law was taken to prove that evolution had proceeded on more or less a straight line from the simplest organisms to its epitome in human beings. We no longer have this view of evolution, but the biogenetic law remains a useful generalization about the way an embryo develops" (Trefil, 1992, p. 23).

Although most current textbooks no longer use Haeckel's fraudulent drawings, some evolutionists still ignore the overwhelming evidence against recapitulation theory and use the often vague similarities found in developing species to argue for Haeckel's theory—or a watered-down version of it. A major reason why Haeckel got away with passing off his theory for so long was because his drawings "became famous and have been repeatedly reproduced by publishers over the course of the last 120 years or so" (Youngson, 1998, p. 176). Haeckel's drawings were even prominently displayed on the cover of one college text (Gerhart and Kirschner, 1997) and reproduced inside (p. 329) as if they were valid. Drawings similar to Haeckel's were also pictured on an advertisement for another college embryology text, but interestingly, Haeckel's ideas were never mentioned in the text itself (Gilbert and Raunio, 1997).

The Human Tail

The putative human embryo "tail" which is gradually reduced until it usually disappears before birth, is also misinterpreted as an example of "recapitulation." It was interpreted as evidence of our tailed ancestors. This "tail" is actually the human spine and the developing vertebra that ends in the coccyx. The "tail" appearance develops because the brain and spinal cord mature very early in development in order to coordinate the rest of the body's development, and, at this developmental stage, the developing spinal column system is longer in the embryo torso.

It is for this reason that humans and most animals possess what superficially *appears* to be a tail during their early development. Tailed animals do not have a spinal cord or spinal vertebras in their tail as do part of the "tails" of embryos. Unfortunately, many older textbooks published highly misleading and often totally erroneous claims about the human embryo spine. A good example is Haupt's 1940 text that claimed embryology teaches that "many structures which are permanent in the lower members of a group appear only in embryonic stages in the case of the higher members, and then either later disappear completely, persist as vestiges, or become modified to form other structures. For example, during an early period of prenatal development, the human embryo has a tail as well developed as that of any of the other vertebrates" (1940, p. 347). A survey of modern biology textbooks indicated that few texts today even mentioned the now disproved tail claims.

The Embryological Down or Lanugo

During the so-called hair stage of human embryo development, an extremely fine soft downy hair known as *lanugo* or *embryonal down* covers most of the embryo (Harrison, 1963). Evolutionists once taught that this hair was evidence of our hairy primate and mammalian stage of evolution and is nonfunctional today. We now know that this fine hair plays an important role in both embryo and fetal development. Lanugo hair is most prominent during the seventh and eighth month of fetal development and, among the functions it serves includes helping hold the *vernix caseosa* in place. The *vernix caseosa* is a sticky white secretion that covers the undeveloped skin to protect the developing embryo from the corrosive effects of the surrounding amniotic fluid (Butler and Juurlink, 1987).

As the embryo develops, the skin thickens and is keratinized. Consequently, around the 36th week the lanugo hair no longer is necessary and, as a result, normally almost all of the lanugo hair is "lost" just before birth (Carlson, 1996, p. 362). Actually it is not "lost," but converted to vellus hair, fine colorless short body hairs. The actual number of hair follicles per square inch of skin is the same throughout life, only the *type* of hair produced changes. The three basic types of hair shafts include lanugo, vellus, and terminal. The terminal hairs are the scalp hairs that cover our head and produce the male beard. When males reach puberty, their face follicles stop producing vellus hairs and begin producing terminal hairs.

Small amounts of this downy hair persist throughout life on certain parts of the body, especially on the face and ears. This fuzz is not comparable to the coarse pelts of hair existing on mammals but still can be very useful in certain situations, even in adult humans (DuPuy and Mermel, 1995). Since the outer layer of skin consists of highly keratinized "dead" epithelial cells useless for tactile sensations these hairs are necessary to increase tactile sensitivity in order to enable the skin to communicate effectively with the outside world. Downy hair also can develop over the entire body during famine or in anorexics to help replace the insulation lost by a decrease in body fat (DuPuy and Mermel, 1995, p. 179).

The Conserved Stage Theory

The biogenic law now is widely recognized as both erroneous and misleading because so many exceptions were found. It is also flawed for the reason that Haeckel based his biogenetic law on the "flawed premise that evolutionary change occurs by successively adding stages onto the end of an unaltered ancestral ontogeny, compressing the ancestral ontogeny into earlier developmental stages. This notion was based on Lamarck's concept of the inheritance of acquired characteristics" (Hickman et al., 1996, p. 161)

A major problem with the "conserved stage" hypothesis of recapitulation is that different organs develop at different times in different species, making it impossible to point to a single conserved stage when all species have the same body plan (Richardson, 1997b). Furthermore, organ development was often contrary to what recapitulation predicted. For example, if the human embryo repeated its assumed evolutionary ancestry as it developed, the human heart should begin with a single chamber and then develop successively into two, then three, and finally four chambers. Instead, the human heart begins as a two-chambered organ which fuses to a single chamber, which then develops directly into four chambers. In other words, the sequence is 2-1-4, not 1-2-3-4 as required by the theory. The human brain develops before the nerve cords, and the heart before the blood vessels, both out of the assumed evolutionary sequence. It is because of many similar contradictions and omissions that the theory of embryological recapitulation has been abandoned by embryologists (Gish, 1995, p. 358).

41

Another excellent example is that the development of similar forms of animals from very dissimilar pathways is common at later stages of development. "Many types of animals pass through a larval stage on their way to adulthood, a phenomenon known as indirect development. For example, most frogs begin life as swimming tadpoles, and only later metamorphose into four legged animals. There are many species of frogs, however, which bypass the larval stage and develop directly. Remarkably, the adults of some of these direct developers are almost indistinguishable from the adults of sister species which develop indirectly. In other words, very similar frogs can be produced by direct and indirect development, even though the pathways are obviously radically different. The same phenomenon is common among sea urchins and ascidians (Wells and Nelson, 1997, p. 16).

Many other examples of organs and structures exist that do not develop in the order predicted by the biogenic law. Examples Rusch gives include the tongue in mammalian embryos develops before the teeth and certain environmental conditions can change the sequence order that embryo differentiation occurs (1969, p. 28). Nor do anatomical evaluations of the developing embryo support recapitulation: "while many authors have written of a conserved embryonic stage, no one has cited any comparative data in support ... the phylotypic stage is [evidently] regarded as a biological concept for which no proof is needed. This has led to many problems, not the least of which is the lack of consensus on exactly which stage is conserved" (Richardson et al., 1997) Furthermore, the biogenetic law has misled researchers who for years were looking for evidence that does not exist and ignored, or tried to explain away, the large body of evidence that contradicted the biogenetic law.

Other Problems with the Biogenetic Law

The biogenetic law originally tried to explain virtually *all* aspects of development. An example is the claim that the human embryo has "gill slits and aortic arches, which undergo exactly the same transformations that take place in other mammals. Man's heart is at first like that of a fish, consisting of one auricle and one ventricle. His backbone begins as a notochord, is next a segmented cartilaginous rod, then each segment or vertebra consists of five separate bones, and finally each fuses into a single bone. He has in the course of his development three different pairs of kidneys, first a pronephros (or forekidney), like that of the lower fishes, then a mesonephros (or mid-kidney), like that of the frogs, and finally a metanephros (or hind-kidney) like that of reptiles, birds, and mammals, which alone survives the adult. His brain, eye, ear, in fact, all his organs, pass through stages in development that are characteristic of lower vertebrates. Even in those adult features that are distinctively human, such as the peculiar form of the

42

hand and the foot, the number of bones in the ankle and wrist, the number of pairs of ribs, the absence of a tail and the relative hairlessness of the skin—in all these respects the human fetus resembles anthropoid apes more than adult man. Why are not these and a hundred other structures made directly? Why this roundabout process of making a man? There is no answer but evolution" (Conklin, 1928, pp. 74-75).

"Another major problem is recapitulation almost certainly would *not provide the animal with any selection advantages* but instead would likely result in many major selection *disadvantages* during the embryonic stage. As a result of the growing knowledge of biology and development in the 1950s, the law was actually largely discredited almost a half century ago, even though it took decades before this new knowledge was reflected in the textbooks. The evidence supports Sir Arthur Keith's statement made three-quarters of a century ago regarding embryology and evolution that biologists expected that the embryo would recapitulate the features of its ancestors from the lowest to the highest forms in the animal kingdom. Now that the appearances of the embryo at all stages are known, the general feeling is one of disappointment; the human embryo *at no stage* is anthropoid in its appearance (1925, p. 867).

"The many flaws that eventually mortally wounded the biogenetic law included the fact that it lost favor due to the rise of experimental embryology and finally became *untenable in theory* (when the establishment of Mendelian genetics converted previous exceptions into new expectations). The biogenetic law was not disproved by a direct scrutiny of its supposed operation; it fell because research in related fields refuted its necessary mechanism" (Gould, 1977, p. 168).

Attempts have been made throughout the years to repair and revise the law, but all have failed. One early attempt offered by XIX century embryologist K. E. von Baer argued that early developmental features were *more* widely shared among "different animal groups than later ones … The adults of animals with relatively short and simple ontogenies often resemble preadult stages of other animals whose ontogeny is more elaborate, but the embryos of descendants do not necessarily resemble the adults of their ancestors. Even early development undergoes evolutionary divergence among groups, however, and it is not quite as stable as von Baer believed" (Hickman et al., 1996, p. 162)

No reason now exists to believe that the recapitulation theory is true except that it appears to support evolution. Harvard's Steven J. Gould (1977) even wrote a 501 page book on Haeckel's biogenetic law documenting its history from its appearance in the pre-Socrates days to its demise in the early twentieth century.

Why the Biogenetic Law is Still Taught

The reason the biogenetic law is still taught in some textbooks is not because of the realization that it is erroneous is recent. Many biologists recognized that the biogenic law was falsified as long as one-half century ago. "For a time during the latter half of the nineteenth century this theory was received with great enthusiasm, and it was predicted that a study of living things in the light of this "law" would revolutionize biology. But, unfortunately, the predictions are not being verified. As a working hypothesis the theory has been a great help, but there are so many exceptions, apparently even contradictions, that its application is frequently misleading" (Hauber and O'Hanlon, 1946, p. 156).

Many scientists at least nominally still accept biogenetic law because it, or remnants of it, are still taught in textbooks and at many colleges for the reason that the "biogenetic law has become so deeply rooted in biological thought that it cannot be weeded out in spite of its having been demonstrated to be wrong by numerous subsequent scholars. Even today both subtle and overt uses of the biogenic law are frequently encountered in the general biological literature as well as in more specialized evolutionary and systemic studies" (Bock, 1969, p. 684).

Many scientists and scholars are either unaware of the criticism of the law, or chose to ignore the evidence against it. Professor Glover claims that the "vast majority" of his medical students believe the human embryo has gill slits even though their medical text on embryology correctly explains that the human embryo does not possess gill slits but pharyngeal grooves (Ham, 1992). And Youngson concluded that while "Haeckel's theory of recapitulation was, for a time, almost universally believed, the debunking of Haeckel was not widely known ... His book *The Riddle of the Universe,* an extraordinary mishmash of real science and imaginary nonsense, was a great popular success and ran into numerous editions. So, although his ontogeny ideas were brushed aside at a fairly early stage by the serious scientists, they continued to be accepted by the lay public" (Youngson, 1998, p. 177). "Another major reason for the continued acceptance of the biogenetic law by some evolutionists is because it is now part of the accepted worldview of scientists, a belief that they were exposed to from the earliest days of their training. Most scientists are influenced by social pressure, and many fear recriminations from their fellow scientists if they do not conform to what is currently viewed as valid. To prove their orthodoxy, many scientists have become unscientific and have embraced the worldview of twentieth century naturalism" (Johnson, 1993).

Biogenetic Law and Racism

This problem is of major interest to creationists and others. The biogenetic law "became extremely influential outside of science" and "caused a great deal of mischief" (Milner, 1990). Gould claimed that the theory was one of the two or three leading scientific arguments for racism (1977, p. 216). One example Milner cites is the idea that the brains of certain races were stuck at a lower, childlike stage of evolutionary development. In Gould's words for "a half century the proponents of recapitulation had collected" evidence that "argued adults of 'lower' races were like white children" Gould notes that proponents of recapitulation asserted the fact that "women are more childlike in their anatomy than men" was proof of their inferiority" (1977, pp. 219-221).

Gould gives many other examples of the use of the biogenetic law to endorse racism, such as the argument that black males are more primitive than whites because the distance between their navel and penis remains small relative to body height as adults in contrast to white children, which start with a small separation that increases during growth. The rising belly button was seen as a mark of evolutionary progress because it could be used to rank the evolutionary "level" of many primates including the apes (1977, p. 218). Gould also noted that *Recapitulation* had its greatest political impact as an argument to justify imperialism. Kipling, in his poem on the *White Man's Burden*, referred to vanquished natives as "half devil and half child" If the conquest of distant lands upset some Christian beliefs, science could always relieve a bothered conscience by pointing out that primitive people, like white children, were incapable of self-government in a modem world. During the Spanish American War, a major debate arose in the United States over whether we had a right to annex the Philippines. When anti-imperialists cited Henry Clay's contention that the Lord would not have created a race incapable of self-government, Rev. Josiah Strong replied: "Clay's conception was formed before modem science had shown that races develop in the course of centuries as individuals do in years, and that an underdeveloped race, which is incapable of self-government, is no more of a reflection on the Almighty than is an undeveloped child who is incapable of self-government" (1977a, pp. 218-219).

Moore even concluded that "recapitulation was a leading argument for racists in the late nineteenth century" (1999, p. 1). Furthermore, some of Sigmund Freud's more radical (and now discredited) ideas came directly from Haeckel's biogenetic law (Milner, 1990, p. 177). Milner even claims Haeckel's views "became a major cultural force in shaping the militant nationalism in Germany" that led to the holocaust which resulted in the loss of over six million lives (1990, p. 205). This view of development has even been used as an argument to justify abortion in the early stages based on the reasoning that it is not wrong to kill life at this development stage because the embryo is not yet human, but is only a fish or less (Major, 1994, pp. 175-177).

45

Implications for Creationism

The ontology law was a major weapon in the arsenal used to attack, not only creationism, but also Christianity, minorities and even the existing social order in favor of communism. Gould noted that recapitulation was Haeckel's favorite argument "to attack nobility's claim to special status—are we not all fish as embryos?—and to ridicule the soul's immortality—for where could the soul be in our embryonic, wormlike condition?" (Gould, 1997, p. 217). Our ability to reason, to determine right and wrong, to live according to a conscience, to exercise domination over plants and animals, to enjoy music and art, and to worship our Creator are all only a small part of the enormous chasm that separates humans from *every other living creature*. The biogenetic law is only one of many hypotheses that Darwinists have used to attempt to support their naturalistic theory that is gradually being proven wrong as new evidence accumulates.

Naturalistic evolution requires faith in an embryological theory that has now been disproved and belief in evolutionism requires a blind, often in credulous faith induced partly by pressure to conform to the world of science that is saturated with naturalism. The history of the biogenetic law should force all persons to look critically at the current lack of evidence for the whole evolutionary model of origins.

Conclusions

The biogenetic law was based on very superficial similarities in developing embryos. As our knowledge of embryology and especially genetics increased, it became increasingly obvious that the "law" was fundamentally in error. The three major early scientific objections to Haeckel's version of the biogenetic law can be summarized as follows:

1. The path of embryological development varies enormously for both organs and body structures. Each ontogenetic stage is an inseparable mixture of organs in different stages of putative ancestral repetition.

2. Larvae and embryos possess many features that help them adapt to their individual mode of life. New characters are often introduced at stages of embryological development that do not follow the biogenic law.

3. Development can be retarded as well as accelerated compared to the expectations of the biogenic law. Embryonic or larval stages of ancestors can become the adult stages of descendants—a phenomenon directly opposite to the recapitulation law prediction (Adapted from Gould, 1977, p. 168).

The stages of embryological development of many animals do show some

46

similarities but the major reason for this fact is design constraints. Likewise, adult organs show much similarity because only so many ways exist to design a heart or lung, and we would expect the *earlier* in development, the *fewer* the design possibilities that exist. All sexual reproducing organisms start out as one-celled zygotes that superficially look remarkably similar and, as development and differentiation proceed, they increasingly look different.

All hearts begin as a single contracting artery tube which, depending on the animal type, develops into a one, two, three or four chambered heart. A human heart does not start as a simple one chambered heart because our ancestor had a one chambered heart, but because embryological development in general mandates simple to complex progression. The same is true for all other organs. Because all life begins as one cell does not prove all life evolved from one cell, but that this is the only way that life can develop by either asexual or a sexual reproduction (Milton, 1997).

Acknowledgements: I wish to thank Dr. Wayne Frair for his comments on an earlier draft of this paper and for his excellent article on this topic (*CRSQ* 36:62-67).

References

Asimov, Isaac. *The Wellspring of Life.* Abelard-Schuman, New York, 1960.

Assmuth, J. and Ernest Hull. 1915. *Haeckel's Frauds and Forgeries.* P.J. Kennedy, New York.

Bock, Walter J. Evolution by orderly law. *Science,* 164:684-685, 1969.

Butler, H. and H. Juurlink. *An Atlas for Staging Mammalian and Chick Embryos.* CRS Press, Boca Raton, FI, 1987.

Carlson, Bruce M. *Patten's' Foundations of embryology.* McGraw-Hill, New York, 1996.

Conklin, Edwin Grant. Embryology and Evolution in *Creation by evolution.* Frances Mason, editor. MacMillan, New York, 1928.

Coyne, Jerry. "Not Black and White" *Nature*, 396(6706):35-36, Nov. 5, 1998.

Darwin, Charles. *The Origin of Species.* Reprinted by D. Appleton. New York, 1897.

De Beer, Gavin. Darwin and embryology in *A Century of Darwin.* Samuel Barnett, editor. Heinemann, London, 1958.

Dodson, Edward O. *Evolution Process and Product.* Reinhold, New York, 1960.

DuPuy, Nancy and Virginia Lee Mermel. *Focus on Nutrition.* Mosby, St. Louis, MO, 1995.

Exalto, N. "Early Human Nutrition" *European Journal of Obstetrics Gynecology and Reproductive Biology*, 61(1):3-6, 1995.

Gerhart, John and Marc Kirschner. *Cells, Embryos and Evolution.* Blackwell Science, Malden, MA, 1997.

Gilbert, Scott F. and Anne M. Raunio (editors). *Embryology: Constructing the Organism.* Sinauer Associates, Sunderland, MA, 1997.

Gish, Duane. *The Fossils Still Say No.* Institute for Creation Research, El Cajon, CA, 1995.

Gould, Steven Jay. *Ontogeny and Phylogeny.* Harvard University Press, Cambridge. 1977. Racism and Recapitulation in *Ever since Darwinism.* Norton, New York.

Gramet, Charles and James Mandel. *Biology: Serving You.* Prentice-Hall, Englewood Cliffs, N.J., 1958.

Grimes, Charles. *A Story Outline of Evolution.* Bruce Humphries, Boston, 1944.

Grigg, Russell. Ernst Haeckel: Evangelist for Evolution and Apostle of Deceit. *Creation Ex-nihilo,* 18(2):33-36, 1996.

Grigg, Russell. Fraud Rediscovered. *Creation Ex-nihilo* 20(2):49-51, 1998.

Haeckel, Ernst. *Natural history of Creation* (English translation, 1870). Appleton, New York, 1906.

Haeckel, Ernst. *The Evolution of Man.* D. Appleton, New York, 1920.

Ham, Ken. A Surgeon Looks at Creation. *Creation Ex-Nihilo,* 14(3):46-49, 1992.

Harrison, Ronald G. *Textbook of Human Embryology.* Blackwell, Oxford, England, 1963.

Hauber, V.A and M. Ellen O'Hanlon. *Biology: A Study of the Principles of Life for the College Student.* F.S. Crofts, New York, 1946.

Haupt, Arthur. *Fundamentals of Biology.* McGraw Hill, New York, 1940.

Hickman, Cleveland, Larry Roberts and Allan Larson. *Integrated Principles of Zoology.* William C. Brown, Dubuque, IA, 1996.

His, Wilhelm. *Unsere Korperform.* c.W. Voegel, Berlin, 1874.

Johnson, Phillip. Science without God. *Wall Street Journal,* May 10, p. AIO, 1993.

Jordan, David and Vernon Kellogg. *Evolution and Animal life.* D. Appleton, New York.

Kaufmann, David. 1985. Further comments on baleen fetal teeth and functions for yolk sacs. *Origins Research,* 8(2): 13, 1908.

Keith, Arthur. The Nature of Man's Imperfections. *Nature,* 116: 867, 1925.

Kent, George C. *Comparative Anatomy of the Vertebrates;* fourth edition. C.V. Mosby, St. Louis, MO., 1978.

Kent, George C. *Comparative Anatomy of the Vertebrates;* sixth edition. Times Mirror Mosby, St Louis, MO, 1987.

Kimball, John W. *Biology.* Addison-Wesley, Reading, MA. 1965.

Lindsay, Daniel ; Ian S. Lovett, Edward A. Lyons, Clifford S. Levi, Xin-Hua

Zheng, Susan C. Holt, and Sidney M. Dashefsky. Yolk Sac Diameter and Shape at Endovaginal US: Predictors of Pregnancy Outcome in the First Trimester. *Radiology,* 183: 115-118, 1992.

Major, Trevor. Haeckel: The Legacy of a Lie. *Reason and Revelation* 14: 68-70, 1994.

McNamara, Ken. Embryos and Evolution. *New Scientist* 164(2208): 1-4, 1999.

Milner, Richard. *The Encyclopedia of Evolution.* Facts on File, New York, 1990.

Milton, Richard. *Shattering the Myth of Darwinism.* Park St Press, Rochester, VT, 1997.

Moore, John A. (Editor). *Biological Science: An Inquiry into Life.* Harcourt Brace and World, New York, 1963.

Moore, Randy. Science, Objectivity and Racism. *American Biology Teacher,* 61(4):1, 1999.

Muller, Fritz. *Facts and Arguments for Darwin* (English translation). John Murray, London., 1869.

Parker, George. *What Evolution Is.* Harvard University Press, Cambridge, MA, 1931.

Pennisi, Elizabeth. "Haeckel's Embryos: Fraud Rediscovered" *Science,* 277:1435, 1997.

Pitman, Michael. *Adam and Evolution.* Rider, London, 1984.

Raven, Peter and George Johnson. *Understanding Biology.* Times Mirror/Mosby, St. Louis.

Richardson, Michael. 1997a. Embryonic Fraud Lives on. *New Scientist,* 155(2098): 23, 1988.

_____. Heterochrony and the Phylotypic Period. *Developmental Biology,* 172: 412-421, 1997b.

Richardson, Michael, James Hanken, Mayoni Gooneratne, Claude Pieau, Albert Raynaund, Lynne Selwood, Glenda Wright. There Is No Highly Conserved Embryonic Stage In The Vertebrates: Implications for Current Theories of Evolution and Development. *Anatomy and Embryology,* 196: 91-106, 1997.

Romanes, George. *Darwin.* Chicago: The Open Court Publishing Co, 1901.

Rusch, Wilbert. "Ontogeny recapitulates Phylogeny" *CRSQ,* 6(1): 27-34, 1969.

Sadler, Thomas W. *Langman's Medical Embryology.* Williams and Williams, Baltimore, 1995.

Shier, David, Jackie Butler, and Ricki Lewis. *Hole's Human Anatomy and Physiology;* eighth edition. McGraw Hill, New York, 1999.

Sillman, Emmanuel I. Further Comments on Baleen Teeth and Functions for Yolk Sacs. *Origins Research,* 8(2): 13, 1985.

Smallwood, William Martin. *A Text-book of Biology;* sixth edition. Lea and

Febiger, Philadelphia, PA, 1930.

Solomon, Eldra Pearl, Linda Berg, and Diana Martin. *Biology*. Saunders, Orlando, FL, 1999.

Spock, Benjamin. *Dr. Spock's Baby and Child Care*, 7th edition. NY: Dutton, 1998.

Taylor, Ian. *In the Minds of Men*. TFE, Toronto, Ontario, Canada, 1984.

Trefil, James. *1001 Things Everyone Should know about Science*. Doubleday, New York, 1992.

Wells, Jonathan and Paul Nelson. "Homology a Concept in Crisis." *Origins and Design,* 18(2): 12-19, 1997.

Wells, Jonathan, "Haeckel's Embryos & Evolution" *The American Biology Teacher*. 61(5):345-349, 1999.

Wilson, Carl. *Botany*. The Dryden Press, New York, 1954.

Youngson, Robert. *Scientific Blunders, a Brief history of How Wrong Scientists can Sometimes be*. Carroll and Graf, New York, 1998.

Author Bio

Jerry Bergman collects degrees like some people collect stamps. His nine degrees, including from the Medical University of Ohio, Wayne State University in Detroit, University of Toledo, and Bowling Green State University, are all in the sciences. Dr. Bergman has taught numerous college level courses including biology, genetics, chemistry, biochemistry, and anthropology for over thirty-eight years. He is listed in six different regional and national Who's Who lists, including in science, medicine and theology. He is a member of Mensa, a association of the intellectual top 2% of the population. In 1998 he received the Edgar Langsdorf award for excellence in writing and was awarded a public health service grant valued at more than $20,000 while still a graduate student. He is a fellow of American Scientific Affiliation (ASA) has been an active ASA member for over thirty years.

Dr. Bergman is currently serving as the Biology Editor for the ***Creation Research Society Quarterly***. Many of his nearly 1,000 publications in twelve languages and twenty books and monographs deal with the creation/evolution controversy. He has taught at the Medical University of Ohio where he was a research associate in the department of experimental pathology, and at the University of Toledo and Bowing Green State University. He is now an adjunct professor at the University of Toledo Medical College in Ohio. You may email him at: jerrybergman@verizon.net or JBERGMAN@north- weststate.edu.

Homology and What It Really Indicates
De-Sanctifying Darwin's Sacred Cow
George Howe, Ph.D.

Editor's note

Dr. George Howe has been my long time friend. He was my Sunday school teacher long ago when I was a lowly doctoral student at the University of California, Los Angeles and we have stayed in touch. He is has long been one of the foremost leaders in the modern Creation science movement having served as president of the Creation Research Society and the science editor for the prestigious *Creation Research Society Quarterly* for decades. In terms of Creation science he is truly a giant among men. I am honored to call him friend and delighted he has chosen to make major contribution to this book. Besides his accomplishments in Creation science, he is also a well published and respected botanist. His field work with yuccas and other desert plants are legendary.

Introduction

What is homology?

Internal similarity between different groups of living organisms is called "homology," a subject regularly studied in biology from high school level upward. An example often given for homology is the correspondence between bones inside a bat wing, a human arm, and the flipper of a seal.

History of "Homology"

The scientific aspects of homology were developed and described by creation scientists, well before Darwin. But Darwin wrote that homology is evidence solely for evolution and that it becomes "hopeless" in the creation model. Later, the religious followers of Darwin, whom I call "Darwinites," completely adopted Darwin's revised definition of homology, making it a proof for evolution alone. They turned it into a "sacred cow," never to be questioned. These and other aspects of homology history will be developed.

It is Advisable to De-sanctify Such Sacred Cow.

The term "sacred cow" is a very useful metaphor, applicable to almost anything that has received undue veneration. Attaching the "sacred cow" moniker to a long-revered but unsubstantiated topic enables it to be discussed, exposed, modified, or even abolished, as needed. In this book the light of logic is focused on many "holy hobbies" found in the scientific community, which shows that scientists have been no better than other people when it comes to avoiding the beatification of unfounded concepts. Homology has been promoted as a unique proof for evolution and has thus become a sacred cow.

Worship of Cows Still Takes Place

In dealing with sacred cows, it should first be recognized that homage is still paid to actual cows and that the origin of humans worshipping bovines harks back to early times. Two instances of cattle being venerated will be studied—one from ancient Egypt and another in modern India. This is done in order to shed light on the worship of "sacred cow" ideas that are now being exalted in science. Some of the other subjects covered in this chapter include: the American origin of the metaphor "sacred cow," Darwin's distorted view of homology, and an in-depth discussion of scientific facts to set the topics of homology and "analogy" free from their adherent evolutionary trappings. In a conclusion suggestions are made regarding what can currently be done to eradicate the Darwinian beatification of homology as an untouchable proof of evolution. A postscript encourages creation scientists to speak about the Creator when reporting on His works. Hopefully this book will help people dispel existing sacred cows, like homology, and enable them to avoid accepting new ones.

Examples of Actual Cow-Worship: Past and Present

In understanding the nature of the sacred cow figure of speech, it is well to demonstrate that worship of real cows occurred historically, with cows being deified in ancient Egypt. The plagues that God sent to Egypt were directed against spheres of influence attributed by the Egyptians to their own deities. Walvoord and Zuck (1983, p. 120) showed that various Egyptian cattle gods were targeted twice in the plagues: once in the Nile River plague and then in the plague involving the death of livestock. Hopi, an Egyptian bull god who had the responsibility of controlling the Nile River, met his match when the river was turned into blood (Exodus 7:14-26). In Exodus 9:1-7, the cow god Hopi was again confronted in a livestock plague, together with an Egyptian goddess Hathor, who had a cow's head. All the cattle of the Egyptians died—the animals belonging to the very individuals who worshipped the two cow gods. Then too, Egyptian cattle worship was probably involved when Pharaoh tried to arrange a compromise insisting that Moses and the Israelites stay in Egypt to slaughter sacrificial animals. Moses refused Pharaoh's offer, perhaps because he knew that the Egyptians would have been angered if they had witnessed cattle being sacrificed this way. Even now, devotees of an ideological sacred cow become enraged when their favorite themes are criticized.

God had many other truths to teach in the plagues. As Smith has pointed out, God manifested a sense of humor in the plagues as well. "It is my understanding that the Egyptians worshipped a frog god. Our Creator was in essence saying, 'You want frogs, I'll give you frogs.'" (Smith, 2009).

Cows are still worshipped in Hindu countries where killing an animal of any kind is a sin because it interrupts the assumed cycle of reincarnation. Above other animals, cows are held to be especially sacred in Hinduism, where it is forbidden to eat beef. This sinful fetish for cows may have developed because Hindu people have relied on cattle historically for milk, tilling their fields, and as a source of dung useful for fertilizer and fuel. "Despite differences of opinion regarding the origins of the cow's elevated status, reverence for cows appears through the major texts of the Hindu religion" (*Wikipedia*, 2009, p. 2). [Realizing that *Wikipedia* is not a highly respected scientific source, I am quoting it only in those portions of this chapter that deal with religion or with subjects of a very general nature. I will not depend on it in the scientific sections.] In Hindu areas ... where there is a ban on cow slaughter, a citizen can be sent to jail for killing or injuring a cow (Wikipedia, 2009, p. 4). Perhaps the most extreme devotion of a Hindus shown for a cow is found in the statements of Ghandi concerning this animal: "I worship it and I shall defend its worship against the whole world. The central fact of Hinduism is cow protection ... [The cow is] the mother to millions of Indian mankind ... Our mother, when she dies, means expenses of burial or cremation. Mother cow is as useful dead as when she is alive. We can make use of every part of her body—her flesh, her bones, her intestines, her horns and her skin (Wikipedia, 2009, p. 4). [Bracketed words in these quotations are added.] Ghandi and other Hindus need to realize that death is death and that after death there is the wise judgment of the loving God (Hebrews 9:27-28). and there is no such thing as reincarnation. Also, the foolish homage like people pay to real animals like cows is very similar to the commitment some scientists make for their own ideological sacred cows.

"Sacred Cow" as a Metaphor in the English Language

To develop an understanding of philosophical sacred cows in science, it will be helpful to learn how that "sacred cow" phrase itself ever arose and became a figure of speech. Martin (2009) stated that the first printed use in the English language of "sacred cow" for actual worship of cows, was in 1854 when a man named Wady Jahed wrote a letter from Wisconsin to the high Brahmin in Benaries, India—a letter that somehow got published in *The Janesville* [Wisconsin] Free Press, January 1854. Jahed was obviously shocked by American culinary practices and he related to the Brahmin that Americans drank their grain from the brewery, as liquor, in honor of "the gods," rather than "... feeding it to sacred bulls and cows …" (p. 1). He added that Americans did other vile things too, such as eating the flesh of animals.

"Sacred cow" was first used as a simile in the New York Herald, March 1890. The author of that article took issue with certain individuals whom he thought had been treating a particular New York waterway, that he called "the great ditch," "… as [or like] a sort of sacred cow" (Martin, 2009, p. 2). In 1909, nineteen years thereafter, "sacred cow" appeared as a full-fledged metaphor in *The Galveston Daily News*. It was reported in that article that some people were treating "raw material" in the fashion of a "sacred cow" not to be subject to tariff reform, and "enjoying the blessing of incidental protection" (p. 3).

During the many years since that Texas article was printed, the metaphor "sacred cow" has become commonplace parlance for almost anything that is placed above change. Each author in this book is showing that even science, which is supposed to be continuously "open" and "objective" on all issues, is itself replete with these protected philosophical pockets that are seldom if ever questioned. Sacred cows in science are particularly dangerous and hard to dislodge because of the respect afforded to the "scientific method"

The Original "Homology": Before Darwinian Distortion

In biology, "homology" designates a similarity between two different creatures, as noted at the onset. In its adjectival form, this term "homologous" is applied to features like the corresponding bones in the wing of a bat and those of the human arm. Each human bone such as the humerus, radius, or ulna is said to be "homologous" to a complementary bone given the same name inside a bat's wing. Where a human has an ulna in its lower arm, a bat also has an ulna as part of its wing. The wing surface is then stretched across the bat's elongated finger bones called phalanges, the name which is also given to human finger bones. Homology should simply specify this kind of underlying similarity between matching organs in different types of creatures. Once upon a time homology was not a sacred cow of evolutionists but a phenomenon reported by early creation scientists, as will be historically documented below.

Carolus von Linnaeus was a creation scientist in the 1700s who set the stage for homology research. "Homology" was actually coined by scientists who believed in creation. Carolus von Linnaeus was one of them, a key eighteenth century botanist who devised and established our modern scheme of classification, called "taxonomy" He prepared the way for the discussion of homologies because he wrote that the Creator, while creating, used various patterns, plans, or "archetypes" (as he called them). Based on similarities and differences within the many archetypes, Linnaeus predicted that living organisms would fit into an outline of classification—and they did. As a result, he described that creation outline and predicated our modern system of "natural" taxonomy on the basis of similarities. Among his many contributions to science, Linnaeus described homologies between different parts of flowers, parallelisms, which are still used in plant classification.

Linnaeus Followed the Facts

In his mature years Linnaeus became aware of biological variation. As a consistent scientist, he modified his early opinion, which was fixity of species, and brought it into conformity with the variation he observed in nature. His final thoughts, after years of study, were that changes in species and even in a genus can occur, but only within the boundaries of the "kinds" originally established by the Creator. He discussed the possibility that certain species in the same genus may have had a common ancestry, and he did this more than a hundred years before Darwin. Unlike Darwinism, however, Linnaeus' view of origins fits well with the facts of biology and with the statements in the book of Genesis where it is stated ten times in chapter one and seven more times after the Flood that God did not evolve life from one kind but had each kind reproduce "after its kind" Linnaeus' outlook on origins is accepted nowadays by all scientific creationists who take the Bible literally. The revised form of Linnaeus' taxonomy is likewise practiced by all biologists.

Linnaeus' Work Was Buried During the Beautification of a Sacred Cow.

Unfortunately Linnaeus' many contributions to systematic biology have been obscured while undue reverence was showered upon Darwinism. Some authors, however, have recognized and recorded the greatness of Linnaeus. Greene (1959, p. 137) wrote: "In the fruiting organs of plants Linnaeus believed he could discern characters written by 'the hand of God' to aid man in distinguishing the genera" Eisley (1958, p. 23) summarized Linnaeus' outstanding scientific work by writing that "... he [Linnaeus] glimpsed, more than his fellows, the wonderful plan of creation, the unities as well as the diversities of form that existed in the mind of God" Linnaeus was able to accomplish all this because his theology was one of openness to God as the Creator. It would be well for scientists to show the respect that is due this forgotten hero, Linnaeus.

Georges Cuvier Was the Founder of Modern Animal Taxonomy.

In the early 1800s another creation scientist named Georges Cuvier devised a systematic view of the animal kingdom and it also was based on homologies. Differences between the groups of animals, according to Cuvier, were: "... merely slight modifications, founded on the development or addition of certain parts, which produce no essential change in the plan itself" (Eisley, 1958, p. 87). Cuvier's zoological taxonomy "... was the greatest advance in classification since Linnaeus, and it has formed the basis of all subsequent animal classification" (Taylor, 1963, p. 138). Creation scientists did the foundational work for describing and utilizing homology.

Like Linnaeus, Cuvier presented opinions on taxonomy and homology, which were formulated on the belief that the Creator had worked with a limited number of underlying archetypes—a helpful scientific concept that was later vilified by Charles Darwin. According to Linnaeus and Cuvier, the Creator modified each archetype slightly when producing the various categories within a biological group.

Cuvier Related Homology to Creation.

Cuvier's knowledge of taxonomy and his monumental contributions to zoology were predicated on his firm belief in creation. The following remark of Cuvier's shows that he scorned materialistic philosophies and agnostic theories of science: "... but we also see how puerile are the philosophers who have given nature a kind of individual existence, distinct from the creator, from the laws which he has imposed upon motion, and from the properties or forms which he has given to the creatures" (Coleman, 1964, p. 26). Dobzhansky (1955, pp. 227-228) was a dogmatic evolutionist who nonetheless acknowledged that Cuvier's useful classification involving archetypes was founded on an understanding of homology: "Cuvier's four types of animal structure were also based on his ability to perceive the homologies of the corresponding parts in different animals"

Cuvier Opposed Evolution.

Cuvier opposed the idea that creatures of one kind changed gradually into those of a different kind. Thus Cuvier's anti-evolutionary stance was a scientific decision. It is still quite possible to present creation in a completely scientific format, as Cuvier did, and thereby squelch the oft-repeated Darwinite sophism that "creation is religion, while evolution is science"

The anti-evolutionist Richard Owen actually coined the word "homology." Owen was a scientist in Darwin's day, who described the biological differences between humans and apes quite vigorously, and who opposed the idea that they were related by evolution. He denounced Darwin's book, *The Origin of Species,* which shall be abbreviated hereafter as *The Origin*. Owen said that Darwin, in writing *The Origin*, had left the question of the origin of the species very nearly where he (Darwin) had found it. Incidentally, this same non-evolutionary scientist, Owen, actually coined the word "homology" in its original, pre-Darwinian sense. Owen, Cuvier, Linnaeus, and others before Darwin were

56

amassing scientific support for what was already written Sacred Scripture: each living form reproduces "after its kind"

Owen's ideological stance proved that not all creationists are Bible-believing Christians. Owen did not take the Genesis creation account literally, rejecting what he called "literal scripturalism" Owen, Louis Agassiz, and several other famous biological scientists make up a significant group of top-level scientists who believe in creation, but are not otherwise very "religious" or "Biblical" in their approach to the topic. While I think these individuals are wrong on that account, their situation nicely refutes the deprecatory assertion that "It has been only Bible believers who adhere to scientific creationism"—a statement that is just not true!

Homology: A "Cow" Rustled, Redefined and Sanctified

Darwin redefined homology and castigated creation at the same time. In sections of his book, *The Origin*, he discussed homology and attached to it his own religious idea that it must have come into being entirely by natural selection, without direct action on the part of the Creator. This "rustling" (which, of course, meant the "capturing" and "stealing" of cattle in "the early West") and redefining of homology can be seem in the following quotation from a section of *The Origin* in which Darwin used the term "analogies" for what we now call "homologies."

In all these respects the species of large genera present a strong analogy with varieties. And we can clearly understand these **analogies** [now called homologies], if species once existed as varieties, and thus originated; whereas these analogies are utterly inexplicable if species are independent creations. [Bold face type here and in other quotations has been added by me.] (Darwin, 1859, p. 75).

In these confusing sentences, Darwin was simply proclaiming that homologies between different species are explicable only if those creatures were related by evolution. He even asserted that such similarities are "utterly inexplicable" within the creation view. This is a peculiar remark because homologies had been first observed, discussed, named, and even explained by creationists like Linnaeus, Cuvier, and Owen. Homology was in fact quite **explicable** in the scientific creation view. Darwin expanded the earlier definition of homology (the simple fact of similarity) and asserted, in paraphrased form, that: "homology is always an indicator of evolutionary descent and cannot be explained by creation"

Darwin even Denied the Usefulness of Front Limb Bones.

Within *The Origin* made the preposterous claim that the various bones in the arms, forelegs, wings, or flippers of different mammals are of no specific "use" to each animal but are simply leftovers indicating evolutionary descent from the same ancestry:

> ... we cannot believe that the similar bones in the arm of the monkey, in the fore-leg of the horse, in the wing of the bat, and in the flipper of the seal, are of special use to these animals. We may safely attribute these structures to inheritance (Darwin, 1859, p. 184).

According to Darwin these bones are merely leftovers, not adapted for any particular advantage in each of the animals, persisting simply because an ancient ancestor possessed them. Many comparative anatomists would shudder at the arrogant ignorance of these remarks because scientists have shown repeatedly that any particular mammal bone, like the ulna, has a specific design in each creature (monkey, horse, bat, or seal) to play unique roles peculiar to that animal's activities. Many of the features of the human and animal bodies originally assumed to be useless vestiges of evolutionary inheritance can be show to play unique functions (Bergman and Howe, 1990).

Darwin Likewise Maintained that "Serial Homologies" in Certain Animals (like Crustaceans) are "Inexplicable" in the Creation View.

In this way he closed the door to scientific investigations of their possible functions. Insects and certain other invertebrate creatures are clearly "segmented," having a series of many well-marked body sections. In *The Origin* Darwin made the claim that these "serial homologies" cannot be explained in the creation view, implying that they too were solely the result of evolutionary heredity, their constant number has no purpose whatsoever:

> How inexplicable are the cases of serial homologies on the ordinary view of creation! ... Why should one crustacean, which has an extremely complex mouth, formed of many complex parts, consequently always have fewer legs; or conversely, those with many legs have simpler mouths? Why should the sepals, petals, stamens, and pistils in each flower, though fitted for such distinct purposes, be all constructed on the same pattern? (Darwin, p. 416)

Although it would take too much space for this chapter, any creation-minded invertebrate biologist could quickly relate examples showing that it is indeed helpful and efficient for there to be a constant number of segments. "Design economy" on the basis of underlying genetic plans is one general rationale for the Creator's keeping the total number of legs plus mouth parts consistent in crustaceans.

Unlike Darwinism, the Creation View Fosters Research.

If there is a similar underlying pattern in a series of different flower parts like sepals and petals, this is no automatic support for evolution. The creation view of mysteries like these is scientifically productive, forcing workers to ask questions like "what function might there be in keeping a constant number of crustacean segments?" Or "what advantage is it that the underlying structure of the sepals and petals in flowers be similar?" Creative questions are ignored when the issues are simply dismissed as instances of useless features proving descent from common ancestry. In this way the evolution model stifles intelligent inquiry regarding possible roles of biological phenomena. Bergman and I (1990) found that the organs in human and animal bodies that had been written-off by evolutionists as useless in fact had one or more useful roles to play.

Darwin Repeatedly Insisted that Any Mention of the Creator in *Origins* is both Hopeless and Unscientific.

In *The Origin*, Darwin declared that it was quite "hopeless" to believe that similarities can be explained as the Creator's intelligent action:

Nothing can be more hopeless than to attempt to explain this similarity of pattern in members of the same class, by utility or by the doctrine of final causes … On the ordinary view of the independent creation of each being, we can only say that so it is—that it has pleased the Creator to construct all animals and plants in each great class on a uniform plan; but this is not a scientific explanation (Darwin, 1859, p. 414).

Darwinism leads to a "dead end" without any room to attempt to find a reason for the common shapes and functions of body parts.

The Cause: Darwin's Religious Conversion

A religious change was at the root of Darwin's downgrading of creation. Darwin's disrespectful statements about the creation view resulted from his own religious disbelief in a "hands-on" Creator. He was promoting a "religion" of his own and trying to present it as "science," while stating that creationist opinions about the same data were not "scientific" The tactic of passing off one's own religion as "science" and calling other origins theories "nonscientific," is a technique still used by some Darwinite academicians. They practice a shameful "bait and switch" routine in their textbooks and lectures—"Promise them science, but then deliver them a different religion in scientific clothing!"

Darwin Abandoned Biblical Christianity for Deism.

Darwin was writing these statements in *The Origin* at a time of life when he had already rejected his earlier Christian faith. Some of his remarks were skilled apologetic statements in support of his new religion, which was a type of deism wherein "god" worked only at the very onset of creation, leaving the rest of origins to natural selection, governed only by chance.

It Was in College that Darwin's Religious DemiseBegan.

This and other facts about the religious abdications of Darwin can be studied on-line by entering some phrase like "Charles Darwin's Religion" Deviations from faith began for Darwin when he was still in college at Cambridge. He told a fellow college student there that he [Darwin] could not accept certain teachings of the Church of England and therefore could not become a clergyman. Yet during his college years, and even later (while he was on his famous post-college oceanic voyage), Darwin still expressed belief in the creation account of Genesis.

After Changing His Religion, Darwin Wasn't even a "Theistic" Evolutionist Anymore.

As years went by, Charles Darwin gradually extracted himself from belief in the Christian gospel, the Genesis creation, and biblical faith. He stated that his own disbelief came upon him so slowly that he felt no pain during the transition. He finally asserted that the God of the Bible is no more to be worshipped than any of the various deities in other religions. The suggestion has been made that a driving force in Darwin's decline of faith was the death of his daughter. Although he wanted people to call him a "theist," meaning one who believes in a personal and active God, Darwin had actually converted into deism, professing that any direct actions of deity in creation occurred only at the start. Darwin was thus not even what we now call a "**theistic** evolutionist" Although he was heralded at his death as a "Christian" man of science, Darwin probably died outside the faith of Christ (Rusch, 1975).

But, Did Darwin Finally Become a Christian?

In some reports it has been claimed that Lady Hope, a woman of Northfield, England was Darwin's final caregiver. According to these sources, Darwin experienced a spiritual revival near the end of his life and likewise rejected his own theory of evolution (Turner, 1979, p. 3; Smith, n.d.; and Rusch, 1975). After examining all the evidence he could find, however, Rusch (1975) decided that the accounts of Darwin's near-death revival and return to Genesis have not been adequately supported. Rusch and Klotz drew this same conclusion in their later book (1988).

The Overall Nature of the Book *The Origin*

Darwin displayed religious motives while writing *The Origin of Species*. Darwin did not actually state what he was attempting to accomplish by railing against creation, creationists, and the Creator in *The Origin*. At several places, however, he remarked that any reader of *The Origin* who had got that far and still doubted evolution, might as well "close this book" Sectarian outbursts like this are extremely rare in science books and demonstrate an uncontrolled religious fervor on Darwin's part. Apparently his desire to promote his own naturalistic conclusions was so desperate that he ignored logic and good manners when writing *The Origin*.

Yet, there is much "science" in *The Origin*, particularly facts about animal breeding and about changes that occur in many different varieties and species of living organisms. But, this same idea of change occurring in living groups of organisms had been acknowledged by others, from the time of Linnaeus right down to Darwin's day. Perhaps this lack of scientific novelty on Darwin's part is why Owen maintained that Darwin had left all questions about origins very nearly where he (Darwin) had found them.

In one section of *The Origin* Darwin discussed the ability of seeds to remain viable after soaking many days in ocean water (1859, pp. 353-360). This work of Darwin's played an important part in my own research on how seeds of many species may have survived Noah's Flood (Howe, 1968, pp. 110-111). Rather than being pure science, however, the science in Darwin's book was usually focused on trying to prove things that cannot be demonstrated by science alone, e.g. how life arose, how it developed across assumed eons of time, and how all species have arisen by natural selection from common ancestry.

The Origin accordingly has the tone of a religious "tract" favoring deism. A friend of mine has aptly classified this type of slanted Darwinian discussion as "imaginary science" Darwin wrote several scientific books analyzing invertebrate animals and insectivorous plants, books replete with real scientific information. But *The Origin* is a book in which the facts Darwin selected were chosen from among many other facts not included and were aligned to "convert" the reader to a deistic philosophy of origins.

It was his personal religious beliefs and not science that prompted Darwin to describe creation science as "hopeless" Darwin thought that to believe the Creator formed animal features for "utility" (usefulness) was a "hopeless" idea. Like his Darwinite successors who still criticize the design model, Darwin supplied no explanation concerning why this "utility" or "design" view is

61

"hopeless" By asserting its hopelessness, however, Charles may have been attempting to divert attention away from his own rather hopeless task of showing that little changes, under the influence of natural selection, across eons of time, produced all the different useful body organs by chance alone from common ancestry. The problems facing Darwin's deistic religion become quite obvious if a creation theory is simultaneously presented as a viable alternative. But neither Darwin nor many of his disciples have been interested in openness or in the fair presentation of alternative origins theories. By dismissing the whole creation argument as "hopeless" or "religious," their task of defending evolutionism becomes easier. In this manner, Darwin tried to discredit Divine design, without disproving it. Advocates of the "evolution only" approach to origins currently use these identical diversionary tactics in their textbooks, class lectures, and research papers. The design model in origins, however, is still alive and well, thanks to God and to the good work He is doing through creation scientists, Intelligent Design advocates, and other critics of the general evolution theory (Bergman and Wirth, 2008).

But if the Facts Fit Design, Why Not Accept It?

Concerning Darwin's remark that creation is "not a scientific explanation," it must be recalled that "science" is a search for truth in technical fields, a search involving the construction of better theories to explain what is observable. If a biological fact, like homology, actually fits best with the theory that an astute Designer used a sagacious plan by which to produce body organs for a whole class of organisms, we ought to adopt or at least openly discuss the design option. Close agreement between scientific data and the idea that a brilliant Planner was responsible for origins **is** both "scientific" and "hopeful" If "it pleased the Creator to construct all animals and plants in each great class on a uniform plan," as Darwin sarcastically wrote in *The Origin*, then, hats off to the Creator, Charles, because this concept has good fit with the biological facts! God left clear clues fostering a creationist origins model. In creating homologous bones, the Creator was possibly using an intelligent engineering principle called "economy of design" By pursuing this concept, biologists face a remarkably "hopeful" task of trying to understand God's plans and purposes in each and every minute detail of life.

What Analogy Really Is and How It Is Explained

Evolutionists explain homology by assuming that the speculative process called "divergent evolution" was the cause. They mean that a branching descent from an imaginary common ancestor took place, and they think that homology demonstrates divergence. Evolutionists believe that similarity (homology) always means kinship by a common ancestry. Creationists, on the other hand, think that similarity resulted because there was one "common" Originator who was responsible.

Analogy Needs to Be Defined and Illustrated too

People who hold these beliefs about homology face additional problems because other parallelisms exist between creatures that are unrelated and are in vastly different taxonomic groups. For example, both bats and butterflies have wings to fly. The bat wing, however, is made of skin stretched over the bat's elongated finger bones, while the insect wing is filmy, boneless, and covered with tiny scales. These two wings are said to be **analogous** (not homologous) in that while they both promote the same function (flight), each has an extremely different composition. Furthermore, the bat and the butterfly belong in two very different phyla—Chordata and Arthropoda respectively. Distefano (2004) stated that analogous structures like these "… **are made** of different materials and originate from different evolutionary paths" (p. 155). In this quotation his statement, I have put his phrase **"are made"** in bold face. Biological systems do look exactly like they would if some intelligent being had actually **"made"** them. Concerning analogous organs, Alders and Alders (2006) said that: "these structures do not have the same embryological origin and do not share the same underlying anatomy" (p. 244).

Here Are Some Other Examples of Analogy.

The marsupial mammals, called Marsupialia, carry their young in pouches, after being born early. By way of contrast, the young of the placental mammals (Placentalia) are attached to the ovary wall by a placenta during the full term of pregnancy. Finally, some other mammals, known as the Monotremata, actually lay eggs, which later hatch. These are all "mammals," of course, in that each nourishes its young with milk. According to Darwinian speculation, these three mammal branches (Monotremata, Marsupialia, and Placentalia) first diverged from each other eons ago. But surprisingly, within each of the three resulting groups, there is an "anteater"! The egg-laying anteater, however, is more closely allied to an egg-laying duck-billed platypus than it is to either the pouched or the placental anteaters. And the marsupial anteater has nearer affinity to a kangaroo (which, of course, also bears its young in a pouch) than it does to either the placental or the egg-laying anteaters.

The Woodchuck and the Wombat Are Analogous.

One placental mammal found in eastern North America is the woodchuck. But in the marsupial group is another animal, a wombat, which has the same lifestyle as the placental woodchuck, from which it is quite widely separated in taxonomy! The wombat, in fact, has a closer tie to the koala than it does to the placental woodchuck, even though it behaves like the woodchuck! Examples of distant ecological parallelisms like these are widespread in the animal world, and each is a clear instance of "analogy"

Squirrels and Phalangers Also Illustrate Analogous Similarity.

In addition to the wombat/woodchuck analogy or the three different anteaters, several other mammal lifestyles are represented in both of the widely separated marsupial and placental groups. The various placental squirrel species have their striking counterparts in the marsupial phalangers, which live much like squirrels. There are even "flying phalangers," parallel to the "flying squirrels." But in taxonomy, a phalanger is closer to a kangaroo than it is to a squirrel! The placental wolf has a carnivorous marsupial parallel too—the Tasmanian devil. One must ask how fortuitous evolution could have ever produced these and many other amazing counterparts from different groups. It is not enough to say that: "… a niche existed and therefore diverse groups converged independently to fill the niche"

Examples of Analogy and Homology Are Numerous in Botany Too.

My own experience is deepest in the plant world, where one can likewise find many analogies. Some of my papers and those of Armitage, on the creation of plants such as lichens, saguaros, and yuccas are available at www.evolutionflunksbotany.org. Numerous growth stages and morphological structures are similar between widely different groups of bacteria (Howe, 1965, p. 14). There are analogous parallelisms found between certain algae and very different algae of other groups. Analogies occur between some algae and widely separated fungi. Parallelisms exist between life cycle patterns of numerous plants that are otherwise very distant to each other taxonomically (Howe, 1965, p. 15). Quite different plants each contain analogous features such as woody tissue, secondary bark, anatomy of conducting systems, guard cells, leaves, cones, appearance of stems, and more (Howe, 1965, pp. 15ff). A bibliography of older sources covering analogy can be secured from Howe (1965), and Berg (1926). Homologies discovered in the study of biochemistry, DNA, and homeobox genes (Hox genes) are examined in Howe (1999, and 2000).

There Is a "True/False" Puzzle That Analogies Pose for Evolutionists.

To the Darwinite, homology is a "true" indicator of close divergent relationship but analogy must then be seen as a "false" index, not proving any close ancestral tie at all. An evolution-minded analyst thus faces the subjective job of deciding which similarities are **true** indicators of common ancestry (homologies) and which ones are false (analogies). Anderson (1998) clearly noted this necessity to distinguish between the two classes in the following quotation where he called analogy "convergent similarity" or "homoplasty": "Convergent similarity (homoplasty) [actually analogy] is common and the ability to distinguish this false homology is an absolute necessity" (p. 423). Mayr (1953, p. 42) succinctly summarized evolutionary thought about analogy versus homology by writing that:

The first step then toward the achievement of a phylogenetic classification is an analysis of the taxonomic characters to determine which of them are derived from common ancestors (homologies) and which are **spurious similarities** (analogies), usually convergent adaptations correlated with similar habits.

In the evolution view it is therefore essential to carry out a "true/false exam" in order to discern which resemblances are homologies (valid indicators of common ancestry) and which ones are misleading superficial similarities derived from parallel evolution in vastly different creatures (analogy). Making such decisions presents a daunting task to the Darwinites.

Being able to distinguish between homology and analogy, however, is quite essential because evolutionists have proposed a different evolutionary history for each of them. Evolutionists explain homology as indicative of divergent evolution from a common ancestry. To explain analogy, however, they gratuitously propose that the vastly separate lines resulting from the divergence subsequently underwent "convergent evolution" to fill available ecological niches independently! They never define this "convergent evolution" nor do they supply scientific support for its occurrence; they simply introduce the term as an unsubstantiated "explanation" Convergent evolution is merely a convenient invention of the Darwinians to account for the origin of analogy, another example of imaginary science.

How do Darwinites Say It Happened?

According to evolutionism, several classes of animals such as mammals, birds, and reptiles each first diverged from one common ancestry. Then by means of convergent evolution, bats among the mammals, birds of flight among the Aves, and pterosaurs among the reptilia each produced flight systems independently by a parallel but quite separate evolution. In this very "fortunate" manner, mammals, birds, and reptiles each supposedly gained access to the realm of flight.

It is highly improbable, however, that so-called analogous features would ever arise in groups that had undergone a wide separation. Leo Berg (1926, p. 158) commented on the extreme unlikelihood of the Darwinian view of parallel convergent evolution:

This explanation seems quite improbable. Since every useful variation according to Darwin's theory arises by chance, it is scarcely credible that such a variation should arise accidentally even in one species; but still more incredible would beits occurrence in different species.

For more information on the unlikelihood of convergent evolution, consult the works of Denton (1985) and of Shute (1961).

By way of contrast, scientific creationists propose a simple and coherent explanation for both homology and analogy. They maintain that homologous structures were originated when the Creator used the same basic archetype (with appropriate design modifications) to fulfill various functions in different animals.

Conversely, analogous structures were produced in animals of quite different archetypes (Howe, 1999, p. 4). To a creationist, the presence of diverse analogous organs having a similar function (e.g. wing of bird versus wing of insect) is a tribute to the Creator's versatility in having worked with very different biological "starting materials" to produce parallel systems. Creationists hold that homology and analogy are each the result of the same Designer, a "common Designer," Who used different engineering techniques.

I developed a biology laboratory exercise (Howe, 1972), which was presented at the National Association of Biology Teachers conference in San Francisco. In it I showed that the idea "homology indicates common design" has a remarkable correspondence in the realm of human manufacturing. Parking garages, bowling alleys, and college classrooms all have light switches on the walls and steel I-beams in their structure. Not for one second, however, does anyone imagine that the light switches and I-beams ("homologies") seen in these three types of buildings indicate their derivation from a "common ancestry" Nor does anyone think that all the buildings "converged" to each produce the switches and beams independently! We know instead that such similarities in diverse structures all derive from the common designer—in this case *Homo sapiens* and not from a common ancestry. Although this illustration is derived from the non-living world of human construction, it shows that the creation view of origins can satisfactorily explain the similarities seen among God's living organisms—they were all created by a single, common Deity. All of this is true even though many workers interpret all these scientific data to support evolutionism. Perhaps this is an instance in which a strong dilusion has come among them "… that they should believe a lie" (2 Thessalonians 2:11, KJV).

The principle of parsimony (Occam's razor) favors the creationist explanation of homology and analogy. The philosopher Bishop Occam once wisely proposed that wherever there are two or more explanations for the same phenomenon, the simplest and most coherent explanation ought to be accepted—a useful philosophical principle now called "Occam's Razor" The creationist theory is simpler and more coherent than the evolutionary scenario in various ways. Likewise, it is free of the Darwinian problems and unsupported assumptions. "Occam's Razor"

The Current Definition of Homology and Analogy

The word homology is used in several other fields than biology. The biological term "homology", discussed in this chapter, appears only rarely in general discourse or ordinary literature. One can discover present attitudes about this word, however, by consulting dictionaries, textbooks, and other sources. In their definitions, are these repositories distinguishing between the fact of similarity and the theory of common ancestry? Or are they still promoting homology as a sacred cow proving nothing more or less than divergent evolution from common ancestry?

Dictionary definitions are instructive as to current usage of the "homology" concept in biology. Here are three dictionary entries in which the words discussing the evolution theory tied to homology are put into bold face: homology: "Likeness in structure between parts of different organisms (as the wing of a bat and the human arm) due to evolutionary differentiation from a corresponding part in a common ancestor" (Merriam Webster's Collegiate Dictionary, 2004, p. 596). Homology: "likeness short of identity in a structure or function between parts of different organisms due to evolutionary differentiation from the same or a corresponding part of a remote ancestor …" (Webster's Third Unabridged Dictionary, 1971, p. 1085). These two show that dictionaries typically slant the definition of homology into an evolutionary pattern, paying homage to the sacred cow. The next dictionary definition was an exception in a small but significant way: homology: "The fundamental similarity of a particular structure in different organisms, which is assumed to be due to the descent from a common ancestor" [Note how this author was pointing out that the evolutionary connection between homology and common ancestry is "assumed!" May his tribe increase!] (Allaby, 1999, p. 255). Would that most evolutionists would recognize, as Allaby did, the evolutionary overtones of homology are really **an assumption!** This same attitude should be found in all evolutionary presentations.

Next, the following are a few definitions of homology gleaned from science textbooks found in nearby local libraries:

Homology and Analogy: "... overall similarity does not distinguish between features that were acquired from a common ancestry (homologous features) from those that arose convergently (analogous features)" (Anderson, 1998, p. 423).

Homology: "... similarity due to common ancestry is called homology" [Then next, Campbell and Reece had this to say about the front limbs of human, cat, whale, and bat] "If these limbs had completely separate origins, we would expect that their basic design would be very different. However, structural similarity would not be surprising if all mammals descended from a common ancestor with a prototype limb" (p. 256). [Note how these authors skillfully fostered the "evolution-only" technique (Campbell and Reece, 2001, p. 256).

Homology: "How similar the structure of two organisms are, can be used to show a common ancestor...Homologous structures have the same evolution and origins but not necessarily the same function" (Distefano, 2004, p. 155). Here is another: "Homologous structures are similar in construction and evolutionary development but dissimilar in function" (p. 331). [Concerning bones in arms, he wrote that:] "...all mammals at one time changed and diverged from a common ancestor"). "... Evolutionary relationships based on comparative anatomy depend on homologous structures" (Rechtman, 2004, p. 280).

It will not be necessary to list them, but several definitions of biological analogy were also found in these same textbooks and dictionaries, and each linked the analogy concept to convergent evolution. The same is true of most on-line definitions, which show that biological homology is still consistently being treated as Darwin's sacred cow. To these writers, analogy always indicates convergent evolutionary descent from common ancestry with no alternatives.

What Can Be Done?

This final question is addressed not only to creationists and intelligent design advocates, but likewise to all evolutionists who desire to distinguish between the true facts of homology and the evolutionary speculations about its origin. All fair-minded evolutionist scholars can and should discriminate wisely between the facts they use and the philosophical origins conclusions they draw. These same wise-hearted evolutionists ought to also defend creation scientists who face job losses for merely expressing views like those presented in this chapter! There is much that can be done by people of all persuasions to liberate homology from its "blessed bovine" status and to bring common sense back to job tenure in the biological science.

Many different forms of communication can be used to promote the truth about biological similarities. Textbook chapters, scientific papers, popular articles, public lectures, letters-to-the-editor, on-line entries, and private conversations can all become venues by which to remove homology and all of evolution theory from the pedestal of "cow worship" Hopefully many balanced treatises and communiqués will be initiated, contacts by which minds of both the young and the old will be brought to see this subject and other evolutionary themes in a much broader two-sided perspective. Films like the classic documentary *Expelled, no intelligence allowed,* produced by the very gifted and insightful Ben Stein, will go a great distance in showing the vast unconformity between real science and origins theories. Bergman and Wirth (2008) have shown that the creation movement and its Intelligent Design counterpart has had great success, even though they have had their freedom abridged and their jobs removed because of their failure to worship the various evolutionary sacred cows.

Together with two other scientists, I wrote a balanced textbook of biological sciences (Lester, Englin, and Howe, 2002, p. 13.6) in which we discussed both the creation and the evolution origins alternatives for homology and analogy. We likewise attempted to accomplish this same bi-partisan treatment for all origins topics. More work is necessary to make textbooks like this one available in both public and private schools, books in which facts are separated from interpretations and where creationist alternatives are included. Let's not allow strident activists to suppress creation science in public education by the false claim that it is "too religious to be treated in the public classroom" Help people see that all discussions of what might have happened in the past (before "science" was practiced) have a religious substructure.

One complete and permanent solution to the homology problem would be to change the vocabulary. Both the words homology and analogy have been so thoroughly blended with evolutionary speculation that abolishing them would be a boon to biology. One word "similarity," with modifying adjectives, could quite nicely serve to replace them both because homology and analogy each designate various types of similarity. Homologies would then be spoken of as "underlying" or "structural similarities" The question of how such structural similarities arose would be an entirely different matter. An analogy like that of the bat wing versus an insect wing could be called a "functional similarity," showing that the function of flight is controlled by wings that differ widely. Or the wings could be classed as a "distant similarity" indicating that the bat and the insect are from distinctly different phyla. Analogies like those seen between the wombat and the woodchuck (or between phalangers and squirrels) could be named "lifestyle similarities" or "ecological similarities" These and other changes in terminology should be enacted to help demolish the worship of homology as a sacred cow.

A Postscript

If the creationists and the intelligent design advocates are correct, and we think we are, then the name "God" or "Designer" ought to be directly used in biological research. To eliminate His Name is a form of negligence, bordering on plagiarism. Our deistic and atheistic evolutionary friends are guilty of this omission, to be sure, but we should not fall into that same malpractice. Making mention now and then of the Creator while studying living forms that He has made, is an intelligent act of "true science" Giving "credit" in biology to the Creator is a scientific way of giving Him "praise" for the work He has so wisely accomplished.

References:

Anderson, D. T. *Invertebrate Zoology.* Oxford University Press, Oxford, NY, 1998.

Allaby M. *Dictionary of Zoology.* Oxford University Press. Oxford, NY, 1999.

Alters, S. A. and B. A. Alters. *Biology: Understanding Life.* John Wiley and Sons. Hoboken, NJ, 2006.

Berg. L. S. *Nomogenesis or Evolution Determined by Law.* M. I. T. Press. Cambridge, MA, 1926.

Bergman, J. and G. F. Howe. *"Vestigial Organs" Are Fully Functional.* Creation Research Society Books. Chino Valley, AZ, 1990.

Bergman, J. and K. Wirth. *The Slaughter of the Dissidents.* Leafcutter Press. Southworth, WA, 2008.

Campbell, N. A. and J. B. Reece. *Essential Biology.* Addison Wesley Longman. San Francisco, CA, 2001.

Cattle in Hinduism. http://en. Wikipedia.org/wiki/Cattle_in_religion, 2009.

Coleman, W. *Georges Cuvier Zoologist.* Harvard University Press, Cambridge, MA, 1964.

Darwin, C. *The Origin of Species.* Dent: Dutton. New York, NY, 1859. (Reprinted from earlier edition. Foreword by W. R. Thompson, noted entomologist who was quite critical of evolution theory. Thompson's foreward is worth the whole price of the book.)

Denton, M. *Evolution a Theory in Crisis.* Adler and Adler. Bethesda, MD, 1985.

Distefano, M. *Biology Homework Helpers.* Career Press, Franklin Lakes, NJ, 2004.

Dobzhansky, T. *Evolution, Genetics, and Man.* John Wiley, New York, NY, 1955.

Eisley, L. *Darwin's Century.* Doubleday. Garden City, NJ, 1958.

Greene, J. *The Death of Adam.* The New American Library. New York, NY, 1959.

Lester, L. P., D. L. Englin, and G. F. Howe. *Designs in the Living World.* SymBioSys Publishing, lane.lester@gmail.com, 2002.

Howe, G. F. Homology, analogy, and creative components in plants. *Creation Research Society Quarterly* 2:11-21, 1965.

Howe, G. F. Seed germination, sea water, and plant survival in the Great Flood. *Creation Research Society* Quarterly 6: 105-112, 1968.

Howe, G. F. 1972. Homology, analogy, and innovative teaching. Unpublished manuscript. georgefhowe@sbcglobal.net, 1972.

Howe, G. F. Homology and origins. *Creation Matters* 4:1-5, 1999.

Howe, G. F. The origin of flowering plants. *Creation Research Society Quarterly* 37:66-67, 2000.

Martin, G. 2009. Wikipedia: Sacred cow. http://www.phrases.org.uk/meanings/309- 250.html, 2009.

Mayr, E. *Methods and Principles of Systematic Zoology.* McGraw Hill. New York, NY, 1953.

Merriam Webster's Collegiate Edition, Eleventh Edition. Merriam Webster. Springfield, MA, 2004.

Rechtman, M. *Biology.* Cliffs Study Solver. Wiley. Hoboken, NJ, 2004.

Rusch, W. H. Sr. Darwin's last hours. *Creation Research Society Quarterly* 12(2):22-102, 1975.

Rusch, W. H. Sr. and J. W. Klotz. *Did Darwin Become a Christian?* Creation Research Society Books. Chino Valley, AZ, 1988.

Schute, E. *Flaws in the Theory of Evolution.* The Temside Press. London, UK, 1961.

Smith, N. E. Personal Correspondence, 2009.

Smith, O. J. *Darwin's Confession.* Peoples Church, Toronto, Canada, (no date).

Taylor, G. *The Science of Life.* McGraw Hill. New York, NY, 1963.

Turner, C. E. A. Darwin's last hours. *The Evolution Protest Movement* 2:3, 1979.

Walvoord, J. F. and R. B. Zuck. *Dallas Theological Seminary: The Bible Knowledge Commentary.* Victor Books. Wheaton, IL, 1983.

Webster's Unabridged Dictionary. G. and C. Merriam. Springfield, MA, 1971.

Author bio

George Howe earned the Ph.D. degree in botany (plant physiology) from the Ohio State University, studying the time-course of photosynthesis in leaves of various flowering plants. He has taught many different science courses in two Christian colleges where he was a professor of biology for over forty-two years.

He has been a member of the Creation Research Society (CRS) since its inception and a member of its board for 41 years, holding such CRS Board offices as president, vice president, and secretary. For five years he served as editor of the *CRS Quarterly* and four years as its managing editor.

He has published numerous technical papers on botany in the *CRS Quarterly*, the *Journal of the Southern California Academy of Sciences*, the *Journal of the Ohio Academy of Sciences*, and *Crossosoma*. Several books and papers published by Dr. Howe are found on his web site www.evolutionflunksbotany.org. He continues research on plants such as the unicorn plant, also known as devil's claw. He is a scientific creationist who believes that general macro-evolutionism will someday lose its "luster" when workers come to see that it has little predictive value and only very meager support in the sciences.

Micro-Evolution Is <u>Not</u> Mini-Evolution
Charles Jackson, Ph.D.

Editor's Note

One of the more persistent sacred cows is that minor genetic changes over time can account for major evolutionary changes. As you will see in this chapter, that idea is simply not true. Something more is needed than minor genetic change even with vast periods of time. We have learned a great deal about genetics during the past two decades and this assumption is not supported by observation nor does genetics provide a mechanism for such major to occur no matter how much time passes. Evolution is without a plausible mechanism. As clearly stated in Genesis the originally created "kinds reproduce after their own kind" This essay addresses this important concept in depth.

Abstract

There is much confusion among Christians about the proclaimed differences and similarities of the concepts of "microevolution" and "macroevolution." It is said that one is acceptable within the framework of Biblical objectivity and the other is not. The truth is neither of them (as so stated/worded) is compatible with a Biblical worldview.

Introduction

Most Bible-believers get confused when it comes to this "micro-evolution" versus "macro-evolution" thing. The *very worst* obstacle to understanding the meaning of these words is the very *words themselves*. They are misleading—at least the first one is. They are misleading because they are misused. That is the plain and simple truth of the matter. I will explain.

Evolutionists are fond of citing things that can happen *without* evolution being true, and then citing those *very* same things as "proofs" of evolution. This is self-deceptive at best, and outright deliberate lying at worst. They'll say that since you look different from your grandparents, that's *proof of evolution*. And who can deny there are over 200 breeds of dogs? *That's evolution*, they'll say. And people are taller now than they were a century ago. Aha! *Evolution!* If these *observable things* … were *all* they were asking us to believe and that's *all there is* to "evolution," *then I'd believe* in evolution right now!

But that's *definitely not* what they mean when they say the word "evolution" Theory *doesn't really* <u>only</u> say that humans have changed or that Collies and Dachshunds are cousins. It says that worms became *bald eagles*, that *fish* became dinosaurs, and that dinosaurs became chickens, and of course, that people were once *monkeys!* The slick lie in everything they say about "micro-evolution" is this: *"if you just stretch this thing out over hundreds of millions of years, then you can easily see how this makes all the rest of the theory of evolution true, too"* No. It doesn't.

The processes that made you look different from your grandparents, that made the different breeds of dogs, that made people today taller, honestly have nothing to do at all with *any* of the processes that would be needed for salamanders to turn into people (which, yes, *is* a part of their theory)! We could pretend that these processes have gone on for millions, billions, or *even trillions* of years of time and we'd *still never* get any kind of an upward movement of any species into the "evolution" of, or the "origin" of ... any new, more complex, and better-functioning species.

The Truth: Micro-Evolution Should Be Called Micro-Variation.

The reason that even billions of years of "micro-evolution" cannot achieve the claims of evolutionary theory (namely, "the origins of the species"), is that "micro-evolution" *is **not*** "evolution," no matter how much *anybody* tries to talk like it is. Micro-evolution really ought to be called "micro-*variation*." That's more like what is *really* happening here. And the word is more honest and not as misleading from the *observable* truth.

Remember that genes are made up of DNA and that they contain all the information for the growth and development of your particular human body, physically. Every generation that survives long enough to reproduce and to make a new generation, must give their DNA to their offspring before they die. You get half your DNA from genes in the chromosomes of your father. You get the other half from the chromosomes of your mother. Since your parents each have two versions of every gene they've got, it becomes a 50:50 chance which of the two genes you'll get of Mom's and which of the two you'll get of Dad's. After that, then you will also be born with two versions of every gene – one you got from your mom and one you got from your dad. Later then, when you have children of your own, your child will get one of your two versions of each gene, and one of your spouse's two versions ... which explain why your children will look a lot like a combination of both you and your spouse. We may say you look somewhat like your mother's side of the family, or maybe somewhat more like your father's side. So, sometimes we see that chance plays out more heavily on one side than the other. I'm sure you can think of examples of this amongst the people that you know and in your own family.

Try to think of the genes in all of your chromosomes like the cards in a full deck of playing cards. All of this mix n' match going on with every new generation, results in a sort of a "shuffling the deck" of genes between generations each time before a new "hand" of genes is dealt out to the next "round of players" (the offspring). This every-generational re-shuffling is called "sexual recombination" It is a mechanism for the species' defense against the dangers of inbreeding from the damages that mutations that have built up in the human genome over the past six thousand years. The re-shuffling explains why we look a little like our grandparents, but usually more like our parents. Families change a little with each new generation. But *this* is not evolution—not in the way the word is used by our anti-Biblical and atheistic colleagues in the evolutionary community!

Yes, we change. But, how is this accomplished? By shuffling and re-shuffling the *already-existing* genes in the human population. For true **evolution** to do its work of taking bacteria "three billion years ago" and working with them and then making them into everything from giraffes to dinosaurs and people—you must have some kind of a mechanism in place that is capable of creating *brand new* genes! You can't get that from just re-shuffling the genes that are *already there!* Bacteria *do not* have the genes for a human brain. Lizards *do not* have the genes to grow bird feathers. Fish *do not* have the genes for horse hooves. But each of these examples that I have given above, *is* a part of the evolutionary "story." *This* whole thing is what they are talking about when they use the innocent-sounding word "evolution."

Don't believe it when they tell you that evolution is the same process that got us poodles and boxers from wild dogs. Don't you believe it when they tell you that "evolution" is the same process that you got a different look from your grandparents. It is most definitely *not* the same thing. The processes that we can see, that are visible, observable and provable, are *not* the processes of the so-called "evolution" from molecules to man! The processes we can see every day are *not* in *any way* what would be needed to change *a salamander into a human being*. No *observable* processes have *ever* been seen to produce new genes containing new genetic information coding new traits for the next generation of plants or animals, never.

Something slightly *akin* to this does happen however, in viruses, and even in some bacteria. Virus DNA mutates about a million times faster than other DNA. So, since mutations are *supposed* to be the *driving force* behind evolution, this means that viruses should be driven to "evolve" a million times faster than we do. Do they? Well, this does make them *change* genetically much faster than we do. But is it evolution? Stop here and think a minute. Evolutionists say viruses can evolve a million times faster than we can. And evolutionists *say* that viruses have been at it for *over a billion* years longer than we have (keep up with me now!). They believe viruses can "evolve" a million times faster than we can. Think. So

75

then, after *all this time* of *super-fast* "evolution," just what have the viruses become by the magic of evolution? They are still viruses. They have not "evolved" into anything else! Why not?—if they've had all this time, and evolution is real?

The point of course is, that the processes we observe in the *real* world – whether given the Biblical time-frame of six thousand years or *even* the evolutionary time-frame of three <u>billion</u> years, obviously **cannot create new species**. Nobody's ever seen them do so. It is therefore logical to assume that no one ever will. Viruses never got out of being viruses. Bacteria have never been able to become anything but bacteria—not in our observation, not in the real world, not in the world of science.

So, what the evolutionists are asking you to do, is to look at processes that we indeed know to be true, and then just *take it on **faith*** that these processes *can* somehow magically produce things that we have *never seen* them to produce while we were watching them. But evo-believers ask you to join them in believing it, anyway, because they told you so … because they told you so. Is that good enough for you? It's not good enough for me. And no point can be proven by the one-word argument: "because" *I learned that in <u>grade school</u>.*

The Lie: Micro-Evolution Produces Macro-Evolution

Let's go back to our analogy of the deck of playing cards. Take a deck of "Uno" cards. Shuffle them. Deal out a hand to the players for the next round of the game. Shuffle and deal, shuffle and deal, *do it all you want!* You'll <u>never</u> deal out a hand with an *"Old Maid"* card in it or a *King of Hearts*, or a *"go to jail"* card. That's obvious, right? But this is <u>exactly</u> what evolution demands you to believe, that you <u>can</u> deal out new and different cards from the *same old deck* that were never there before! But you can't. It doesn't work that way. It can <u>never</u> work that way. They'll <u>tell</u> you *"mutations can make the changes"* of one kind of a card into another kind of a card. No they can't. How *could* they? Do evo-believers ever tell us the "how?" No. That's because they <u>can't</u>. There is no "how." There is no mechanism for this thing they say can happen to happen! None has ever been verified. What is the process? No process has ever been demonstrated that is known to do all this that is needed by their theory. What principles of science are involved? None can be named. Can you observe this to happen? That answer too is "no" Upward changes in species have never been observed. And even the sideways changes, have only been observed in bacteria or viruses, which are a thousand times simpler than even the one-celled organisms that you may have looked at under a microscope in high school biology class … like amoebas, or paramecia or euglena.

76

Not *only* are the evolutionists expecting you to believe in something that they *themselves* have never seen, they are asking you to believe in something which they themselves cannot even *imagine* the process by which it may happen! Yet, they *expect you* to believe that it "somehow" *did* happen. This doesn't have just a "slim chance" of happening. It's not a matter of *"overcoming the odds."* There *are* no odds. It has a **zero** *chance* of happening! As far as the known laws of science in nature and in biology, in light of all of the observations we have made of living things in the history of the enterprise of science, there is not even any way *to imagine* how any truly novel DNA instructions in new genes could ever be created. New cards just cannot "appear" in the old original deck. New genes cannot just appear in an old original chromosome, to be dealt out to the next generation and to make them "more evolved" than their parents' generation was. All of that part of evolutionary thinking, the main part, must forever remain under the heading of *pure* **make-believe**. Yet it is still presented as a "proven fact" in most all science textbooks, by teachers and professors, and on every science cable TV channel that you can name! This is bad science. This is bad logic. It is bad thinking. It's, well, this is **lying**.

The Solution: Expose Macro-Evolution as Faith-Based.

So how do you *deal with* evo-believers who cannot even *acknowledge* the *known facts* of science? How can you *reason with* someone who has become most certain—that things which *cannot be seen* are really there, that processes *without proof* are really happening—*in spite* of all **evidence** to the contrary? How did the Creator Himself deal with such questions and problems, when he walked and talked and debated the truth among human beings, as one of us on this Earth? In other words, "what *did* **Jesus** do?"

So, we must begin by looking at how Jesus dealt with the ill doctrines and unbelievers of His days on Earth. First of all, he never tried to bring the Pharisees over to His side. Nicodemus came to Jesus, not the other way around. But "the common people heard Him gladly." Jesus *never* argued with any of the Pharisees or the Scribes or Sadducees *in private*. It was always in public—and there was always an audience of "people" standing by as witnesses, as Jesus showed openly and plainly what were all of the flaws in the arguments of those who opposed His own words of life.

You may be tempted to think that Jesus won his public encounters with His opponents, because He was so much smarter than they were, or by virtue of being the Word of God made flesh. He didn't. If that *were* the case, then He couldn't have been setting *an example* for us to follow (since none of us are God incarnate, though He does live in us by His Spirit) and neither are any of us as smart as Jesus clearly was. "Greater works than these, shall ye do," He said to His disciples. And that word comes down to us—to you and me in today's world. What was Jesus really doing in His example of answering the naysayers of that day?

In every response, Jesus was going straight to the flaw in His opponent's thinking, not just the flaws in their words, but the actual flaws *behind* those words. In the "giving tribute to Caesar" encounter, He drew the crowd's attention to His opponents' lack of understanding about the place for the authority of man versus the authority of God (which He also did in His question to them about the Baptism of John). With the woman caught in adultery, He showed up their lack of understanding in matters of "the letter of the law versus the spirit of the law" (as He also did when His disciples were criticized for gleaning wheat on the Sabbath, and His healings on the Sabbath).

So what is the flaw of understanding with our evolution-believing friends, when it comes to this question of "micro-evolution" versus "macro-evolution?" And what is, by the way, the flaw in many Bible-believers' minds on the exact same subject? In both cases, it is the taking of something that we can see as being true and applying it as though it is some kind of proof that something which we cannot see is true. Now there is nothing wrong with this, on the surface. Actually, that's what "faith" is (see Hebrews 11:1), and there's nothing wrong with a person holding onto their faith. But, as the evo-believer will be most quick to tell you … they claim that their position does not depend on any faith of any kind. They will tell you that their position does not arise from any "presuppositions" or "presumptions" that bear any resemblance to "faith" in the slightest. This is not true. And this is what must be seen by the people, even if it can never be seen by those who are trapped in the clutches of the dark Darwinian deception themselves. Their position is based solely upon faith—and certainly not on any science, logic, data, or evidence viewed honestly.

They will say they are "slaves to the evidence" or that their position is "driven by the data only" This is not the case. But how can you show up their lack of understanding in this matter? Be ready to be verbally abused when you do this. But their faith *can be* shown up for what it is, much in the same way that Jesus showed up the ill-conceived tenets of the doctrines of the Pharisees. Just ask a simple question. Ask "how?"

This is a question that they cannot answer. There are many questions that they have been prompted with replies, but not this one. How? How did that happen? How do you know that's true? How did they prove that? Where are you getting this information from? How do you know it is reliable, or do you just put that much trust in someone else's integrity and/or authority? And, you really want to rattle their New-Age Post-modernistic cage? Then just ask them the question "why?" This is a question that even the framing of which creates a horror in the deepest pits of their souls. Do not ask the "why" question, unless you feel urged by the Spirit of God to probe deeper down inside them, all the way right down to where the source of their hurt resides, into their very soul. And God may give you that job, just be open to it.

You may not know much about science. Or, you may know a lot. That doesn't really matter as much as you might think it does. Remember; their theory is not true. So when you merely ask them "how" it works, they merely will not be able to answer you. And do not let them give you the "what happens" answer to your "how it happens" question, and that's what they'll try to do to save face and to camouflage their desperation. If they answer you with just more "information" on the details of their theory, then go back to your original question about the original part of the theory and ask again "yeah, but how?" You may get a violent response on cornering them with the "how" question, since they are only used to the "what" question, for which they have much love (because it gives them a chance to recite and burp back up what they have been taught) and for which they have many words. But keep to it. Do not back down. Do not let them pass over your question. Just keep asking "how" it supposedly works. If you get far enough before they freak out and storm away, you will get them to the point where (when you've asked "how" enough times in a row) they will be forced to admit that— *yes*, it is a ***faith-based*** position that they hold indeed. At that point, you can tell them you're okay with that, and you too understand about faith. But now there's no justification for them to look down on you, just because you do not choose to hold the same faith-position that they have chosen. Truth, reason, and logic are the great equalizers here. Science and the evidence are on our side. These things are not our enemies, but friends. Don't listen to the propaganda about Bible-belief being "anti-science" It's just not so.

And when it comes to micro-evolution versus macro-evolution, just remind them that micro doesn't lead to macro. It can't. The processes that give the result which we attribute to "micro-evolution" are processes that do things ***other than*** what would be necessary to eventually produce the results which would be attributed to "macro-evolution." There is a great wall between them. It's like expecting to drink apple juice from a carton of orange juice, if you'll just have the "patience to wait longer and longer enough" until the ***miracle*** happens. At least creationists believe in miracles with a Miracle-maker. Evo's believe in miracles, and yet deny the existence of a miracle-maker of any kind. Think about which one makes more logical sense.

Author bio

Charles Jackson has impressive credentials. He has earned the following degrees: EdD, Science Education at University of Virginia, an MS Environmental

79

Biology, Medical Science in Education, Science Education and a BS degree in biology. He has been a science teacher and professor since 1980 and is a lifetime member of Creation Research Society, lifetime member of American Mensa, charter member of East Tennessee Creation Science Association, international creation lecturer/blogger/evo-debater, approved speaker by Institute for Creation Research, science editor for Answers in Genesis *Answers* magazine, science editor for Answers in Genesis *The Zoo Guide,* internet creation science teacher for Liberty University, director of college ministries at Creation Truth Foundation, Noble OK. You may email him at the following email address or either of the two websites: DrJ@CreationTruth.com, www.PointsofOrigins.com or www.Creation-Truth.com.

Are Mutations Random

Through him all things were made; without him nothing was made that has been made. (John 1:3, NIV)

Jonathan Bartlett

Editor's note

The conditions on the earth are not static. God in His infinite wisdom knew the conditions on the earth would change over time and that the living things would need to change in order to survive those changes. Genetic variability allows the kinds of organisms that God created to survive and change, and continually produce new variations of the themes that He originally designed.

Abstract

In evolutionary theory, mutations are thought to be haphazard events, with the primary direction of change being provided by natural selection. Recent discoveries in molecular biology are showing that organisms actually have internal mechanisms which generate mutations in a surprisingly directed manner.

Introduction: Mutations and Evolutionary Theory

Evolution by natural selection is currently the most popular scientific explanation for the origin of biological adaptations. Natural selection is a two-part process. The first part of the process is the generation of new variations, and the second part is the weeding out of less-fit members of the population, so that only the variations which are more fit in the current environment are left.

Notice, however, that in the theory there are two parts, but only one part is named. It is termed "evolution by natural selection," not "evolution by natural selection of variations." The reason for this is that it is believed that the variations produced in the first part of the process are produced haphazardly. Therefore the specific mechanisms for producing variations doesn't need any special attention, because the part of the process which truly shapes the direction of evolution is natural selection. Therefore, it is "evolution by natural selection" because natural selection is the only part of the mechanism providing the direction for the change—everything else is essentially accidental.

All of this was before the discovery of genes and DNA. DNA is a class of chemicals that provides a storehouse of information within each cell of an organism. It works by taking for different kinds of DNA, specified by the letters A, T, C, and G, and strings them together in a sequence which stores the information the cell needs to create and regulate the production of proteins for the

81

cell's functioning. DNA provides information about how to create proteins from their building blocks, which are called amino acids. DNA regulates the production of proteins through a series of promoters and inhibitors (which are themselves part of the DNA code) which make sure that proteins are produced in the right conditions. DNA even contains information on different adjustments which can be made on the proteins being made. Each protein-coding segment of DNA is called a gene, and the entirety of the information encoded by DNA in a cell is called the genome.

Every generation of organisms receives a copy of the genome, but not exactly the same one as its parents. The genome is organized into sections called chromosomes. Most organisms have two copies of each chromosome—one from each parent. However, the organisms do not inherit exact copies of the chromosomes of its parents. The DNA of an organism contains many changes from the parent chromosomes. These changes are then part of the DNA which are then passed on to its children. These changes are known as mutations.

The modern theory of evolution focuses on changes which occur within the genome because these changes can be passed on through the generations. Evolution *by natural selection*, then, views evolution as a process of generating variations in the genome, and then keeping the beneficial ones in the population through natural selection. Therefore, in modern evolutionary theory, the sources of new variations for natural selection to act on (to keep or throw away) are the mutations which occur in the genome.

The current theory of natural selection says that the mutations which occur within the genome are random with respect to the fitness needs of the organism or population, and hence are often termed as "copying errors." Therefore, according to the standard theory, for the evolution of new types of organisms, natural selection is the primary directing force. For proponents of natural selection, the mechanisms by which mutations occur may be interesting, but they are evolutionarily insignificant - it is natural selection which provides the direction, not the mechanisms of mutation.

Recent data from molecular biology, however, is turning this notion on its head. It turns out that mutation is not the haphazard process it was formerly thought to be. In fact, it turns out that organisms have very tightly controlled mechanisms for producing mutations. Thus, as we will see, in many cases the direction of change for a population of organisms may be directed more by the organism's own internal mutational mechanisms than by natural selection. This gives us a dramatically different picture of the character, causes, and possibilities within natural history.

What is Meant by Random, and Why is it Important?

There are several different types of randomness, and each of them has slightly different meanings and sometimes drastically different implications. All of them involve some sense of unpredictability, but that is as far as they are similar. We will look at three different kinds of randomness.

Probably the simplest form of randomness is non-correlation. Think of your favorite type of food. Now think of your favorite book. Now, there are a lot of reasons for someone to have a favorite food, and a lot of reasons for someone to have a favorite book, but there is rarely a relationship between the two. Knowing someone's favorite book doesn't tell us anything about what they are likely to have as a favorite food, and vice-versa. This is non-correlation. It simply means that, given two quantities, while they are each likely to have a perfectly good cause on their own, they have no relationship between each other.

Another form of randomness is statistical randomness. Statistical randomness is like a slot machine. Slot machines are designed so that each possible outcome will be achieved at a very steady pace, but the outcome of each individual pull of the lever will not be knowable. So the percentage chances of each possibility are essentially fixed (at least over the long term), but the specific sequence of events cannot be determined ahead-of-time.

The final form of randomness I will discuss is philosophical randomness. Philosophical randomness is an event which occurs outside the control of a system. When you think about a slot machine, even though the outcomes are statistically random, the system is built so that the owners of the slot machine have a guaranteed amount of money they will earn. Because the chances are fixed, the long-term behavior of the system is very reliable. In fact, for certain processes, statistical randomness can be utilized so that the long-term behavior of a system is actually more reliable than if deterministic (non-random) means were used. So, using our slot machine example, if statistical randomness describes its normal operation, philosophical randomness would describe what happens if an angry customer beat on it with a baseball bat.

In evolutionary theory, all of these come into play at some point or another. The problem is that, often, evidence for one type of randomness will be used as proof for another type randomness. As you can see, non-correlation does not imply statistical randomness, and statistical randomness does not imply philosophical randomness. These are each very different, even though they are often confused.

There are several common claims made by evolutionists about the general nature of mutations:

1. Mutations are copying errors made by the cell when it replicates DNA.

2. Mutations are not correlated with the fitness of the organism or the population.
3. Mutations are not correlated with the future needs of the organism.
4. Mutations occur with a fairly reliable (though very small) frequency.

Claim #1 is one of the more common descriptions of mutation given by biologists. By calling them "copying errors" it is clear that the type of randomness being referred to is philosophical randomness. Claims #2 and #3 are claims of non-correlation. Claim #4 is a claim of statistical randomness. In addition, even though claims #2 and #3 are technically claims of non-correlation, they are used as claims of philosophical randomness because *survivability* is the variable they are not correlated with. Claim #4 has almost no bearing on the question of philosophical randomness, because, as we have shown with the example of the slot machine, some systems not only use statistical randomness to achieve their goals, they rely on them.

Now, as mentioned earlier, evolution by natural selection asserts that natural selection is the main director for the path of evolution, and that the mechanisms of variation are essentially haphazard. If the mechanisms of variation were not haphazard, then natural selection would no longer be the force that directs evolution—instead it would be the mechanism which *produced* the variations which were most important. Therefore, claims of philosophical randomness are essential to the theory of natural selection.

Therefore, the remainder of the essay will focus on current evidence that contradicts the first three claims of randomness for mutations. Obviously, some mutations are in fact philosophically random. Exposure to radiation, or chemicals, or honest-to-goodness copying errors actually do occur within the genome, and they are genuinely philosophically random events. However, as the evidence from molecular biology is teaching us, the mutations which are interesting for evolution have turned out to be parts of an exquisite mutational machinery, not haphazard changes.

Generating Diversity in the Immune System

One of the many awe-inspiring systems within the cell is the immune response system. One part of the system are antibodies (also known as immunoglobulins). Immunoglobulins are used to attach to foreign invaders within the body. Your body has millions of different immunoglobulins, but they are all coded from a relatively small set of genes. The way this works is through cut-and-paste DNA. The genome has several different batches of interlocking antibody parts. As immune Cells mature, the cells take one part from each batch and put them together. In at least some cases, if the pieces don't fit right to make a functioning immunoglobulin, the cell can actually patch them slightly to make them function.

By having an assortment of parts, the immune system can have millions or even billions of immunoglobulins, just based on choosing different parts from each batch. This allows the immunoglobulins to attach to a wide variety of differently-shaped invaders. The part of the immunoglobulin that undergoes this batch assembly is called the variable region. At the base of the immunoglobulin is the constant region, which determines the class of immunoglobulin, and attaches to immune cells. It is a very interesting pattern - the part of the protein which interacts with the cells internal systems is very regular and static, which allows the antibody to function predictably. However, the part of the protein which interfaces with foreign substances has a huge, dynamic variety.

In any case, this is not the whole story. When the immune system undergoes a challenge, not only does it have millions of immunoglobulins available to make a fit onto the foreign antigen, it then refines the fit of the variable region of the immunoglobulin so that it has a higher affinity towards the antigen. The process which does this is called somatic hyper-mutation, often abbreviated SMH. "Somatic" means that the process is happening in cells other than germ (i.e. sperm or egg) cells. "Hyper-mutation" means that the cell's DNA is undergoing rapid changes. And this is where it gets interesting.

Out of all the genes in the cell, the gene which undergoes hyper-mutation is precisely the gene which needs refinement—namely, the immunoglobulin gene. Within the immunoglobulin gene, nearly all of the mutations take place in the variable region of the gene (which affects antigen binding) and nearly none of them take place in the constant region (which affects the type of antibody function). So the mutations are not only targeted at the correct gene, but they are targeted at the correct area of the correct gene. These are all controlled by promoters which are in the proper spot for all of this to occur.

Now, the cell does not know *exactly* which parts of the DNA it needs to modify, nor should we expect it to. However, it is clear that the cell does know which *areas* of the genome is likely to produce beneficial changes, and mutations are focused on those points. In the case of the SHM process described, out of the human genome's approximately 3,000,000,000 DNA letters, only a few hundred are targeted for mutation. Therefore, it seems that the likelihood of mutations in given regions is in fact correlated with the fitness of the organism. The cell is concentrating its mutations in the areas which are likely to produce benefits. As such, the claim that such mutations are "copying errors" seems to be a nearly laughable category error for this system. Clearly, in the case of the immune system, the directions that the mutations are taking are primarily directed by the cells mutational machinery, and selection, while it is operating, is taking a far more subordinate role.

What does Evolution in Action Look Like?

Single-celled organisms are interesting creatures to study. Many single-celled organisms, such as *E. Coli*, reproduce quickly, take up a small amount of space, and everything that there is to see happens within one cell. The generation time for bacteria can be as small as 20 minutes (compared with 20 years for humans) and you can have trillions of them in a tiny space. Therefore, for watching evolution happen, they are quite ideal. In 57 years, you can have the same number of generations for evolution that you would have if you watched the evolution of humans for three million years.

Because of these properties, many experiments have been done on single-celled organisms to discover the character of evolution. What we are finding over and over again is that the most interesting changes we observe are primarily based on one of two types of evolutionary mechanisms:

* Mechanisms which produce diversity ahead-of-time
* Mechanisms which produce diversity in response to stress

In the early days of genetics, two experiments were developed which demonstrated that many mutations took place *in advance* of the selection event. If you took a bacteria like E. coli and subjected it to a lethal stress (simulating natural selection), you can determine, using the methods of either Lederberg or Luria-Delbrück, whether or not the surviving bacteria had the gene required to survive ahead-of-time (by a mutation) or if it developed it in response to the stress. For numerous cases, it was found that the organisms which survived already had the gene *before* the selection event took place. In addition, the mutations usually followed a fairly steady statistical distribution.

It is understandable why many biologists used this to assume that the mutations were therefore the results of haphazard copying errors. They were, in fact, not correlated with present fitness (since the mutations pre-existed the selection event), and therefore fit the first type of randomness—non-correlation. In addition, since the mutations seemed to reliably follow a statistical pattern, they also fit the second type of randomness - statistical randomness. It was (incorrectly) assumed that these two types of randomness imply that the mutations are haphazard. However, there is another, better, way of looking at the data, and that is to think like a financial advisor.

Financial advisors cannot see the future. Therefore, for a large portfolio, the portfolio is usually divided up between different types of investments. The type of investment that the financial advisor thinks will make the highest dividends will get the largest amount of capital. However, additional capital will be invested in other investments which are not ideal in the current climate, but will keep their clients from losing everything if the market changes drastically. Now, these investments are just as carefully chosen as the primary investments, but the goal is to keep the client financially healthy in case of a drastic change in the market. Likewise, it seems that these mutations that exist in the population before the selection event are there as a similar type of hedge. If the environment changes drastically, there will be a certain percentage of the population which has an alternate metabolic configuration which may be able to survive. By having the mutations be statistically random, it guarantees, without the overhead of any communication between organisms, that a specific percentage of the population will have the different mutations.

So how do we know if these are haphazard changes or part of a future risk reduction system within the cell? One way is that environmental stress, rather than magnifying the deleterious effects of mutations, often alleviates them. This indicates that rather than being errors, they are alternate configurations which are adaptive to non-standard situations. Another interesting fact is that many of the mutations are reversible. If mutations are haphazard, then it would not be likely that a mutation would eventually be able to reverse itself - we would expect instead that a genome would simply accumulate mutations because the chances for the exact reverse mutation to occur would be small enough that it would not be of any significance. But in fact mutational reversions are common. This is accomplished through several mechanisms in the cell.

One mechanism for producing reversible mutations is turning a promoter on or off by reversing a small segment of DNA. Another mechanism is through adjusting the length of repeatable segments of DNA, where different repeat lengths cause different functions to occur, or different levels of functioning to occur. Another mechanism is by inserting and removing mobile DNA elements into a gene or promoter. All of these mechanisms are much more likely to produce a stable modification which could potentially be beneficial in the future than a modification which was haphazard. We can then conclude that many mutations which occur in the absence of selection are not due to mistakes by the cell, but rather good planning.

But organisms are not restricted to only mutating ahead-of-time. Many mutations are the direct result of the stresses being applied. Barry Hall showed that E. coli possessed the genes to metabolize a certain set of sugars, but even with the genes, it could not use them. However, if the cell was subject to starvation, *and was also in the presence of those sugars*, the cell would induce a mutation which inserted a segment of DNA into the genome at the right position to activate those genes. If the sugars were not there, the insertion would not happen. Only with both the starvation and the presence of the sugars would the insertion occur. Thus, the cell was able to modify its genome in direct response to its environment.

Most environmentally-directed mutations are not quite so specific, and nor should we expect them to be. The environment provides an unpredictable source of stresses, and it is unreasonable for the cell to know precisely which response will work best for which stress. However, what is often seen is that, like the example from our own immune system, the cell has marked the locations and types of mutations which are likely to help out under different types of stresses, and the cells try different ones hoping to achieve a workable result.

Environmentally-directed mutations can operate on a number of different principles. One of them is the mobilization of a fragment of DNA either into or out of a gene or its promoter region. Another one is the initiation of breakage in DNA at specific locations, which are then repaired by a mutagenic repair mechanism. In E. coli, for instance, when DNA breaks, what happens next depends on whether or not the organism is undergoing a crisis. If the organism is doing fine generally, then it repairs the breakage without inducing mutations. If the organism is under stress (starvation or other problems), then it attempts to repair the breakage with a mutagenic repair mechanism. Thus, if the presence of a substance causes DNA breaks to occur at or near favorable sites, mutations can be generated which are likely to help the cell process the substance. The exact pathways linking foreign substances to the location of DNA breakage has not been elucidated, but the evidence so far points to a correlation between the foreign substance needing processing and fortuitous DNA breaks which induce beneficial mutations.

So, when we look at evolution in action, what we see is a highly controlled process, not a haphazard one. We see a system geared to produce hedging (potentially beneficial) mutations ahead-of-time, coupled with a system that can generate adaptive mutations when needed. When we first discovered mutations, they seemed haphazard at the time, precisely because we did not adequately understand them. But as we dig deeper into the mechanics of what is happening, we are seeing that evolution, at least as far as we can observe it, seems to be a controlled process.

Applying the Results to Multicellular Organisms

The previous section focused entirely on single-celled organisms, because so many of their generations of evolution can be easily observed in a short timeframe. However, one major additional problem comes up when applying these ideas to multicellular organisms.

In multicellular organisms, there is a difference between reproductive cells (called the germ cells or germ line) and the non-reproductive cells (called somatic cells). All reproduction is done through germ line cells. Therefore, changes to somatic cells do not automatically result in the changes being passed on to the next generation, since they are not part of the germ line. In single-celled organisms, there was not a difference between germ line and somatic cells, so this was not an issue.

The separation of the germ and somatic tissue was proposed in what was called the Weismann barrier, which said that changes to somatic tissue could not be passed on to the germ tissue. While the Weismann barrier is still applicable for the majority of changes which happen to organisms, current research is pointing to some systems and mechanisms which can penetrate that barrier.

For instance, in the case of the immune system described above, there is evidence that the mutated V regions of immunoglobulins find their way back to germ line cells. The specifics of how this works is still unknown, though some researchers investigating it think it is possibly occurring through retroviruses which move the transcripts from the somatic cells to the germ line.

Another system which may be able to communicate stresses to the germ line cells for directed mutation is the epigenetic methylation system. Cells can mark parts of the genome for different functions (as well as different amounts of mutagenesis) by applying different patterns of methyl groups onto DNA. By selectively marking the parts of the genome which need changing, the organism may be able to determine which parts of the germline need changing and which ones do not, and therefore influence where mutations are likely to occur.

These and many other possibilities are under active investigation. What we know is that mechanisms are available to direct mutations to the locations that are most likely to need them. There is also evidence that at least some of these mutations are making their way back to the germ line cells. Given how fascinating the mechanisms of mutation are that have been discovered over the past decades, I am eager to watch and see what mechanisms for transmission back to the germ line are discovered.

Conclusions

It should be reiterated that not all mutations are part of a planned system of mutation. It is certain that there exist mutations which are, in fact, haphazard. The important point, however, is that the mutations which tend to be the most interesting for evolutionary adaptation also tend to be the ones with the most interesting mutational machinery, and are much less arbitrary than has been historically supposed.

It is quite amazing, then, how much the notion of "random mutations" is still being pushed, even by professional biologists. In the book *The Plausibility of Life*, Kirschner and Gerhart glibly dismiss the whole notion of directed mutation, saying "there is in fact no evidence for facilitated genetic variation [their term for directed mutation] and there is conclusive evidence that it does not exist." This is coming from professors at Harvard and Berkeley. It is difficult to tell if they are unaware of the evidence, dismissive of the evidence, or are simply not philosophically adept enough to understand the implications of the molecular biology research. Given my interaction with other biologists, I would lean towards the latter option. For instance, the mutational machinery that we have discussed regarding immunoglobulins is touted by immunologist Edward Max as evidence of the power of random mutation! How such a targeted mutational system could be described as random mutation is totally beyond my comprehension.

A similar theme comes in the writings of University of Chicago professor Jerry Coyne, who writes that "On the basis of much evidence, scientists have concluded that mutations occur randomly. The term 'random' here has a specific meaning that is often misunderstood, even by biologists. What we mean is that mutations occur irrespective of whether they would be useful to the organism. Mutations are simply errors in DNA replication." Unfortunately for Coyne, the problem is not that the term is being misunderstood. The problem is that terms like "random" (except, possibly, for its statistical meaning), "indifferent," "haphazard," and "copying errors" are not only inadequate to describe the mechanisms discussed, they are simply irreconcilable to the evidence.

The technical literature is also replete with the same types of remarks. In Templeton's Population Genetics and Microevolutionary Theory, Templeton writes "Although many environmental agents can influence the rate and type of mutation, one of the central tenets of Darwinian evolution is that mutations are random with respect to the needs of the organism in coping with its environment."

It is expected that within a field there will be a diversity of opinion. However, it is completely surprising that a whole branch of discovery with such amazing implications would be so blithely disregarded by other biologists. Those who are dismissing the idea of directed mutation do so not by engaging the ideas of biologists who disagree and showing where they are wrong, but rather by simply ignoring them and pretending that neither they nor their evidence exist.

After the dust settles, I think that the existence of directed mutational mechanisms will change the way that we think about evolution and adaptation. Rather than wondering where evolution is going to go next, perhaps we will be able to look inside the cells and see the direction that the cell is taking itself. Rather than endlessly trying to find the origins of novelty in random mutations, we can see the beauty of directed diversity in parameterized, directionalized mutations. If we are looking for the way organisms generate endless forms most beautiful, perhaps looking at the way in which organisms are built to produce such forms is a better method of investigation than merely assuming that such machinery does not exist.

Meet the Author

Jonathan Bartlett is the director of The Blyth Institute (blythinstitute.org), a nonprofit organization dedicated to research and education in new paradigms of biology.

Jonathan began his career in the field of computer science, working on web and e-commerce applications for a variety of companies. Jonathan wrote the book *Programming from the Ground Up* based on his dismay about the low level of education that he was seeing in many recent computer science graduates. *Programming from the Ground Up* was then picked up by classes in a wide variety of schools, including Princeton, DeVry, and Oklahoma State University. IBM invited him to do a series of papers for their DeveloperWorks site which bridged the gap between theoretical and practical computer science issues. Later, IBM invited him to write a series of introductory papers on their most recent processors, including the Cell Broadband Engine which was introduced in the PlayStation 3.

Jonathan's interest in biology and genetics came from his family's battles with genetic illnesses, which required him to stay on top of the current medical and biological literature. During this period of study, Jonathan realized the tremendous overlap between computer science and biology, and how much computer science could contribute to understanding the cell's inner-workings. Bioinformatics had already emerged as a new field, however, bioinformatics is mostly restricted to using computers to process data *about* cells and genes. What was needed was taking the cell's similarity to computer information systems seriously, and using constructs from computer science to help understand biology better.

Towards this end, Jonathan has published a variety of papers and conference presentations. His topics have included the similarity between computer metaprogramming and the cell's recombination events in generating antibody genes, how Wolfram's complexity classes can be used to evaluate genetic regulation and evolvability, how to apply Active Information metrics to cells, developing a classification of mutations beyond their immediate selection benefit, and how Alan Turing's *oracle* concept can be used to enhance cognitive models.

During this time he continued to produce software, cofounding Docvia.com and developing its first release product, a patient-doctor interaction system. He also led the teams to build a web-based tool for evidence-based medicine research, and an SMS-based communication system for patient-doctor notifications in 3rd world countries.

Jonathan is now focusing on biological research and education, and founded The Blyth Institute to coordinate research in uncovering new models of biological adaptation.

Jonathan is married to his wife Christa, and they have had five children together - Daniel, Andrew, Nathan, David, and Isaac. He recently graduated with honors from Phillips Theological Seminary with a Master of Theological Studies degree. He can be contacted at jonathan.bartlett@blythinstitute.org.

References

Bartlett, J. L. 2008. Statistical and Philosophical Notions of Randomness in Creation Biology. Creation Research Society Quarterly 45(2):91-99.

Bartlett, J. L. 2009. Towards a Creationary Classification of Mutations. Answers Research Journal 2:169-174.

Caporale, L. H. 2003. Natural Selection and the Emergence of a Mutation Phenotype: An Update of the Evolutionary Synthesis Considering Mechanisms that Affect Genome Variation. Annual Reviews of Microbiology 57:467-485.

Caporale, L. H. (editor). 2006. The Implicit Genome. Oxford University Press.

Coyne, J. 2007. The Great Mutator. The New Republic (June 12, 2007).

Daev, E. V. 2007. Stress, Chemocommunication, and the Physiological Hypothesis of Mutation. Russian Journal of Genetics 43(10):1082-1092.

Eagle, A. 2005. Randomness is unpredictability. British Journal of the Philosophy of Science 56:749-790.

Eble, Gunther J. 1999. On the dual nature of chance in evolutionary biology and paleobiology. Paleobiology 25:75-87.

Elena, S. F. and J. A. G. M. de Visser. 2003. Environmental stress and the effects of mutation. Journal of Biology 2:12.

Galhardo, R. S., P. J. Hastins, and S. M. Rosenberg. 2007. Mutation as a Stress Response and the Regulation of Evolvability. Critical Reviews in Biochemistry and Molecular Biology 42(5):399-435.

Hall, B.G. 1999. Transposable elements as activators of cryptic genes in E. coli. Genetica 107:181-187.

Hoenigsberg, H. 2003. Cell biology, molecular embryology, Lamarckian and Darwinian selection as evolvability. Genetics and Molecular Research 2(1):7-28.

Kirschner, M. W. and J. C. Gerhart. 2005. The Plausibility of Life: Resolving Darwin's Dilemma. Yale University Press.

Max, E. E. 2001. The Evolution of Improved Fitness By Random Mutation Plus Natural Selection. The Talk.Origins Archive. http://www.talkorigins.org/faqs/fitness/

Papavasiliou, F.N., and D.G. Schatz. 2002. Somatic Hypermutation of immunoglobulin genes: merging mechanisms for genetic diversity. Cell 109 (2, Supplement 1):S35-S44.

Ruden, D. M. et al. 2008. The EDGE hypothesis: Epigenetically directed genetic errors in repeat containing proteins (RCPs) involved in evolution, neuroendocrine signaling, and cancer. Frontiers in Neuroendocrinology 29(3):428-444.

Sanz, I. and J.D. Capra. 1987. V(k) and J(k) Gene Segment of A/J Ars-A Antibodies: Somatic Recombination Generates the Essential Arginine at the Junction of the Variable and Joining Regions. PNAS 84(4):1085-1089.

Templeton, A.R. 2006. Population Genetics and Microevolutionary Theory. John Wiley and Sons.

van der Woude, M. W. and A. J. Baumler. 2004. Phase and Antigenic Variation in Bacteria. Clinical Microbiology Reviews 17(3):581-611.

Author bio

Jonathan Bartlett has worked and taught in computer science for over fifteen years. His textbook on computer science, *Programming from the Ground Up*, has been required reading at schools and universities as diverse as Princeton, DeVry, and Oklahoma State University. His current research is on the overlap between computer science principles and the informational principles which are at play within the genome.

Dinosaurs and Man
Steve Kern, ThD

Editor's note

As a non-scientist, Pastor Kern brings a different perspective and insight to this project. He and I served on the board of directors for Oklahomans for Better Science Education and I was always impressed with his depth of understanding of science. Perhaps because his wife is currently serving as an Oklahoma State Representative, he was always keenly aware of current events and how God' Word applies to them. It seems all children love dinosaurs and he has written a series of adventure books set prior to the flood when man and dinosaurs lived side by side. The church he pastors is located near the site of the Oklahoma City bombing and they provided food and shelter for the many firemen and other emergency responders involved with cleaning up after that disaster. I have long felt the description he cites below clearly refers to a dinosaur. Enjoy his fresh insight. Your children will not hear this from our public schools.

Can you pull in the leviathan with a fishhook or tie down his tongue with a rope? Can you put a cord through his nose or pierce his jaw with a hook? Will he keep begging you for mercy? Will he speak to you with gentle words? Will he make an agreement with you for you to take him as your slave for life? Can you make a pet of him like a bird or put him on a leash for your girls? Will traders barter for him? Will they divide him up among the merchants? Can you fill his hide with harpoons or his head with fishing spears? If you lay a hand on him, you will remember the struggle and never do it again! Any hope of subduing him is false; the mere sight of him is overpowering. No one is fierce enough to rouse him. Who then is able to stand against me? (Job 41:1-10, NIV)

Of all the "sacred cows" evolution holds to so tightly, dinosaurs have to be number one on the list. They are the super star icon of evolution theory's indoctrination scheme. Everywhere you turn dinosaurs are presented as the great attention getters that are used to draw children and adults into the evolution worldview. They dominate natural historical museums. The movies and documentaries on television draw huge ratings. Children's books and television programs incorporate dinosaurs as characters in their story telling and programming because of their huge popularity. It seems our society cannot get enough of dinosaurs. The evolution spin is that these fossil remains of very large and some very vicious animals are evidence of a world long since passed in earth's evolutionistic history. Their fossil remains are all that are left of the age when reptiles ruled the earth but have now been replaced by the more adaptive mammals that have come to dominate recent natural animal history. The dinosaurs have been used as an effective tool to indoctrinate our society into thinking in terms of millions of years. So much so that when we see a dinosaur our minds have been programmed to think they died out 60 or 70 millions years ago long before man came on the scene.

Why does the evolution crowd work so hard to keep dinosaurs and man separated by millions of earth history years? They know that if it can be shown that man and dinosaurs lived together at the same time then the theory of evolution is dead. The most important ingredient to the theory is time. The more time they have to work with the more credible the possibilities of random selection become. Given enough time, as far as the evolutionist is concerned, anything is possible through the process of natural selection. Darwin's *Tree of Life* that is based on the idea that all life forms evolved from one simple cell into more and more complex life forms becomes erroneous when time is reduced to a matter of only a few thousand years. According to the evolutionistic based geological column life began some 530 million years ago. The dinosaurs were supposedly in existence beginning around 200 million years ago and dying out about 60 million years ago. Modern man is supposed to have come on the scene only about 25 thousand years ago. If that is the case then there should not be even the remotest connection between dinosaurs and man in the fossil record. Nor should there be any hint of dinosaur creatures expressed in man's history described in archeological discoveries.

But what does all the evidence really tell us about dinosaurs? The Bible tells us in Genesis 1 that God created all the living animals during the fifth and sixth days of creation. It also tells us that the earth is only about six thousand to no more than ten thousand years old, if you accept its chronological time frame as a literal historical account. The Bible also tells us that God destroyed the original created world with a worldwide flood that covered the whole earth about five thousand years ago. If all this is true then man and all the animals found in the fossil record, including the dinosaurs, would have been created at the same time and also would have been buried together in the sediment of the great cataclysmic flood as fossils. It also means that dinosaurs would have been on Noah's ark and would have survived for a period of time after the flood like all the other animals on the ark. If this is true then there should be evidence that man and dinosaurs have lived together on the earth for a period of time after the flood. That is exactly what the evidence shows us when we take off our clouded evolution colored glasses in order to see that evidence clearly.

Let's begin with the biblical evidence that suggests man and dinosaurs were created in the same six day creation week and then survived on the ark to live after the flood to live together on earth for a period of time before becoming extinct. The first bit of evidence is found in Genesis 1:21 where the Hebrew word *tannin* is translated "whales" in the King James Version and "great sea monsters" in other translations. The point we want to make here is that the word *tannin* is translated dragon in other parts of the Bible. That is very interesting because we find the description of a dragon like sea monster in Job 41 called *leviathan*. We will look at this creature more closely a little later but suffice it for now to say that apparently the *leviathan* of Job 41 and the sea monster of Genesis 1:21 are the same sea dragon. This dragon of the sea we must remember was the first of all the living creatures that God created and, according to Job 41, was reptilian in form.

The other word we need to look at in the Hebrew, *behemoth* is found in Genesis 1:25 where it is translated "beast" This same word is also used in Job 40:15 to name the animal described there by transliterating the Hebrew into *behemoth*. Job 40:15 says, "Behold now behemoth, which I made as well as you …" Some translations try to say this animal was a hippopotamus or an elephant but the description of this animal in Job 40:15-24 we will look at latter is obviously neither one of those two creatures. The fact that God says He made behemoth as well as man on the sixth day makes it clear that the animal described in Job 40 was the first and foremost of all the beasts created, and thus the ultimate behemoth of all the large animals on land. There are no other larger land animals known to man than the long necked, long tailed Apatosaurus dinosaurs. Some grew to be as large as the blue whale in the sea, one hundred feet long and weighing one hundred tons.

Why is linking these animals described in Genesis 1 to the book of Job so important? The book of Job is considered by most scholars to be a product of the time of Abraham, Isaac, and Jacob. That means it was written just a few hundred years after the flood of Noah's time. These animals, described by God to Job, had to have survived for a time after the flood. If they were dinosaurs then the Bible establishes the fact that man and dinosaurs lived together at the same time. The creation account and the mention of these two dinosaur type animals is a wonderful of example of how God's word is consistent with itself. The creation account makes it clear that all the animals were created in the fifth and sixth days which would have to include all the animals in the fossil record, including all the dinosaurs. The Bible also tells us that all the land animals were represented on Noah's ark. Then we find animals thousands of years later as fossils that are described in Job to help us know that the Bible has been right about natural history all along. Thus the bible as a historical record is consistent with what it records and what we observe. So, let's look at the *behemoth* and *leviathan* a little more closely.

Behemoth is described in Job 40:15-24. When you read this passage you learn that behemoth was made at the same time as man on the sixth day. He eats grass like an ox and is huge in size. The defining part of its description is in verse 17 where we are told that, "He bends his tail like a cedar" Now a cedar is a large tree not a short stubby bush. Alligators and other reptiles have large long tails in relationship to their bodies that could be compared to trees but there are no mammals that have that kind of tail, especially the Hippopotamus and Elephant. But in the fossil record we find several huge dinosaurs that had very large and long tails that could be readily compared to a cedar tree. Brachiosaurus and Apatosaurus are the two most likely examples. These two dinosaurs also stood very tall with long necks. The Job account goes on to tell us that the behemoth could stand in the Jordan River when it was at its deepest season and the water would only rise to this animals nose. Only an animal with long thick legs and a long neck could accomplish this feat. The description given here is obviously of an animal that was very large and now extent, but fits the description of some dinosaurs we find in the fossil record. The important thing we need to remember is that this was a real animal that God had created and the man Job had seen one.

The second dinosaur like animal described in all of chapter 41 of Job is called Leviathan. God takes several more verses to describe this ferocious sea creature then He did with the behemoth. This animal is large and has a scale like covering that is thick and impenetrable by any kind of man made weapon. It has large jaws filled with terrible teeth and breaths fire. It cannot be captured let alone tamed by man. There is no other animal that can compare to the majesty of this animal in the animal kingdom. It is obvious that this is a dragon like creature described in ancient cultures from all around the world. It surely provides the living reality behind much of the dragon myths that survive until now. It is also clear that this animal was in existence during of Job's time and Job was familiar with it time and fits the description of a dinosaur like animal rather than an alligator, as most commentaries try to infer.

This creature was not an alligator. Alligators have tough scaly hides and large teeth, but they do not breathe fire. Alligators can also be caught and taken into captivity or killed by man. There is no animal known by man today that fits this leviathan description. Just because it does not exist today does not mean it did not exist in the near past. The fossil record is full of extent animals while many animals today are in danger of extinction. Dragons may be mythological today but that does not mean they did not exist in the past just like we know dinosaurs did. Some of those dinosaurs probably are the reality behind the dragon myths.

The big problem with dragons for most modern minded folks is their capacity to breathe fire. But why is that so hard to believe. There are many animals still in existence today that have physical capacities that we would have a hard time believing they could do if we had not seen them. What about flying bugs that have a blinking light on their back side or electric eels that produce dangerous electric charges or exotic squids that can change their coloring instantly to fit their environment? These are only a few of the many animals that still exist that we would have a hard time believing their existence possible if we had not seen one. Really, is the capacity to breathe fire any more fantastic than what these animals I just mentioned can do? All it would take is for an animal like a cow that eats grass and belches methane gas to have a spark producing mechanism and you would have a fire breathing cow. Apparently there was at least one animal in the past that had that capacity. It was a large ferocious reptile like creature that lived mainly in the sea and breathed fire. In Job 41, God claimed to have created it and Job must have been aware of its existence as God described a leviathan to him. It is also, more than likely, the sea dragon (*tannin*) described as the first living animal God created in Genesis 1:21.

We have now established that scripture supports three important realities. First, all animals alive today and those found only in the fossil record were created in the fifth and sixth day of creation. Second, all those animals were represented on Noah's ark, including the dinosaurs. Third, they all survived for a time after the flood as testified by Job. We can now look to see if there is evidence outside of the Bible that man and dinosaurs lived together at the same time in recent history. The fact of the matter is that a great deal of evidence can be used to show that man and dinosaurs lived together after the flood. Due to space limitations, we will only look at a portion of that evidence.

In his book, *The World After the Flood*, Bill Cooper recounts the part of the Beowulf story where Beowulf kills the monster Grendel. He shows how the description of the monster fits that of a juvenile T. Rex. To kill the animal Beowulf presses himself under the monsters large jaws to keep himself from being torn apart by its mouth full of sharp teeth. While under those fierce jaws and teeth capable of swallowing its victims in few large "gobbets", Beowulf wrenches off the monsters small upper forelimb. Grendel then runs away on its large bipedal legs in a panic with great pain only to bleed to death in its cave. Cooper also gives a picture of a stone carving from the Saxon era in a church on the British mainland that depicts a Grendel like animal chasing a herd of long necked quadrupeds. He also shows a picture of a Saxon shield that is obviously a depiction of a flying reptilian Pterodactyl. Dr. Coopers point in all of this is that these dinosaur animals were still alive and well known by humans after the time of Christ. That would be more than two-thousand years after the worldwide flood.

Another good source for evidence that man and dinosaurs lived together before and after the flood is Dennis R. Petersen's book, *Unlocking the Mysteries of Creation*. One such evidence is called Malachite Man. This is a group of modern human skeletons buried in Dakota Sand Stone supposedly 140 millions old which would pre-date most of the dinosaur ages. The stones containing the remains of ten humans, four of which were female and at least one was an infant, are found in Moab, Utah in the same 100 million year old rock formations that entomb many dinosaur remains not far away in Dinosaur National Monument not far from Vernal, Utah.

Dennis Petersen also records pictures of human tracks mixed with dinosaur tracks taken in the Paluxy River bed not far from Glen Rose Texas. The human tracks are obviously genuine as well as the dinosaur tracks. The tracks actually cross each other's path in very same sand stone. These tracks have been known to exist for several years but get little notice outside of creationist circles. This is true because such evidence if given wide public disclosure would destroy the old earth evolution paradigm. Petersen only covers the Paluxy River tracks but other similar tracks have been uncovered in other parts of the world. Another source for this information is the book, *Dinosaur*, by Dr. Carl Baugh who has led some of the excavations done in the Paluxy river bed. This evidence makes it clear that man and dinosaurs lived together at the same time.

Some of the most convincing evidence I have ever seen is pictures of Dr. Javier Cabrera's collection of the Peruvian Inca culture burial stones. These stones date from 500 to 1500 A.D. Over 1100 of these stones have been uncovered and several have pictures engraved on them of men and dinosaurs interacting together. There are pictures of Triceratops, T. Rex, Stegosaurus, and Diplodocus as well as others with men either riding on them or battling them with spears. The important thing to remember is dinosaurs were not discovered in the fossil record until the early 1800s. These stones were carved 300 to 1500 years before these dinosaurs were rediscovered in modern times. The point is very clear that the Incas in South America lived with these dinosaurs over a thousand years ago long after Noah's flood.

There are documented pictures of dinosaurs drawn on cave walls, rocks in the Southwest United States, on Roman soldier's weapons, even a grave in a church in England. I could go on and on about evidence that shows how man and dinosaurs lived together before and after the biblical flood. The evidence is all well documented. The reason why this evidence has not been given wide spread attention is it would totally destroy the evolution worldview and establish the Bible as the solid historical record that it actually is. That would totally undermine every philosophical, political, social, and science ideology that has taken control of every base of power in the new world order. Those in power cannot let that happen. The truths of creationism are just as dangerous today to the ruling classes of the world as Jesus was to the Sanhedrin two thousand years ago. He became a threat to their power and so they killed Him. But Jesus, through His resurrection, made it clear that you can never kill the truth, and the truth about man and dinosaurs living together will continue to gain acceptance as the evidence continues to grow. That means that in time dinosaurs will no longer be evolutions number one icon.

Author Bio

I have pastored Southern Baptist churches for the last thirty years in Texas, Idaho, and presently in Oklahoma City. I am married to Sally Rogers Kern who is presently an Oklahoma State Representative. We have two sons who are grown and making lives of their own. I am a graduate from Palm Beach Atlantic University in West Palm Beach, Florida with a bachelor's degree in Psychology/Sociology. I also have Master of Divinity and Doctor of Ministry degrees from Southwestern Baptist Theological Seminary in Fort Worth, Texas. I have been a student of Creationism since my college days back in the '70s. I have taught literal creation and creation science in college classes, conferences, and seminars for twenty years. I have authored four books. *Judgment's Greatest Question* is my life story and philosophy of inner city church ministry where I have pastored for the last fourteen years. *No Other Gods* is a chapter by chapter, verse by verse apologetic of the first eleven chapters of Genesis. It is the product of twenty years of study and gives a literal historical interpretation of those chapters. It also includes evidence from science and archeology. *Eden's Veil* and *Eden's Son* are sequels of two fiction adventure novels that take place in the environment before the flood. They tell the story of a man who seeks out the truth about the God of creation as he goes on a quest to find the Garden of Eden. The story includes dinosaurs and much of the other characteristics of the pre-flood world. My email address is: sdkern@sbcglobal.net.

Passive Fear. Alternative to Fight or Flight

E. Norbert Smith, Ph.D.

Editor's note

For nearly a hundred years zoologists and others have been aware of the classic fight or flight response. It is a common response for frightened animals to flee from danger or to fight for survival when attached or cornered by a predator. This response has been studied extensively and is a sympathetically dominant response and results in a marked increase in heart rate, metabolism and the rate of respiration. Body temperature may even increase due to prolonged strenuous activity. Each encounter with a predator is a life and death situation. For nursing mammals the survival of their young is also determined by the outcome. The predator must also acquire prey for its survival and that of its young.

While very important, the fight or flight response is only one reaction an animal may display when approached by a predator or other threat. An alternate passive hiding response is just as widespread, but not nearly as well known. I was the first zoologist to study this response in detail using a variety of wild animals under natural free ranging conditions. In sharp contrast, the passive fear response is parasympathetically dominate and results in reduced metabolism, heart rate, respiration rate and in some cases body temperature also drops. The most extreme example is death feigning by the American opossum and other animals including man and as recorded in scripture in the scripture quotation below. To some, this opposite response is still seen as an anomaly, another sacred cow in science. Following is an attempt to expose this sacred cow and finally put her out to pasture.

There was a violent earthquake, for an angel of the Lord came down from heaven and, going to the tomb, rolled back the stone and sat on it. His appearance was like lightning, and his clothes were white as snow. The guards were so afraid of him that they shook and became like dead men (Matt 28:2-4, NIV).

Abstract

The passive fear response is a common response to fear of wild animals when they have a safe hiding place. It is opposite the classic fight or flight response and is a parasympathetically dominant response with reduced metabolism reflected by a marked reduction in respiration and heart rates. In prolonged responses, even the body temperature may drop. This passive fear response had been reported on a wide variety of animals including, swamp rabbits, cottontail rabbits, fox squirrels, gray squirrels, box turtles, sea turtles, sloths, chipmunks, deer fawns, fish and many other species. Death feigning as seen in the both the American and South American opossums is the extreme case with a 98% reduction in heart rate. During death feigning they are totally unresponsive and do not even exhibit the normal blink response when the cornea of the eye is touched. Yet, they were found to be fully conscious and aware of their surroundings including the position of the attacking predator.

Introduction

Misconceptions have surrounded our understanding of the behavioral and physiological responses of animals to fear since the groundbreaking work of Walter Cannon, Ph.D. at the Harvard Medical School over eighty years ago (Cannon, 1929) He documented profound cardiovascular and other physiological changes associated with fear and was the first to use the term "fight or flight." It has now been well established the fight or flight response is only one of two options animals may exhibit when frightened (Smith, 2006). Nevertheless, the response continues undaunted as a sacred cow in science in spite of overwhelming empirical evidence to the contrary.

The classic fight or flight response, also known as the hyper-arousal or acute stress response is familiar to life science students and others. When approached by a predator or other threat, many vertebrates including man respond with a complex sympathetically mediated autonomic nervous system response resulting in the release of norepinephrine (adrenalin), an increase in metabolic rate accompanied by simultaneous increases in heart and respiration rates. Mental acuity is heightened and the frightened animal or human is mentally and physiologically prepared for prolonged intense physical activity such as fleeing or fighting. Examples of superhuman strength and endurance abound during life threatening emergencies (Dozier, 1999). This should not come as a surprise to anyone familiar with wild animals under natural conditions because any encounter with a predator is a life and death event. The stakes could not be higher as life of the animal and perhaps the life of its nursing young are at risk as well.

The alternative passive fear response is an equally important and perhaps is an even more widespread response to fear. In sharp contrast, it is mediated by the parasympathetic portion of the autonomic nervous system and results in increased release of acetylcholine. Early detection of an approaching predator or other threat often elicits an orienting response first. The animal looks toward the disturbance in order to collect more information. Next, the well known freeze response is triggered if the threat appears menacing. The frightened animal remains motionless and may avoid detection by the predator. A wide variety of animals use this approach including many insects.

Alternately, if the threat continues or increases and a safe hiding place is available the animal may retreat and hide. A woodchuck retreats to the safety of its underground borrow; a squirrel runs up a tree and hides on the side of the tree away from the threat. The freezing and hiding responses are accompanied by a reduction in metabolism and slowing of the heart and respiration rates. If prolonged, a drop in body temperature may result. In extreme cases death feigning may occur.

This response is best demonstrated in the American Opossum, *Didelphis virginiana (*Gabrielsen and Smith, 1985). Opossums are known for their uncanny ability to "play dead" as a way of surviving an attack by a predator. Death feigning apparently frustrates the normal chase-kill sequence of the attacking predator and it quickly loses interest and moves on in search of other more responsive prey. A similar response is seen in man and is even mentioned in the Bible. The Roman guards are considered by many to be the best trained warriors at the time yet upon seeing an angel and witnessing the bodily resurrection of Jesus Christ, ***"The guards were so afraid of him that they shook and became like dead men"*** (Mathew 28:4, NIV). In light of hard evidence to the contrary, the erroneous view that animals always respond to threat with the fight or flight needs to end. It is time for this sacred cow to be dispatched.

Consider the following pictorial example of threat and the typical responses animals often demonstrate. My discovery enabled us to see these various responses in a new light with greater understanding as illustrated in the following drawings.

Disturbance and orienting response

Hiding or death feigning response

Fight or flight Response

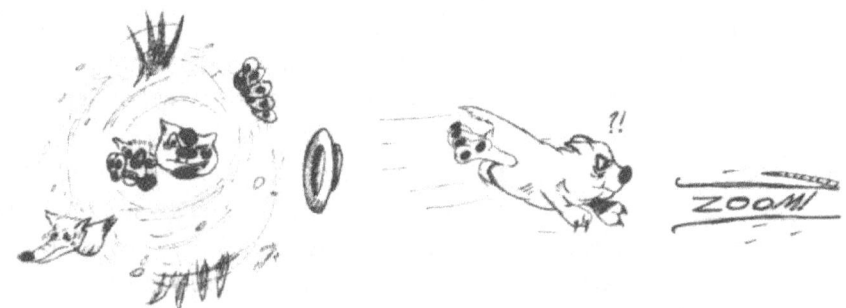

Death feigning in an American Opossum

Opossums should be awarded an Oscar for their excellence in acting. One can touch the cornea of the eye without the normal blink reflex, yet they are totally aware of their surroundings and the position of the threatening predator. Before returning to the research details about death feigning in opossums, let me provide a bit of context for my accidental discovery of this widespread and important response.

Accidental Discoveries in Science

The importance of accidental discoveries in science is well known. Serendipity played a major role in the discovery of x-rays, penicillin, Teflon, Velcro, nylon, various plastics, safety glass, certain sugar substitutes, Newton's description of gravity and DNA to mention a few (Roberts, 1989). My own discovery of the passive fear response of animals also occurred by accident, but first let me provide a bit more background.

In 1929, P. F. Scholander, Ph.D. of Norway, began studying the heart rate response of seals to forced diving. That was the same year Cannon first described the fight or flight response. It was a good year for animal physiology. Scholander did not publish his definitive work until over a decade later (Scholander, 1940). Since that time, other scientists have studied the response of many animals to diving and most show a marked reduction in heart rate while submerged (Anderson, 1966). The term "diving bradycardia" was coined to describe the response. That term is also misleading and has become yet another sacred cow in science. Let me recounting my own accidental discovery. I will discuss some aspects of this discovery in more detail in the next chapter devoted to diving bradycardia as yet another sacred cow.

I designed, built and surgically implanted a sophisticated multichannel radio transmitter that enabled me to measure heart rate and three different temperatures inside free ranging alligators. After five years of preparation, I released a large telemetered alligator near Big Lake at the Welder Wildlife Refuge in South Texas. The alligator's heart rate dropped precipitously upon submergence under the water. This was expected from the literature on the diving response of animals and demonstrated everything was working flawlessly. I was on cloud nine as I would finally reap the benefits of my years of intensive preparation. My excitement was short lived.

My assistant and I spent most of the night constructing a nearby makeshift blind and setting up the telemetry and recording equipment, directional radio antenna and meteorological devices to monitor the lake and water temperatures, wind direction and velocity and other necessary monitoring gear. The alligator remained in the area. The next morning the alligator was seen intermittently swimming nearby, but something was wrong. It submerged often, without slowing its heart rate. It seemed my alligator had not read the books. It was not doing what it was supposed to do when diving underwater. It failed to exhibit diving bradycardia. It's heart rate remained unchanged. The alligator remained nearby and I collected a large amount of data that would later shed light on how alligators regulate their body temperature. This was the major thrust of my study, but the seemingly anomalous cardiovascular response to submergence continued to perplex me. I was able to monitor the telemetered alligator for several days. It moved around the lake and I followed in boat, trying to stay out of sight. After several days and nights without much sleep, I briefly dozed off and when I awoke the familiar sound of the telemetry signal was gone. The alligator had moved to another lake and it took several days of relentless searching to find it. Once again I was monitoring the animal from a makeshift blind. Then it happened. I made an important accidental discovery that forever changed the direction of my research and caused countless old studies to be repeated under more natural conditions. That singular event shed new light on the fight or flight response as well as the cardiovascular response to diving.

As I now expected, the telemetered alligator was off shore and occasionally diving without any significant slowing its heart. Its body temperature was five degrees warmer than the lake surface water and its heart was beating a normal 32 beats per minute which was normal for an alligator that size and temperature. I heard an approaching pick-up truck and recognized the owner as Will Reagan. He was another graduate student that often fished the lakes. He had a canoe and his German shepherd dog, Napoleon with him, but he was not here to fish today. He was studying the hatching success of Least Bitterns, a common shore bird that nested among the reeds at the lake. He and Napoleon were slowing coming my way as he photographed each nest and counted the eggs. He found a new nest and flagged it. Certainly he had as much right to be in the lake as I did, but I was afraid he would frighten my alligator. I tried to wave him away. He thought I was just being friendly and waved back. He was headed for a marked nest less than ten yards from my submerged alligator and there was nothing I could do to stop him.

Suddenly, as his canoe passed directly above the submerged alligator its heart slowed from 32 beats per minute to less than 2 beats per minute. At first I thought my telemetry transmitter had failed. Gradually, as Will moved away the alligator's heart rate increased. In ten minutes its heart was beating at 5 beats per minute. After thirty minutes the heart rate returned to normal. After reliving that event over and over for several days and nights, it finally made sense to me. The alligator was safe underwater. When Will approached in the canoe, the alligator became frightened and slowed its heart in order to prolong its dive. It has long been known alligators can hold their breath for over an hour if necessary. They even have a special shunt so blood bypasses the lungs while they are submerged, must as the human circulatory system by passes the lungs until birth.

Over the next few days I repeated the experiment. My assistant approached the submerged alligator and beat noisily on the bottom of the boat. The alligator's heart slowed. At the time I was driving a large Honda motorcycle and I popped wheelies near the water's edge and the disturbance resulted in marked slowing of the alligator's heart rate. By the end of the summer I had confirmed the results with over two hundred experiments using four different alligators. The response was real. Scientific orthodoxy was wrong. Another sacred cow was in trouble. Still, there was a question as to if the scientific community would take seriously the observations of a lowly graduate student. In time they did and many subsequent studies with a variety of other animals confirmed the results of my accidental discovery at Big Lake when a telemetered alligator was frightened by an unwanted intruder in a canoe.

As happens with many accidental discoveries, the results had nothing to do with alligator thermoregulation which is what I had set out to study. This accidental discovery forever changed my life, helped me earn two graduate degrees, publish several scientific papers and even enabled me to publish a book on the topic. It also earned me an opportunity to be the keynote speaker at a major telemetry conference at Oxford University in England. My research was even highlighted in at PBS TV documentary, *A Smile for the Crocodile*. That accidental discovery changed the direction of my future research. More importantly it gradually changed to way others look at both the fear response of wild animals and the diving response.

Let me say one more thing about accidental discoveries. It is not enough to find something in science is awry from what was expected. One must also follow up on the event in order to understand it. No doubt many others had bacterial cultures contaminated with penicillium mold as it is ubiquitous in households and laboratories around the world. No doubt most were disappointed at the contaminated culture, trashed it and tried again under more sterile conditions. Only Alexander Fleming noticed the bacteria-free region around the contaminated area, understood the significance and had the curiosity to discover the reason. His accidental discovery ushered in the modern age of antibiotics saving untold millions of lives. My accidental discovery also unlocked some previously closed doors. The accidental discovery is but one element. It also takes follow-up to grasp the significance and application of the discovery.

Some Related Physiology

As all biology majors understand, the heart has its own pacemaker and will beat long after it is removed from its owner if suspended in the proper nutrient saline solution. In a time before animal rights activists were so vocal; I would often allow students to take the still beating heart of a turtle from the lab in a beaker of saline so they could show others the miracle of the beating heart removed from its owner. Although the isolated heart of all vertebrates will indeed beat without neural influence, the heart rate in intact animals is influenced by both the sympathetic and parasympathetic portions of the autonomic nervous system. Let me explain this rudimentary cardiovascular physiology in a bit more detail.

Parasympathetic neural activity slows the heart rate by releasing acetylcholine near the sinoatrial node or pacemaker region of the heart. In contrast, the sympathetic portion of the autonomic nervous system releases norepinephrine into the body which increases heart rate. Both systems are normally active all the time and both can speed up or slow down the beating heart. Let me give a simple example to help you understand this process. Compare the situation to a not-so-smart automobile driver with one foot on the brake and one foot on the accelerator. He can accelerate two ways, but taking his foot off the brake or by pressing harder on the accelerator. Conversely, he can slow down by pressing harder on the brake or by removing his foot from the accelerator.

In summary, animals often have several choices when frightened. They may fight or flee with an increased heart rate as has been long established in the literature. Things are seldom as simple as they seem and that is certainly the case here. Alternately, frightened animals may freeze or hide with a marked slowing of their heart rate.

As is often the case in science, additional studies are needed, especially with humans. Thus far the cardiovascular response to fear in humans is unclear. Perhaps it is because of differing emotional responses and other unique traits including greater intelligence, complex language and many other unique traits. The same stimulus might frighten one person while making another angry with contrasting physiological effects. More studies are needed to discover these important differences and to better understand our own response to threat or fear.

References

Anderson, H. T. 1966. Physiological adaptations in diving. Physiol. Rev. 46:212-243.

Cannon, W. B. 1929. *Bodily changes in pain, hunger, fear and rage.* 2nd edition, Appleton and Co. New York and London.

Causby, L. A. and E. N. Smith. 1981. *Control of bradycardia in the swamp rabbit, Sylvilagus aquaticus.* Comp. Biochem. Physiol. 69C:367-370.

Dozier, R. W. 1999. *Fear Itself.* Thomas Dunne Books.

Gabrielsen, G. W. and E. N. Smith. 1985. Physiological response associated with feigned death in the American Opossum. Acta Physiol. Scan. 123:393-398.

Roberts, R. M. 1989. *Serendipity: Accidental Discoveries in Science.* John Wiley and Sons, Canada and the United States.

Scholander, P. F. 1940. *Experimental investigations in respiratory function in diving mammals and birds.* Hvalrad. Sk. 22:1-131.

Smith, E. N., K. Sims and J. F. Vich, 1981. Oxygen consumption of frightened swamp rabbits, *Sylvilagus Aquaticus.* Comp. Biochem. Physiol. 70A:533-536.

Smith, E. N. and E. W. Tobey. 1983. *Heart rate response to forced and voluntary diving in swamp rabbits, Sylvilagus aquaticus.* Physiol. Zool. 56(4):632-638.

Smith, E. N. 2006. *Passive Fear: Alternative to Fight or Flight.* Published by iUniverse New York, Lincoln, NE. and Shanghai.

Smith, E. N. 2006. *Do you sing while you work?* Creation Research Society Quarterly 43:261.

Smith, E. N. 2007. *Endothermic Skunk Cabbage*. Creation Research Society Quarterly 44:153-155.

Smith, E. N. 2010. *Which prey do predators eat?* Journal of Creation 24(2):75-77.

Author bio

E. Norbert Smith grew up on the family farm southeast of Weatherford, Oklahoma. He became interested in electronics while still in high school and obtained an amateur (ham) radio license. His novice call sign of KN5PH.D. was prophetic of things to come for he had a Ph.D. from the FCC. He joined the Air Force five days after high school graduation in 1959 with a promise of electronics school and graduated top of his class. Following his honorable discharge, worked three years first as an electronic technician and then as an electronic design engineer in the Dallas, Texas area.

He wanted to do biological research and become involved in the Creation/evolution debate. He knew he needed a graduate degree to do both of these things effectively. In 1967 he returned to his home town and enrolled in biology at Southwestern Oklahoma State University. While attending college he designed a radio telemetry system for tracking rattlesnakes and worked with Dr. Hobart Landreth at the Wichita Mountains Wildlife Refuge and graduated with his BS degree three years later. He earned a MS degree from Baylor University in 1972. While at Baylor he designed and used a sophisticated multichannel radio telemetry system to monitor three different temperatures and heart rate of free ranging alligators. He continued his study of alligator thermoregulation and earned his doctorate in Zoology is from Texas Tech University in 1975.

He has published over one hundred technical papers on such diverse topics as the design and application of radio telemetry devices, thermoregulation of free ranging reptiles, the cardiovascular response of wild animals to fear and scientific evidences of creation and the survival of salt water plants and animals during the freshwater flood. The BBC filmed a documentary about his alligator research called *A Smile for the Crocodile* and he was invited as keynote speaker to a major radio telemetry conference at Oxford University and took three research students with him to the event. He was given the outstanding teacher award two years, yet in spite of these accomplishments he, like thousands of other professors was denied tenure for rejecting evolution. With his teaching career ended he became a truck driver until his retirement five years ago. He still lives on the family farm and remains active in the Creation Research Society and other Creationist organizations. He now spends most of his time writing books and gardening. He has published eight books including a series of children's books about Al-the-Gator.

He supports two children in Uganda through Compassion International and has paid for the university education of other young people in Uganda. He is also helping a young girl finish high school and will send her to the university. Dr. Smith feels strongly that education is the best way to break the poverty cycle for families. His email address is docgater@aol.com.

Diving Bradycardia.
Another Sacred Cow Is Slain

E. Norbert Smith, Ph.D.

Editor's note

In 1929 P. F. Scholander, Ph.D. of Norway began studying the heart rate response of seals to forced diving. They were restrained on a wooden frame and lowered under water resulting in a reduced heart rate. Other studies followed and a host of diving animals were reported as having diving bradycardia when submerged underwater. Seals remain the animal studied the most for their ability to endure long deep dives. Diving bradycardia was described as a parasympathetically mediated response with the vagus nerve playing a major role in the response. It has long been thought to be an important response enabling diving animals to remain alive under prolonged submergence.

My study with the fear response of wild animals gave me insight into this misunderstood response. I was the first to demonstrate that diving bradycardia not a response to submergence, but instead is a variation of the fear response. It was found to be yet another sacred cow. What had long been thought to be a response to diving was in fact a fear response resulting from poorly designed experiments. When an investigator forcefully submerges an animal underwater it is a terrifying ordeal over which the animal has no control and does not know when or if it will return to the surface of the water and be able to breathe again. It is a life threatening event. The resulting cardiovascular response is a response to fear and NOT to submergence. Following is an attempt to expose yet another sacred cow and set the record straight.

It is dangerous to be right in matters on which the established authorities are wrong. Voltaire, 1751

Abstract

Diving bradycardia has been described for a wide range of diving mammals and birds as well as many other species. In each case animals were restrained and forced underwater while the heart rate was being monitored. Working with free ranging telemetered alligators I found they often went underwater without a reduction of heart rate. By accident I discovered what has long been called diving bradycardia was a response to fear and not submergence. Later I found the same results with swamp rabbits. When forcibly submerged they too showed the classic diving bradycardia response. In sharp contrast, when they voluntarily submerged underwater there was no slowing of their heart. As a result of this study scores of earlier studies had to be repeated with using radio telemetry of unrestrained animals and what had been called diving bradycardia was instead a response to fear and not to submergence. Another sacred cow was slain.

Introduction

The importance of accidental discoveries in science is well known and was discussed in the previous chapter. In 1929, P. F. Scholander, Ph.D. of Norway began studying the heart rate response of seals to forced diving. That was the same year Cannon first described the fight or flight response. It was a good year for physiology. Scholander did not publish his definitive work until over a decade later (Scholander, 1940). Since that time, other scientists have studied the response of many animals to diving and most show a marked reduction in heart rate while submerged (Anderson, 1966). The term "diving bradycardia" was coined to describe the response. That term is misleading and has become yet another sacred cow in science. In the previous chapter I detailed my own accidental discovery of the fear response in a submerged alligator. I will now discuss my work with swamp rabbits.

I encountered my first swamp rabbits, *Sylvilagus aquaticus*, while studying alligators in east Texas. Although somewhat similar in appearance to the more common Eastern cottontail rabbits *Sylvilagus floridanus*, swamp rabbits are excellent swimmers and divers and prefer wetlands to forest and prairies. Swamp rabbits proved to be an excellent choice.

Swamp rabbit, *Sylvilagus aquaticus*

116

Nine mature swamp rabbits were collected from the Rockefeller Wildlife Refuge in Grand Chenier, Louisiana. ECG radio telemetry transmitters were surgically implanted and the rabbits were studied in the laboratory and under free ranging conditions on a 3 acre island in Lake Tenkiller in Eastern Oklahoma. As with other animals I had studied using radio telemetry, free ranging and captive telemetered swamp rabbits, *Sylvilagus aquaticus* exhibited a marked bradycardia (-63%) while hiding.

As mentioned in the previous chapter the heart rate in mammals is influenced by both the sympathetic and parasympathetic portions of the autonomic nervous system. Parasympathetic activity slows the heart rate by releasing acetylcholine near the sinoatrial node or pacemaker region of the heart. In contrast, the sympathetic portion of the autonomic nervous system releases norepinephrine into the body which increases heart rate.

When the neural blocking drug atropine was given to swamp rabbits in order to inhibit the slowing action of the parasympathetic outflow, resting heart rate increased as expected, but fear still resulted in the heart slowing. However the heart rate slowed much more slowly. Conversely when propranolol was given to block sympathetic influence the resting heart rate was reduced as expected, but was farther reduced when the animals was frightened. When both atropine and propranolol were given the change in heart rate with fear was abolished. Of course this must be the case as all connection to the central nervous system was blocked by the simultaneous administration of these two drugs. These data clearly indicate that both portions of the autonomic nervous system are important for the normal rapid and marked slowing of the heart rate when frightened swamp rabbits hide. Free ranging and captive telemetered swamp rabbits, *Sylvilagus aquaticus* exhibited a marked bradycardia while hiding. Oxygen consumption is also reduced significantly when swamp rabbits were frightened (Smith, et. al. 1981). I investigated the heart rate response to both voluntary and forced diving in swamp rabbits (Smith and Tobey, 1983). For voluntary dives I trained telemetered swamp rabbits to dive underwater for a food reward. They did so repeatedly without slowing their hearts as was the case with free ranging undisturbed telemetered alligators making volunteer dives.

In order to study forced diving I lowered telemetered swamp rabbits underwater in a small cage for ten seconds. On average their heart rates dropped from 213 to 44 beats per minute. Surprisingly this response is totally caused by increased parasympathetic activity as evidenced by its abolishment with atropine treatment. Forced diving frightens animals and is not at all like voluntary or purposeful diving. Both fear and forced submergence results in marked slowing of the heart rate.

When the neural blocking drug atropine is given to swamp rabbits it inhibits the release acetylcholine and thus inhibits the slowing action of the parasympathetic outflow. As expected the resting heart rate increased, but fear still resulted in the heart slowing. However the heart rate slowed at a reduced rate. It took longer to get the full effect of the stimulus. Conversely when propranolol was given to block sympathetic accelerating influence, the resting heart rate was reduced as expected, but was farther reduced when the animals were frightened. When both atropine and propranolol were given the change in heart rate with fear was abolished. Of course this must be the case as all connection to the central nervous system was blocked by the simultaneous administration of these two drugs. In this case the heart rate is determined by the pacemaker alone with no central nervous system influence. Similar results have since been found with a host of other animals. The bradycardia had become yet another sacred cow, but has finally been put out to pasture. Others need to follow.

References

Anderson, H. T. 1966. Physiological adaptations in diving. Physiol. Rev. 46:212-243.

Causby, L. A. and E. N. Smith. 1981. *Control of bradycardia in the swamp rabbit, Sylvilagus aquaticus.* Comp. Biochem. Physiol. 69C:367-370.

Scholander, P. F. 1940. *Experimental investigations in respiratory function in diving mammals and birds.* Hvalrad. Sk. 22:1-131.

Smith, E. N., K. Sims and J. F. Vich, 1981. Oxygen consumption of frightened swamp rabbits, *Sylvilagus Aquaticus.* Comp. Biochem. Physiol. 70A:533-536.

Smith, E. N. and E. W. Tobey. 1983. *Heart rate response to forced and voluntary diving in swamp rabbits, Sylvilagus aquaticus.* Physiol. Zool. 56(4):632-638.

Smith, E. N. 2006. *Passive Fear: Alternative to Fight or Flight.* Published by iUniverse New York, Lincoln, NE. and Shanghai.

Author bio

E. Norbert Smith grew up on the family farm in western Oklahoma and earned a BS degree in Biology from Southwestern Oklahoma State University. While attending college he designed a radio telemetry system for tracking rattlesnakes and worked with Dr. Hobart Landreth at the Wichita Mountains Wildlife Refuge. He earned a MS degree from Baylor University. While at Baylor he designed and used a sophisticated multichannel radio telemetry system to monitor three different temperatures and heart rate of free ranging alligators. His doctorate in Zoology is from Texas Tech University where he continued studying thermoregulation of alligators. He has roughly 100 technical papers on such diverse topics as the design and application of radio telemetry devices, thermoregulation of free ranging reptiles and the cardiovascular response of wild animals to fear. The BBC filmed a documentary about his alligator research called *A Smile for the Crocodile* and he was invited as keynote speaker to a major radio telemetry conference at Oxford University and took three research students with him to the event. In spite of these accomplishments he was denied tenure for rejecting evolution and became a truck driver. He is retired and still lives on the family farm, but remains active in the Creation Research Society and other Creationist organizations. He now spends most of his time writing books and gardening. See the author bio in the previous chapter for additional information about Dr. Smith. His email address is docgater@aol.com

Animal Rights. Animals Having Rights Is a Big Lie!
J. Y. Jones, M.D.

Editor's note

Even as a high school boy, I remembered reading about the fundamental dietary change for people before and after the Genesis Flood, yet few preachers mention it and few Bible scholars write about it. At the end of Creation God said, I ***"give you every seed-bearing plant on the face of the whole earth and every tree that has fruit with seed in it. They will be yours for food"*** (Genesis 1:29 NIV). This is in sharp contrast with God's words after the flood. ***"The fear and dread of you will fall upon all the beasts of the earth and all the birds of the air, upon every creature that moves along the ground, and upon all the fish of the sea; they are given into your hands. Everything that lives and moves will be food for you. Just as I gave you the green plants, I now give you everything"*** (Genesis 9:2-3, NIV). It is obvious man changed from a vegetarian diet to a diet that included meat. Life on earth was forever changed.

Let me also add, I have taught a university course about the use and care of laboratory animals for research and as Dr. Jones says, there are actually many more laws regulating the use of animals in research than there are laws regulating the use of humans as research subjects. Most people are unaware of this fact. There is another reason animals used for research are well cared for and healthy: Results from sick or frightened animals would not be nearly as useful.

I have long found it telling that during particularly harsh winters when deer, elk and other wild animals are starving, it is never the environmentalists or animal rights activists that come to the rescue. It is always hunters who pay to provide hay and other food substances, sometimes even air-dropped from airplanes, to get the creatures through the winter. This important fact seems to always be missed by the media. As rock star Ted Nugent has said: if you want to help animals, buy a hunting license!

Dr. Jones discusses these things and more in a way I have not seen before, providing much food for thought. He is also an avid big game hunter and professional hunting guide. As a medical missionary and ophthalmologist he has given sight to thousands of blind people. Perhaps due to his medical training, he writes in a forceful and provocative way that I find both uncommon and stimulating. Read and ponder what he has to say.

Introduction

This chapter is an effort to capture the essence of my recent book, *Worship Not the Creature: Animal Rights and the Bible*, for a very good cause. Of necessity it cannot be as complete as that well-documented work, but it is worth the effort to utilize a moderate amount of the same information to augment and increase the influence of *Sacred Cows in Science, no Objectivity Allowed*. We will look at the animal rights movement from several angles in this brief treatise.

First, we will inspect the movement from the standpoint of evolution, a completely erroneous and easily disprovable concept that nevertheless dominates academia and our educational institutions, as well as being the official state religion of most nations around the world, including our own. It is easy to demonstrate that evolution, even were it true (which is impossible mathematically and otherwise), does not allow for special treatment of animals in any way, and that "survival of the fittest" is distinctly at odds with animal rights dogma.

Second, the animal rights position is indefensible from the standpoint of its historical perspective. Animals have never had rights and in fact such a ridiculous position has never before been proposed until the prosperity, urbanization, and disconnection from the land that occurred in the twentieth century.

Third, the concept of animal rights makes no sense from a medical, physiological, and nutritional standpoint. The human race cannot exist without consuming animals and/or animal products in order to obtain essential Vitamin B12, available only through animal products. Several vital amino acids are almost impossible to obtain, as well, except from meat or other animal products. To seriously propose otherwise is to put the entire human race in jeopardy. It is an abrogation of authority and responsibility unprecedented in human history, whether one believes in the evolutionary fairy tale or the record left by the Creator in His own handbook for living, the Holy Bible.

Fourth, animal rights activism is a bogus concept because it flies in the face of simple logic. No living being can have rights unless it is able to exercise responsibility. Animals are, in the words of one prominent author of Scripture, "unreasoning animals, to be captured and killed." We will look into this argument in some detail.

Fifth, the animal rights movement is primarily about developing and augmenting a legion of deceived donors, and this greedy truth has little or nothing to do with the welfare of animals. The only manner in which there exists any direct relationship can be summarized with the phrase "donor deception," which means keeping their fat cats in the dark about how their donations to the animal rights movement are actually spent. Any good that accrues to any animal or population of animals is purely accidental in practically all cases, and instances of animals destroyed or damaged by the animal rights movement far outnumber the times actual benefit has been achieved. We will discuss this deception as fully as possible.

Sixth, the economic impact of following the radical animal rights philosophy would be devastating. Most of the world, contrary to allegations emanating from big cities on the east and west coasts of the USA, still is agrarian and highly dependent on the animal resource for their most basic economic needs. It is prudent to expose this state of affairs.

Seventh, and most completely covered in my work referenced above, the very concept of animal rights is biblically absurd. I cannot make all the arguments that prove this point in such a brief chapter, nor can I cover the terrible consequences predicted in the Bible that are sure to result from following such a deviant philosophy, but I will hit the high points. For more information please read my recent publication, *Worship Not the Creature: Animal Rights and the Bible*, since practically the entire book is on this particular subject.

Evolution

More than 24,000 actual scientists have rejected evolution (www.ridgecrest.ca.-us/~do_while/-sage/v5i1of.htm) in favor of creation or "intelligent design," the latter of which is little more than a euphemism for creation. Dr. Russell Humphreys, a physicist at Sandia National Laboratories in New Mexico, is quoted as stating there are some ten thousand practicing scientists in the USA alone who believe in a six-day creation! (Ashton, 1999). This figure is more than ten years old, and the number doubtless continues to grow as serious deficiencies in evolutionary theory are increasingly exposed, and the overwhelming evidence for a young Earth and universe are brought to light. This astounding number of converts to the most logical conclusion has occurred not because of faith in the God of the Bible nearly as much as the immense volume of contradictory evidence one can observe by simple application of the time-honored scientific method.

Evolution is the epitome of non-science, the capstone of a godless philosophy that is totally disinterested in finding truth. The only goal of the evolution movement is to eliminate Creator God from the equation, and that mainly for reasons of sexual freedom (fornication, adultery, homosexual activity, abortion, etc.), since the record left for us by that all-powerful Being specifically forbids such acts. Evolutionary theory is well-countered by several other chapters of *Sacred Cows*, so I won't belabor the point. Suffice it to say that the evolutionary underpinnings of the animal rights movement rest on a shaky and unreliable foundation of shifting sand. There are many excellent resources to prove this point beyond doubt, of which I have found one to be especially outstanding, a book developed by people with mind-boggling credentials (Stack,

2009). My own book reveals my personal journey as a dedicated evolutionist who, through the study of the human eye, simply changed my mind.

Historical Perspective

Today's animal rights movement is an outgrowth of empty lives, plain and simple, a futile effort to find meaning apart from the great Creator God. The only fulfillment that is deep, lasting, and satisfying is found by accepting Jesus Christ as Savior and Lord and living according to His perfect Word, which is the progressive revelation of the all-powerful Being who created all that exists. Until this past century, and to some degree in the nineteenth century, there existed no mechanism to avoid the concept of Creator God as described in the Holy Bible. Virtually all mankind sought something greater than themselves, though only those who encountered the living Christ found the all-important Truth that transforms the human heart from sinner to saint.

The movement to embrace evolution as the new god of the universe was almost completely foreign before Darwin published his ill-advised and erroneous book, *Origin of Species*, in 1859. Even Darwin may have had doubts about his improbable theory, though accounts of his "death-bed confession of faith" in Christ appear to have been exaggerated or even fabricated, according to most sources. It really doesn't matter, because such a confession from him is surely unnecessary in view of the vast body of truly scientific evidence that evolution has not occurred, and cannot occur.

Animals have always been considered the servants, workers, companions, and source of needed goods for the human race. Even the new "theory" didn't result in the animal rights movement, however. For this radical change to occur, a fall from faith (or its possibility) was required, a fall into total hopelessness by the adherents of such a philosophy. I quote in my own previously-mentioned book, "Empty lives are easily filled with foolishness." I believe this is the only plausible basis for the concept of animal rights and the radicalism that often accompanies it. A person filled with the Holy Spirit of God through the knowledge of Jesus Christ cannot be so easily deceived, because they live full lives brimming with purpose and meaning. Until Darwin, most of the human race had this same attitude, though naturally skeptics and self-willed individuals have always been a prominent aspect of human society. In many instances whole societies reject the truth and choose to live in error. This, too, is covered fully by the Holy Bible, and the outcome well-predicted. The animal rights movement is destined to fulfill many end-times prophecies, all of them highly destructive and costing the lives of many millions of people, as well as innumerable animals. This will be covered more in the final section of this chapter.

The Nutritional Necessity

Even people who believe the Bible is God's Word can nevertheless be poorly schooled in its basic concepts, an increasing fact in today's world. However, with only a little digging, major findings emerge that square perfectly with today's real scientific knowledge. We must actually go back to the very beginning to understand some of these well-known facts.

It is a truth that animal and human life cannot exist without Vitamin B12. A deficiency of this vitamin can lead to several problems, the most devastating being pernicious anemia, a form of blood disease in which there are not enough red blood cells in the body. A severe lack of oxygen-carrying ability then leads to all kinds of symptoms, including unusual tiredness, shortness of breath, and other lethargic symptoms. A deficiency which is not corrected in a timely fashion leads to permanent nerve damage, as well as heart disease, coronary disease, deterioration of the entire nervous system, and irreversible brain damage. When inadequate amounts of B12 are present in the body, regeneration of the myelin sheath that covers all nerve cells in the body is severely retarded or absent. This hinders the brain's ability to function properly and the ability to handle stress; (www.naiaonline.org/body/articles/archives/arterror.htm).

There is absolutely no plant source from which one can obtain this essential nutrient. This was apparently not the case when God created the Earth and all its diverse life forms, however. He gave specific orders in the Bible that green plants were to be the food for all creatures, including man (Genesis 1:29-30). Apparently Vitamin B12 was easily available from these plant sources before the worldwide flood that occurred in the time of Noah. However, the Vitamin B12 plant was apparently destroyed or rendered impotent by that worldwide cataclysm. God therefore gave immediate instructions after the people and animals emerged from Noah's ark: Mankind was now permitted to eat meat.

The only living things that can today manufacture the intricate molecule cyanocobalamin, or Vitamin B12, are microorganisms such as cyanobacteria and blue-green algae. These tiny animals exist in huge numbers in the intestines of today's true herbivores, and these higher animals absorb and store B12 in their tissues in large quantities as it is produced. In fact, the main difference between a vegetarian animal and a predator (including omnivores like bears and swine) is the presence or absence of sufficient B12-producing bacteria in the intestines to sustain life, and the ability to absorb and store this essential nutrient.

Human beings are firmly in the omnivore category, although contrary to failed evolutionary theory, humans are in no way animals. However, we do have a common Creator and thus certain biological and physiological similarities to the animal kingdom. Thus it is no surprise that God, in His absolute knowledge, was intimately aware that changed circumstances would require human beings to eat meat in order to survive. He also, in almost the same statement, placed a supernatural fear of man on every animal on the face of the Earth, protecting and preserving what I call His image-bearer (man) from even the fiercest of predators. Even today that illogical fear of man pervades the animal kingdom, causing even the most vicious of predators almost always to flee at the mere scent of a human being.

Proof of the veracity of the Holy Bible comes in many forms, but its absolute accuracy in the area of science is one of the strongest of these. Matters of which only God could know are stated in matter-of-fact terms, even though the human writers could not have known the underlying science, only to have even outrageous ideas vindicated millennia later by science. There are numerous examples, but God's order to "now eat meat" is an absolute proof. There is no other way Moses could have known of the necessity to eat meat except by divine revelation. In the same way, the entire Law of Moses contains not a single premise or order that has failed to stand up to any amount of scientific scrutiny. This fact alone proves the integrity of the Holy Bible, since the only educational background Moses had was the faulty science of ancient Egypt.

The bottom line is that no human being can live and prosper without ingesting animal products. Supplements of Vitamin B12 are invariably the work of animals, although very small ones. Thus, vegetarianism for human beings is a myth—a myth that must be dispelled completely if the scourge of "animal rights" is to be abolished from the world, as it rightly should be.

Unreasoning Animals

The eminent apostle and Bible author, Simon Peter was first instructed by the resurrected Christ to "*Arise, Peter, kill and eat*" in Acts 10:13. This was repeated three times to emphasize that all animals were now acceptable as food, with none of the constraints imposed on the Jewish people prior to the time of Christ (constraints designed to set them apart from other people, more than for any other reason). Peter later recounted this incident for his associates (Acts 11:5-10), indicating that any animal should now be considered clean, and could be killed for food if needed.

Anthropomorphism is attributing human characteristics to animals or caricatures of animals. It is entertaining when it comes in the guise of Mickey Mouse or Donald Duck, but it is taken to the extreme by animal rights activists, who consider human beings and animals as moral equivalents. In truth, animals cannot even be considered moral agents in the slightest degree. Peter's later writings compare non-humans to false teachers of the Scriptures, referring to them as *"unreasoning animals, born as creatures of instinct to be captured and killed ..."* (2 Peter 2:12).

This summarizes the essence of God's attitude toward animals. They are to be used for human good whenever and wherever such usage is appropriate or even expedient. Down through the centuries this has always been the case, until the Darwinian movement arose to erroneously equate man with animals. In centuries past, it was perfectly permissible for a cavalryman to shoot his horse and use its carcass for cover if he were under heavy enemy attack. Today he might be prosecuted for animal cruelty, if radical animal rights advocates have their way.

Even creatures that might be on some "endangered list" have been eaten to meet human needs in times of famine. One major example occurred during the Boxer Rebellion in China, where every Père-David's deer in the country was killed to feed hungry soldiers. These creatures survive today only because of the efforts of an eminent sportsman, the Duke of Bedford, who had a few surviving specimens in England.

As if to punctuate the warped thinking being perpetrated on society today, a member of President Obama's cabinet believes animals should be permitted to litigate against human beings for mistreatment of any kind, using human lawyers as their representatives in court. Such twisted thinking is not a new phenomenon, but for its proponents to have the ear of the President of the United States of America is certainly a radical and ominous development.

Since an animal cannot testify in court, it certainly should not be allowed to sue in court. In order to respond to the queries of lawyers, an animal would necessarily be required to reason out its responses. Even a talking parrot only repeats a sound it has heard hundreds or thousands of times. Peter astutely and accurately puts his finger on three major differences between animals and humans: Animals can't reason; taking their lives to meet human needs is God-ordained; and animals are good for use as human food. None of these characteristics applies to normal human beings.

There are copious animal cruelty laws on the books, and all compassionate people are against outright animal cruelty of any kind. Utilization means to kill or harvest in the most humane fashion possible, whether killing an animal for meat, shearing it for wool, or using it to produce milk or eggs. The hard task becomes to assure that such laws aren't used to prevent legitimate and time-honored human utilization of animals, something that is being done in wholesale fashion even today. If this line of thinking ever prevails in its most radical form, the human race is doomed.

Animal Scam

I borrow the title of this section from my well-known friend, Kathleen Marquardt of the American Policy Center, who penned a best-selling book by that name. It was from her that I learned the real truth about how deep mendacity runs in mainline animal rights organizations. Her book is still a valuable resource for those who want to know the truth about this insidious and dangerous movement.

The animal rights movement has nothing to do with (or is only incidentally related) to traditional animal welfare or "being kind to animals." I remember as a boy sympathizing greatly with animals whenever someone brought up the subject of neglected pets, making animals fight one another for human amusement, failing to feed penned animals, and the like. In those days the American Society for the Prevention of Cruelty to Animals, the American Humane Society, and many other such organizations, actually *cared* how animals were treated. They were funded by like-minded individuals who knew their doctrinal positions and supported such groups financially. These organizations unashamedly supported traditional uses of animals such as meat production, dairy farming, rodeos, circuses, zoos, hunting, fishing, and medical research. Their concern was that animals be treated in a humane fashion and be sacrificed as necessary in as painless a way as possible. Theirs was, in fact, very close to a true biblical position on animal stewardship. The times and the population were different back then, too, though—society in general was biblically literate, whereas today most of American society is illiterate in this regard (as well as functionally illiterate in many other areas, in what amounts to a total failure of whole educational system under the heavy hand of centralized government power. This should be the subject of another chapter of this book).

My how things have changed today! Almost all the old mainline animal welfare organizations have been infiltrated and hijacked by radicals of the "animal rights" bent, and many newer, more powerful groups have arisen as well. In fact, traditional animal welfare organizations are now the exception rather than the norm, as competition for big-dollar donors has required ever more outrageous stunts and radical steps toward animal equality with humans.

One huge difference today is that most of these groups play one set of donors against others, with contributing members being given the impression their particular set of priorities is being pursued, while at the same time a different goal is followed by the organization. Anyone who believes that contributing to some animal rights group will advance the welfare of pets is totally in the dark regarding the official policy of that organization toward "animal slaves" of all kinds. The more radical groups have myriad pet owners as major donors, but they don't trumpet their opposition to pet ownership to this group. That tidbit is reserved for truly radical donors who want to change the whole human/animal relationship. As far as pet owners are concerned, their giving goes toward doing away with medical research involving animals, the cruelty of hunting, or some other acceptable goal. Usually opposition to medical research is disguised as opposing the use of animals for research on cosmetics!

Another changed circumstance is the total disregard mainline animal rights groups have for the *actual* welfare of animals. They should be ashamed of their opposition to removing enough deer from a herd to make the rest healthier, better fed, and more able to survive and thrive. Instead, they so oppose hunting that even when it benefits a population of animals immensely (as regulated sport hunting *always* does), they are opposed to it on so-called principle. Their principles have nothing to do with the health and welfare of animals, however; the overriding principle is the amount of publicity they can generate for their cause and their organization, and how much money they can raise in the process.

The third major change that has occurred is the actual encouragement and support given to acts of terrorism on behalf of their "cause" According to the National Animal Interest Alliance, *Animal rights and environmental extremists do more than demonstrate and push radical legislation. They also use physical assaults, intimidation, vandalism, harassment, theft, property destruction and terrorism. During the past two decades these attacks have escalated ... this is an international movement and major acts of terrorism and petty acts of vandalism and harassment occur virtually every day ...* (www.nass.usda.gov).

This group maintains a running list of animal rights terrorist acts going back to 1983, with close to one hundred such incidents occurring most years. Interestingly, practically all these occur in highly developed countries, mainly in the USA and Western Europe. This underscores the fact, noted in the "Nutritional Necessity" section above, that it is an impossible fact for most of humanity to survive without products from higher animals (i.e., they don't have available the advanced supplements made of bacterial products to provide certain nutrients, especially Vitamin B12). This thought leads us into the next section.

Economic Impact

In developing cultures, it is suicide to go without meat and/or dairy products or eggs, so the animal rights movement has nothing to offer except death. Even in countries that would be considered modern, utilization of animals is a huge part of the economy. In the USA, over nine million dairy cattle produce 190 *billion* pounds of dairy products (some 95 million tons) each year. Some 30 billion pounds of broiler chickens are produced in the USA each year, having a value of some $35 billion. This figure is matched by beef production, as some 100 million cattle produce 30 billion pounds of beef each year worth about $35 billion

129

(www.nass.usda.gov). These statistics show only a small part of the value of animals as a food source in only one country.

In developed countries with higher education rates and more available resources, it is *possible*, though by no means best for one's health, to get by without products from higher animals. It is *not possible* to be a pure vegetarian at all, if one considers bacteria to be members of the animal kingdom (which they are). In countries where high-tech supplements from bacteria and yeast are not available, the animal resource *must be used* in order to sustain life. This is the explanation why animal rights activism is not tolerated outside of the developed Western world and its offshoots, such as Australia and New Zealand.

Herding and domestic production of numerous animal products have been going on since before the flood of Noah, as evidenced by the fact that the son of Adam and Eve, Abel, was a herdsman. He was approved by God in this practice, and in fact his animal offering was accepted by God in the biblical account in Genesis 4. His brother Cain's vegetarian offering was rejected by God, leading to a fit of jealousy and the first human death when he murdered Abel.

Animals have always been used by the human race for myriad products, from clothes to dairy products to meat. Northern peoples even used animal products exclusively, and were even able to construct dog sleds and homes out of animal materials. In virtually all cultures around the world, including our own, animal products is a priceless commodity that simply cannot be replaced with vegetarian products. These domestic animals are able to eat and transform into essential nutrients plants that grow naturally on otherwise nonproductive ground. It is a fact that more than two-thirds of the land mass in the world is too cold, too rocky, or too mountainous for growing plant crops.

Were some world dictator to arise who mandated vegetarianism for all, it would be a catastrophe. Not only would economies around the world collapse, but starvation and disease would kill millions or billions of people. Just such a scenario is predicted in the Bible, which brings us to the final point of this chapter.

The Bible Tells Me So

As already mentioned, my book, *Worship Not the Creature: Animal Rights and the Bible,* covers this topic in depth, and I must reference that work again as a recommended source of information. The short story on this subject is that the Bible in *no way* supports the concept of animal rights. The concept of animal rights, as stated previously, is firmly based on evolution, not creation by an all-powerful, all-knowing Creator God as revealed in the Bible. Therefore, it is not surprising to me that the Bible refutes completely the weird, late-arriving doctrine of animal rights. The Bible does firmly support good animal stewardship and humane treatment of animals, in a manner similar to the philosophies of old-time animal welfare groups.

From Adam and Eve (clothed in animal hides after God Himself performed animal sacrifices to temporarily cover their sin) to Jesus Christ, utilization of animals is prominent and is presented positively. Jesus Himself ate fish, fed fish to thousands, and apparently caught and quite obviously cooked fish (John 21:1-14). He definitely ate red meat in order to participate in the Passover meal and thus fulfill the Law of Moses perfectly. He stated that he came to fulfill the Law (Matthew 5:17), and He had to do it perfectly in order to become the sufficient sacrifice for the sins of all mankind. To somehow skip the highly significant Passover meal, which consisted of roasted lamb, would have left Jesus as an incomplete keeper of the Law and thus unworthy to be the all-sufficient sacrifice.

Virtually every writer of the Bible, from the Old Testament prophets to the apostles who wrote the New Testament, refers to humane animal utilization of all kinds, and they give very specific and direct approval of the practice. The references are too numerous even to include all in my comprehensive book on the question, but suffice it to say God unhesitatingly approves animal utilization.

In what is known in eschatology (the study of future things) as the End Times, a world ruler will arise who will wreak havoc on the human race. I believe he will be much like Adolph Hitler: an animal rights activist and vegetarian, intent on world domination, and determined to impose his personal values and ideals on the whole human race. He will lead billions to destruction, and much of the suffering will come about because of four great disasters predicted in Revelation 6:8, which are: war, famine, disease, and *the wild beasts of the earth*. Many people miss this last phrase, even when they read it. War will come about because there will be some resistance to the warped new world order; famine will occur because the animal resource will be denied people; and disease will occur because of malnutrition and the impossibility of animal research to develop and manufacture vaccines and medicines. The role of wild beasts needs a little more clarification.

This final antichrist will create a scenario which allows wild beasts to kill at will. Of course, anyone who opposes God and Christ is an antichrist, by definition, but none will measure up to this "beast", as he is called in Revelation. When the "rapture", or removal of the Church universal occurs (1 Thessalonians 4:15-17), all who have received Christ as Savior and Lord will be suddenly gone. I believe God at that very instant will also remove the unnatural fear of man that affects all animals now, the fear imposed when Noah left the ark. We are seeing an increase in wild animal attacks around the world as a premonition of this huge catastrophe. Add this to the specter of disarmament of all people in the world, and you have a recipe for just such disaster. Even in cities, the antichrist will mandate freedom for all zoo animals and pets. Large packs of vicious dogs and fearless lions, tigers, and hyenas will feast on human flesh at will.

There is far more in the Bible about animal rights than I can cover here, but suffice it to say that the role of the animal rights movement as portrayed in the Bible is not that of "good guy" It is the role of destroyer and accomplice of Satan as the "serpent of old" plays out his own role to its conclusion.

Research the subject for yourself. Read the Bible on your own and see if what I'm saying isn't truth. First, though, get the Holy Spirit to help you. Ask Christ into your life, trust Him with your life, your hope, your all. He promises to indwell you by His Holy Spirit, who will help you to understand the Holy Bible (John 14:26). No amount of animal rights activism could bring such fulfillment!

References

www.ridgecrest.ca.us/~do_while/sage/v5i1of.htm (accessed November 2009).
Ashton, John F., *In Six Days*, Master Books, 1999, p. 284.

Brown, Walt, *In the Beginning*, 8th edition, available at
www.creationscience.com.
Stack, Carol, "Vitamin B12—Necessary for Your Heart and Brain," http://ezinearticles.com/?VitaminB12---Necessary-for-Your-Heart-and-Brain&id=404175, accessed December 2009.
www.naiaonline.org/body/articles/archives/arterror.htm, accessed December 2009. www.nass.usda.gov, accessed December 2009.

Author Bio

J. Y. Jones is a native of Georgia and was educated in its public schools and colleges. He is a graduate of North Georgia College, where he was inducted into the Nu Gamma Scholastic Honor Society. He graduated from the Medical College of Georgia and did an ophthalmology residency at Walter Reed Army Medical Center in Washington, DC. He is board certified by the American Board of Ophthalmology, has been in private ophthalmology practice since 1976, and has done almost twenty-thousand eye operations. He is a Vietnam veteran, and was awarded the bronze star medal, three air medals, and the army commendation medal for his service there. As a writer, he has achieved success in the outdoor press, where he has authored almost three hundred articles and ten books (*Impossible to Fail*, (1999). Providence House Publishers, Franklin, TN; *One Man, One Rifle, One Land* (2001—now in its second printing); *Ask the Elk Guides* (2005)*, Ask the Whitetail Guides* (2006), *Ask the Black Bear Guides* (2007), *Ask the Mule Deer Guides* (2008), and now in press *Ask the Brown Bear Guides* (2010), all these by Safari Press, Huntington Beach, CA, www.safaripress.com. He is under contract to write *Ask the Wild Sheep and Goat Guides* for this publishing house for release in 2011. In 2009, his book, *Worship Not the Creature: Animal Rights and the Bible* was published by Nordskog Publishing, Inc (www.nordskogpublishing.com). He is also working on a major book on Eurasian big game hunting, where he has hunted some thirty countries, to be completed in 2011.

His fiction works are *Lightspeed to Babylon* and *Crossing the Burn*, available from Hoy Publications (www.booklocker.com). Dr. Jones is active in numerous outdoors, wildlife, and conservation organizations, and regularly contributes to several outdoor publications, including *Sports Afield* and *Safari* magazines, as well as *The Hunting Report*. Most importantly, he is also a committed evangelical Christian who has traveled to Honduras and Jamaica twenty-three times to do eye surgery for disadvantaged people and to use his surgical skills as an evangelistic outreach. He financially supports more than a dozen Christian missionary and relief organizations. He is a frequent speaker around the country at church wild game suppers and other events. You may email him at: jyj@jyjones.com.

The Endangered Species Act:
Over Extended and Flat Broke
J. Y. Jones, MD

Editor's note

D r. Jones discusses an important contemporary topic few have the courage to address. Again as an avid big game hunter and physician he brings new insight into this discussion. Certainly we must be good stewards of all of God's creation and care should be taken to avoid the extinction of plants and animals. As Christians we see all living things as evidence of the Creator ... the work of His hands and all of creation and especially living things must be protected from extinction, but at what cost? Again Dr. Jones documents how the protection of some species costs and in some cases is far more than most people realize. There will be opposition to some of his views on this important topic, but we must be informed about not only this issue, but also what it is costing all of us.

Introduction

Two California condors were released in January 1992 in native habitat in California.[1] The cost of this program so far (over thirty years in the making up to this point) is some $30 million dollars. There were only a handful of these rare, giant soaring birds when the few survivors were captured in 1987 and placed in a captive breeding program. Today some 150 fly wild and free over the Far West, at a cost of some $200,000 per bird. The conservation community is not against spending tax dollars in situations where the expenditure is warranted. The bird is back from the brink of extinction.

As expensive as these birds have been, this type of scenario is exactly what the public expected when the Endangered Species Act (ESA) was passed in 1973. I was a young man then, and to me it seemed eminently sensible to protect such animals with a Federal law. Little did I know that sinister forces of which I was unaware would not let the California condor remain a typical case study.

Moving Rapidly From Function to Abuse

The snail darter, which was used to hold up construction of Tennessee Valley Authority's (TVA) Tellico Dam, was known to exist in other areas and yet the US Fish and Wildlife Service (FWS) listed it under the ESA, claiming it was nearly extinct. Actually, opponents of the dam had gone all out to find some species, any species, to stop the project. This small, unimportant fish was used to this end, and the ploy almost succeeded.[2] Most Americans were completely unaccustomed to such an impact from a Federal law, but it took years of litigation to finally break the logjam and complete the project. Estimates of economic losses are staggering, and the tiny fish was eventually found to be common in many other waters nearby. End of episode, but it was only the beginning of the story of the much-abused ESA.

It was a naïve American public that assumed we were going to protect valuable plant and animal resources from decline and possible extinction with the ESA. It was even more adolescent to believe that an expanding government could in any way be cost-effective in saving endangered animals. This law, which has turned out to be the epitome of the word *draconian*, has spawned a huge, power-grabbing bureaucracy that continues to burgeon. It has also distracted FWS from its primary mission of managing the nation's wildlife resources. This agency alone now spends some **1.4 billion** dollars each year on ESA issues.[3] At least three dozen additional Federal agencies spend many millions more dollars on ESA-related costs, with estimates of total expenditures running as high as high as four billion dollars and more. Our nation can no longer afford this illogical and usually counterproductive extravagance.

Most people think of eagles, woodpeckers, grizzly bears, and other warm-blooded animals when they think of endangered species. The fact is that of the top ten most costly animals on which ESA funds are spent, nine of the ten are fish[4]. The other is the Stellar's sea lion, of which there exist more than 100,000 today. There has been a decline in these animals which environmentalists attribute to commercial fishing activity (mainly, as usual, to shut down human utilization of virtually **any** natural resource), even though only about thirty sea lions per year are victims of fishing nets, a highly acceptable mortality factor in a population this size. These animals were placed on the ESA as threatened in the early 1990s.[4] (This animal is also protected by another unnecessary and costly act that needs repeal, the Marine Mammal Protection Act. The fact of the Stellar's sea lion's decline in numbers even before that act was passed is simply evidence of this act's similar ineffectiveness.)

The desert tortoise is listed on the ESA even though there are an estimated three million tortoises in Western deserts of the US and another 100.000 in captivity. Government reports state the decline of the desert tortoise is due to raven predation,[5] stating, "Raven predation will lead to the extirpation [total extermination] of the tortoise population" In an eminently logical response, the Bureau of Land Management (BLM). decided to kill 1,500 ravens. An environmental group immediately sued the BLM. The BLM settled out of court and agreed to kill only 56 ravens providing it could be shown that the ravens killed were "habitually preying on tortoises." What's going on here, is it tortoises or ravens that are endangered? Or is it either? And quite obviously for the radical "environmentalist" who cares so long as human progress is stymied?

A pond weed, which was found on private land in Texas, was petitioned for listing on the ESA without the knowledge of the landowner.[6] FWS workers simply trespassed to conduct their research studies. This is a pattern of abuse that has occurred time after time as government zealots seek more and more control over private property.

A snail prevented a retired veteran in Kanab, Utah from building a recreational vehicle park and tourist site on his own property, even though experts disagree as to whether this snail is endangered or not, and even whether or not it is only a subspecies.[7] (Most government officials involved with the ESA apparently believe the acronym stands for "Endangered Subspecies Act.") FWS alleged an endangered snail (Kanab amber snail). was found on this individual's property. An "emergency" listing of the snail was obtained. The veteran was told he could not use his property and has no option but to sell it to the government.

Shrimpers in the Gulf Coast states of Texas and Louisiana are being harassed by the National Marine Fisheries Service (NMFS) over a non-native sea turtle, the Kemp's Ridley sea turtle.[8] This animal does not naturally nest in the United States, but in Mexico. Government biologists bring the eggs to the US, hatch them on South Padre Island in Texas, and release the hatchlings into the Gulf of Mexico. The NMFS has conducted over 1,100 indiscriminate searches of shrimp boats under the guise of turtle protection. Not one turtle has been found on these boats. Fines for failing to have a functioning Turtle Excluding Device (TED) on shrimp nets range from $8000 to $25,000.

A $240 million ($240,000,000) dollar world-class telescope project in Arizona was stopped by a small squirrel.[9] The University of Arizona wanted to build this observatory on Mount Graham in southeastern Arizona, where an observatory already exists, but the Mount Graham red squirrel (genetically not a distinct species at all but a subspecies of the plentiful common red squirrel) is listed as endangered on the ESA. It was claimed that if the telescope were built, the squirrel would become fixated with human activity, forget to collect nuts, and ultimately fall prey to its prime predator, the goshawk.

The sockeye salmon is listed as endangered on the Colombia, Snake and Salmon Rivers of the Pacific Northwest, even though it is exactly the same species (not even a separate subspecies) that is plentiful and at historical population highs in many other river systems.[10] This ESA listing thus only affects 900 miles of river systems in the Pacific Northwest. This recovery program will cost between $200 million and $1 billion dollars in the next five years - a price to be paid by electric rate-payers, farmers, river operators and commercial fishermen (as well as those who dine on fish).

Clarkea australia is a small one- to two-inch high plant, but has shut down logging in parts of California. It's not because the plant is in the logging areas per se, but because the logging areas are located in areas where the Forest Service said the plant "might" be located. This tenacious little plant is in no way endangered, fends well for itself, and will even grow in skidder tracks left by intense logging activity.

Some sample FWS expenditures in fiscal year 1990 on threatened and endangered species include: The northern spotted owl ($9.7 million); least bell's vireo [small, green bird similar to the warbler] ($9.2 million); grizzly bear ($5.9 million); red cockaded woodpecker ($5.2 million); Florida panther ($4.1 million); Mojave Desert tortoise ($4.1 million); bald eagle ($3.5 million); ocelot ($3 million); jaguarundi [slender, short-legged wildcat] ($2.9 million); and the American peregrine falcon ($2.9 million). The highest costing bug was the valley elderberry longhorn beetle ($952.000) and the highest plant expenditure was the northern wild monkshood ($226.000). Note: These figures are not the total amount spent on endangered species that year, only what can be itemized by species.[11]

Four trash fish in Colorado—the squawfish, two types of Chub, and a sucker, which, until a few years ago were systematically poisoned by FWS, are now listed under the ESA. Their recovery is estimated to cost $60 million dollars and impacts costs are ten times that amount, over $650 million. Meanwhile, in the state of Washington, anglers are paid $3 by FWS for every squawfish caught that measures over eleven inches.

A Maryland couple could not prevent erosion from destroying their home, which was built on a 60 foot cliff overlooking Chesapeake Bay, because it might harm an insect known as the Puritan tiger beetle. An official from the Maryland Natural Heritage Program stated that protecting bug habitat was more important than protecting the couple's home.

A $100 million golf course and resort complex was stopped by a butterfly (the Oregon silverspot butterfly). A man sought to turn a fenced cow pasture and an area strewn with beer cans and vehicle tracks on the Oregon coast into a world class golf course. To help save the butterfly, he met with FWS personnel and even hired a butterfly expert. All his good-faith efforts were to no avail. Construction could not commence unless the man could guarantee he could build the golf course without killing a single endangered silverspot butterfly.

FWS wants to save hybrid animals as well as purebred animals (never mind the fact most animals listed on the ESA are actually **not** species at all, but subspecies of much more common animals). One examples of a costly hybrid is the dusky seaside sparrow. When efforts involving millions of dollars spent failed to save the sparrow, a hybrid program was tried, and it also failed. The red wolf (a hybrid cross between the grey wolf and coyote) is another costly example.[11]

In a well-known boondoggle, government officials designated 6.9 million acres of forest in the Pacific Northwest to be set aside for a subspecies of owl genetically identical to the common spotted owl (the reserved area for this subspecies, which many argue was never endangered—or even threatened—is larger than the states of Massachusetts and Rhode Island combined). The government's estimate of job loss is 33,000 jobs. Private sources say 60,000 jobs is a more accurate figure. Additionally, landowners of some three million more acres in the Pacific Northwest were told they could not harvest timber on their own land due to the presence of this northern subspecies of spotted owl, even though seven million acres of government-owned wilderness had already been set aside for the owl.

An elderly couple in Georgia, needing money for medical expenses, sought to sell timber on their private land only to be stopped by an arguably common bird, the red cockaded woodpecker, which many deer hunters in Georgia has seen from time to time from their deer stands, often pecking the very tree the hunter is leaning against. The bird was not found to live on the couple's land, but FWS and Georgia Forestry Commission officials reportedly discovered seventeen trees with "possible" abandoned red cockaded woodpecker nests. Nobody, including the FWS, had ever seen this woodpecker on the property belonging to this couple. The family has lived on the land for eighty years, yet a temporary ban on timber harvest was imposed. This bird requires pine trees eighty years old or more for nesting habitat, so naturally Georgians never allow their pine trees to attain this age, lest their property be seized if the birds move in.

The Columbia white-tailed deer has been listed as endangered for years and millions of dollars have been spent on its recovery. The same can be stated for the Florida Keys white-tailed deer. Studies reveal that both subspecies are "genetically plain-Jane white-tailed deer."[11]

An Ongoing Pattern of Ineffectiveness and Costliness

Some 1,600 species have been listed under the ESA, but only a small handful have been "recovered" and removed from the list. In some years hundreds of species and subspecies are listed by FWS, never to see any progress in their "recovery" despite massive infusions of hard-earned (and increasingly scarce) taxpayer money. By all accounts, the reasons for delisting the few "recovered" species had little or nothing to do with the ESA's efforts to recover them. Seven species were delisted because they are extinct—the Tecupa pupfish, the longjaw cisco, the blue pike, the Santa Barbara song sparrow, Sampsons pearly mussel, the Amistad gambusia, and the dusky seaside sparrow. Sixteen species were taken off the list because they were listed in the first place due to data errors: the Mexican duck, the Pine Barrens tree frog, the Indian flap-shelled turtle, the Bahama

139

swallowtail butterfly, the purple-spined hedgehog cactus, the Tumamoc globeberry, the spineless hedgehog cactus, the Mckittrick pennyroyal, the cuneate bidens, the eastern brown pelican, the Palau fantail, the Palau dove, the Palau owl, the American alligator, the Rydberg milk-vetch, and the gray whale. The Arctic peregrine falcon, which was decimated by the pesticide DDT, was delisted because it recovered on its own after the 1972 ban on DDT. Australia's eastern gray kangaroo, the red kangaroo, and the western gray kangaroo were delisted as a "response to Australian policies"[12] Having hunted in Australia, I know that the eastern gray kangaroo has never been threatened or endangered; landowners are issued permits to kill thousands of them because of overpopulation problems, and the fact that kangaroos pull up grass by the roots while grazing, leading to a totally barren landscape if the animals are not culled frequently.

The ineffectiveness of the ESA in helping species that are, or may be, actually endangered is legendary. According to a 1997 Hoover Institution report, most of ESA's activity involved listing species instead of instituting methods that would help listed species to recover. For example, of the $171,811,000 that federal and state agencies spent to "protect" the 639 endangered domestic species listed in 1991, 50% went to just seven species, and 90% was spent on 54 species. The remaining species had to fend for themselves.[13] This same pattern continues today, although the amount of wasted public funds has climbed astronomically in amount during the interim. Suffice it to say that the taxpayer is being ripped off regularly by practically all efforts involving species listed under the ESA.

Consequences of Gross Abuse

The extent of destructiveness to the U.S. economy by enforcement of the ESA cannot be overestimated. The instances cited above are but the tip of the proverbial iceberg. Vast areas of land, both public and private, have been locked up for commercial and recreational use, the northern spotted owl used to shut down timber harvest in the Pacific Northwest being one of numerous examples.

The ESA has also been a prime method government attorneys have used to deprive individuals of their constitutional right of protection of private property. For example, there would be the distinct threat of a $100,000 fine and mandatory prison sentence for killing an "endangered" wolf that is in process of attacking a farmer's livestock. Such an action would likely be deemed a felony, as well, and the fine could thus be raised to a maximum of $250,000 in that case.[14]

The "takings clause" of the US Constitution is the portion of the Fifth Amendment that says "nor shall private property be taken for public use without just compensation." This clause is one of the few parts of the Bill of Rights that authorizes the government to violate individual liberty, since under the takings clause, the government, exercising so-called eminent domain, can compel a person to sell his property. But although the framers unfortunately believed the government would sometimes have to compel the sale of property, they also were aware of the potential for abuse that government power always represents, so they inserted the "just compensation" phrase. Increasingly, the ESA and other similar measures have been used by the government to get around this inconvenient obstacle to mandating what uses shall be allowed for private property, and to thus deprive individuals of their property without just compensation, i.e., restriction on use and/or hostile acquisition of land as critical habitat for endangered species. Such "regulatory takings" are different from the concept of eminent domain, where private property is taken for a project beneficial to the public at large, and where a fair market value must be paid for the property. Numerous examples could be cited where the value of private property (and usually the timber or other resources found on the property) has plummeted by ninety percent or more when an endangered species is found on the land, resulting in marked restriction by the government of its use.

According to one well-informed source, "It should be understood that a landowner's right is in the physical property, not its value. The value—or market price—is the result of other people's estimation, which cannot be owned. If someone's land falls in value from $1 million to $500,000 as a result of voluntary action (changes in consumer tastes or in the surrounding area, for example), no rights have been violated. One cannot own a particular set of market conditions. Thus, when government restrictions on the use of private property cause the value to fall, the government's offense is not theft of value. That is merely the result of the government's real offense: the restriction on the owner's peaceable actions, which is enforced, ultimately, at the point of a gun"

"If we apply the principle of liberty to the takings issue, we easily see that, in fact, taxpayer compensation for regulatory takings is not the right solution to this problem. It may have the beneficial effect of restraining government activity. But there is no justice in forcing taxpayers to compensate property owners for theft committed by bureaucrats. That merely substitutes one act of theft for another. The real source of the crime is the takings clause itself. It violates individual liberty, and the only corrective is to amend the Constitution and get rid of it. The government should never be able to compel a person to sell his property"[15]

As a hunter and a conservationist, an aspect of the ESA that bothers me perhaps more than any other is how it is used by FWS to thwart legitimate conservation initiatives, not only in the U.S. but especially overseas. FWS agents systematically use the ESA to ignore or interfere in the conservation and wildlife programs of smaller and poorer nations. In unashamed violation of the spirit, if not the letter, of international treaties such as the Convention on International Trade in

Endangered Species (CITES), we systematically issue and observe, or choose to ignore, quotas and conservation programs that are given full authority by the rest of the world. This particularly applies to hunting trophies, though it spills over into numerous other areas as well. It becomes clearer every year that FWS, which was at one point in our history a major advocate of hunting as a conservation tool, has now been infiltrated by anti-hunters and preservationists to the point it is now probably the U.S. Government's most blatant example of anti-hunting sentiment.

It should be emphasized that the ESA is totally redundant in every one of these international actions of interference, since international law (CITES) is already applicable almost everywhere, and is accepted as authoritative by the rest of the world. While the U.S. participates in CITES, FWS mostly ignores quotas set by the organization, insisting on either totally ignoring such quotas or else setting their own quotas. As an alternative, sometimes FWS chooses to apply other US laws (such as the Marine Mammal Protection Act) to shut down imports.

The Cure

In view of these many abuses of the ESA, plus the fact our nation is facing bankruptcy and cannot afford to offer full government protection to every imaginable species of plant and animal, along with expensive and redundant enforcement and permitting processes, the following actions must be taken to abolish the ESA in its present form, and to reform and improve an entirely new ESA.

In applying the provisions of the new ESA, patriotic Americans should insist that the protection of the US Constitution for individual citizens must take precedent over any endangered species issue. We should also demand that any species to be listed must be in imminent danger of worldwide extinction, not extinction in some specified portion of its range.

Virtually all the economic damage done by the ESA has been inflicted by listing subspecies, not species. We must be adamant in all reform language that subspecies that are genetically identical to a plentiful species inhabiting another location cannot be listed under any circumstances (scientific name = genus + species *only*). This simple action alone would eliminate whole categories of abuse and save billions of dollars while effecting no loss whatsoever to the biosphere as a whole.

The new, revised ESA must include only vertebrate animals and higher (flowering) plants, and specifically exclude lower plants, insects, and animals which should be targeted for elimination as pests (such as rats and mice). The new ESA must have provision for evaluation of any proposed listed species as to economic impact of such listing, and no listing shall occur if a positive economic impact on the overall health of the U.S. economy cannot be demonstrated.

It is a fact that there have been millions of "natural" extinctions in past Earth history. Therefore, since extinction is inherent to the natural world and in some cases is even a desirable phenomenon, we must insist that a precondition for listing under the new ESA be that man's activities are the primary reason for a species' population decline. Naturally occurring population declines must be specifically excluded as candidates for efforts to reverse that decline by listing such species under the ESA.

In view of the damage done worldwide by misguided efforts to mandate certain "conservation behavior" in other sovereign nations, we must insist that provisions of the new ESA be applied only to qualifying animals and plants in the United States or its territories. It must be prohibited for US bureaucrats to hinder, hamper, interfere with, or destroy the conservation programs of other nations, most of which adhere strictly to the CITES treaty recognized as authoritative by practically all nations. CITES, and CITES alone, should be the authority when it comes to import/export regulations and procedures, quotas, and enforcement.

It is perhaps even more important that U.S. landowners be given positive incentive to have endangered or threatened species on their lands. The status quo, under which the U.S. Constitution is trampled underfoot in the name of whatever species or subspecies of questionable abundance can be found on a citizen's land, must not be allowed to prevail.

References

Jurek, R.M., California DFG Nongame Wildlife Section, http://www.dfg.ca.gov/wild-life/nongame/t_e_spp/condor/ (accessed March 2010).

Murchison, K.M., "The Snail Darter Case: TVA Versus the Endangered Species Act," Lawrence University Press, Vol. 17 No. 8, August, 2007, pp.720-723. http://www.libraryindex.com/pages/3033/Endangered-Species-ActENDANGERED-SPECIES-ACT-SPENDING.html (accessed March 2010). http://www.marineconservationalliance.org/press/pr20091116.pdf

Kristan, W.B. III, Boarman, W.I., "Spatial Pattern of Risk of Common Raven Predation on Desert Tortoises," *Ecology,* 84(9), 2003, 2432–2443.

Whittall, J.B., et al, "Cryptic Species in an Endangered Pondweed Community (*Potamogeton,* Potamogetonaceae), *American Journal of Botany.* 2004;91:2022-2029.

Littlefield, J., "Endangered or Not? Taxonomy of the Kenab Ambersnail," http://www.snr.arizona.edu/files/shared/person/Culver_Kanabambersnail.df (accessed March 2010).

NOAA Fisheries Office of Protected Resources, "Kemp's Ridley Sea Turtle," http://www.nmfs.noaa.gov/pr/species/turtles/kempsridley.htm (accessed March 2010).
http://www.mountgraham.org/WhitePapers/squirarticle.html (accessed March 2010)
http://www.nmfs.noaa.gov/fishwatch/species/sock_salmon.htm (accessed March 2010).

Much of the above information comes from: http://www.aws.vcn.com/flawed.html.

Annett, A. F., "Reforming the Endangered Species Act to Protect Species and Property Rights," November 13, 1998, The Heritage Foundation.
Simmons, R. T., "Fixing the Endangered Species Act" in Terry Anderson, ed., *Breaking the Environmental Policy Gridlock* (Stanford, CA: Hoover Institution, 1997), p. 82.18 U.S.C. § 3571(b) (3), (5).
Richman, Sheldon, "Takings: The Evils of Eminent Domain," *Freedom Daily* The Future of Freedom Foundation, July 1995.

Author bio

J. Y. Jones is a native of Georgia and was educated in its public schools and colleges. He is a graduate of North Georgia College, where he was inducted into the Nu Gamma Scholastic Honor Society. He graduated from the Medical College of Georgia and did an ophthalmology residency at Walter Reed Army Medical Center in Washington, DC. He is board certified by the American Board of Ophthalmology, has been in private ophthalmology practice since 1976, and has done almost 20,000 eye operations. He is a Vietnam veteran, and was awarded the bronze star medal, three air medals, and the army commendation medal for his service there. As a writer, he has achieved success in the outdoor press, where he has authored almost 300 articles and ten books (*Impossible to Fail*, (1999) Providence House Publishers, Franklin, TN; *One Man, One Rifle, One Land* (2001—now in its second printing); *Ask the Elk Guides* (2005)*, Ask the Whitetail Guides* (2006)*, Ask the Black Bear Guides* (2007)*, Ask the Mule Deer Guides* (2008), and now in press *Ask the Brown Bear Guides* (2010), all these by Safari Press, Huntington Beach, CA, www.safaripress.com. He is under contract to write *Ask the Wild Sheep and Goat Guides* for this publishing house for release in 2011. In 2009, his book, *Worship Not the Creature: Animal Rights and the Bible* was published by Nordskog Publishing, Inc (www.nordskogpublishing.com). He is also working on a major book on Eurasian

144

big game hunting, where he has hunted some thirty countries, to be completed in 2011.

His fiction works are *Lightspeed to Babylon* and *Crossing the Burn*, available from Hoy Publications (www.booklocker.com). Dr. Jones is active in numerous outdoors, wildlife, and conservation organizations, and regularly contributes to several outdoor publications, including *Sports Afield* and *Safari* magazines, as well as *The Hunting Report*. Most importantly, he is also a committed evangelical Christian who has traveled to Honduras and Jamaica twenty-three times to do eye surgery for disadvantaged people and to use his surgical skills as an evangelistic outreach. He financially supports more than a dozen Christian missionary and relief organizations. He is a frequent speaker around the country at church wild game suppers and other events. You may email him at: jyj@jyjones.com.

Section Two

Physical Science

Physical Science Introduction

Charles W. Imes

The Physical Sciences are under attack; and, irrespective of what you hear, it is not from the religious right. The beneficial results of the Physical Sciences to this current world and America in particular have been spectacular. There are very few Americans who are not directly benefited by these incredible results every single day. Scientists, especially those in the Physical Sciences, are probably esteemed above every other major profession in this country.

The success and the esteem of this sublime profession have NOT been lost on the politicians and the ideologues of this country and world. Every political agenda must have some "justification" for their policies, especially destructive policies. Many in the major environmental groups have used the facade of caring for the environment as a smokescreen for their "real" agenda. This leftist agenda is so strong that the "results" of their scientific agenda are more important than using "objective" scientific methodology.

Obviously, no one is truly objective. So, why should we expect scientists to be any more objective than men and women in other vocations? This is an important question and needs to be answered.

Thomas Sowell in his book *Intellectuals and Society* says this, "Verbal virtuosity has enabled many intellectuals to escape responsibility for filtering reality to create a virtual reality more closely resembling their vision. Some among the intelligentsia inflate to insoluble levels the problem of choosing between filtering and non-filtering, and then dismiss critics as expecting the impossible—namely, perfect objectivity or complete impartiality. "None of us are objective," according to the *New York Times'* public editor. Of course no one is objective or impartial. Scientific *methods* can be objective but individual scientists are not—and need not be. For that matter, mathematicians are not objective, but that does not mean that quadratic equations or the Pythagorean Theorem are just matters of opinion. Indeed, the whole point of developing and agreeing to objective scientific **methods** is to seek reliable information not dependent upon the subjective beliefs or predilections of particular individual scientists or on any hope that most scientists would be personally objective. If scientists themselves were objective, there would be little need to spend time and effort to work out and agree upon objective scientific methods.

Even the most rigorous scientist is not objective as a person or impartial in scientific pursuits. Scientists studying the growth of cancer cells in human beings are clearly not impartial as between the life of those cancers and the lives of human beings. Cancers are not studied just to acquire academic information but precisely in order to learn how best to destroy existing cancers and, if possible, prevent new cancers from coming into existence, in order to reduce human suffering and prolong human life. There could hardly be any activity more partial. What makes this activity scientific is that it uses methods devised to get at the truth, not to support one belief or another. On the contrary, scientific methods which have evolved to put competing beliefs to the test of facts implicitly recognize how ill-advised it would be to rely on *personal* objectivity or impartiality among scientists. Although J. A. Schumpeter said, "The first thing a man will do for his ideals is lie," he also said that what makes a field scientific are "rules of procedure" which can "crush out ideologically conditioned error" from an analysis. Such rules of procedure are an implicit recognition of the unreliability of personal objectivity or impartiality.

Why isn't there a huge political controversy about the quadratic formula? Why isn't there a huge political controversy about the formula for gravitational pull? Why isn't there a huge political controversy about Einstein's Theory of Relativity even though there are competing theories?

Competing theories in science are a mainstay of good science. Bad theories have been replaced by good theories since the beginning of science. The "good" push out the "bad" But, when the "end result" or "agenda" is more important than scientific procedures, then the nucleus of what made the physical sciences so successful has been undermined. So, the reason why the quadratic formula, the formula for gravitational pull, and the theory of relativity have not had huge political controversies is that they are either independently provable or there is NO agenda.

Arguably, the two most "controversial" scientific theories in this modern era are Darwinian evolution and global warming. Why these two? The Theory of Darwinian evolution must be defended at all costs because the alternative is unthinkable by its advocates. If the evolution of life did not happen by chance (and the advocates of this theory know it is mathematically impossible), then you are compelled to believe in a designer. So, how do the advocates of this theory retain a virtual monopoly in academia for this theory? The agenda was really quite easy and very successful. When in positions of authority, especially in higher education, propagate yourself. When in the majority, suppress competing ideas and de facto, this has the same effect as filtering data.

The global warming agenda is part and parcel of this same strategy and tragedy. The agenda of nefarious governmental officials, both in individual countries and the New World Order, require complicity with science to further their agenda. If one simply said, "I'm going to tax everything that is living in this world simply because we want to control every individual in every country, there might be an adverse reaction.

150

Because of this, they needed a highly esteemed "objective" segment of society to give scientific reasons for their true agenda. However, the "facts" started to be problem. Therefore, the seasoned methodology that was used by the advocates of Darwinian evolution to suppress and filter data was implemented.

If I could make a comparison, science has become compromised in an analogous manner to how the free market has been compromised in the last two years. Progressive and liberals look at the free market system with distrust and look to the government as the savior of everyone. What happened to the housing market was not a free market issue. Government undermined the natural process of the free market system by eliminating risk. If there is no risk, the free market system cannot properly function.

If competing ideas in science are not allowed and if the physical sciences do not require a methodology to mitigate the natural bias of scientists, then the sciences which we have seen to be so beneficial to mankind, could be used by reprehensible governments and institutions to enable the implementation of tyrannical and despot governments.

The Nefarious Results of Scientific Ideologues
Charles W. Imes

Editor's note

Charles is a relatively new friend. We served together on the board of directors for Oklahomans for Better Science Education and have several interests in common. We often had our meetings in a guest house near his beautiful home in Oklahoma City. He is one of the most widely read individuals I know and an extensive personal library with over four thousand books. He has actually read all of them. His background in engineering gives him a somewhat different and perhaps more critical view into various science issues and he addresses several topics many more traditionally trained science professor type seem to miss. He also has a refreshing writing style and clarity I find uncommon ... must be due to all the reading he does or perhaps his engineering background.

> *Realists do not fear the results of their study.*
> Fyodor Dostoevsky

> *There is nothing so powerful as truth, and often nothing so strange.*
> Daniel Webster

Introduction

Science is not good and scientists are not good people. I know what some of you are thinking. You are thinking, "I know a lot of scientists and they are 'good, ethical people and I have seen the tremendous benefits of science to mankind. So, how can you say that science is not good and scientists are not good people?"

Obviously, I know that there are scientists that are "good people" and I especially know as a heart transplant recipient that science can be "good", i.e. wonderfully beneficial to mankind. But, I'm stating this proposition because there seems to be a naïve and childish acceptance of science and scientists being ethically "good" irrespective of the science that they are advocating or the nefarious consequences of their ideas. Properly understood, science does not make good people, but there are good people who are scientists.

Definition of Nefarious: Evil, Utter Immoral or Wicked

In and of itself, science is not good or evil in the same way that money is not good or evil in and of itself. Many people improperly quote 1 Timothy 6:10

153

which states: ***For the love of money is the root of all evil: which while some coveted after, they have erred from the faith, and pierced themselves through with many sorrows.*** That is much different than saying that money is the root of all evil. And yet, we all know that money has been and can be used for great evil. But, money is ethically neutral. Many "good" people use their money to take care of their own lives and then sacrificially help others with their discretionary money.

Scientists are usually a subset of intellectuals, albeit a more practical intellectual. Now, lest someone thinks I am anti-intellectual, I love to be around intelligent people. I crave their friendship and the mutually beneficial dialogue that I have with them. And, I really don't like to be around obtuse and willfully ignorant and uneducated people. But, being an intellectual does not require that a person be "good" History, especially the twentieth century, is dominated by the political implementation of destructive ideas of intellectuals.

Marxism and its political child communism murdered at least 100 million people in the last century. Add to that the step child of National Socialism and the death count increases by another 50% or more. The important thing to understand about this most tragic century is that these 100+ million deaths were the result of the nefarious ideas of intellectuals and, to a lesser degree, scientists. And yet, these self-anointed and self-appointed elitist took no responsibility for arguable the most tragic century in human history.

Not only is it naïve to think science and scientists are intrinsically "good", it is dangerous. Everyone needs to be held accountable for their actions and for their ideas. This is not to say that if a man reads a post-modern existentialist and then concludes that everything is meaningless and vain; and, then goes out and acts on that nihilistic worldview by committing a "meaningless" act of murder, that the existentialist should be tried for murder.

However, responsible humans need to be assertive in educating our children (and even adults) on the dangers of irresponsible ideas. Responsible parents warn their children about the dangers of over drinking, the dangers of careless driving, the consequences of capricious sex, and the dangers of irresponsible fiscal actions in their personal finances. Why don't we warn our children about the dangers of destructive ideas?

Many parents make statements to their children which not only do not prepare them for this onslaught of caustic ideas, but actually foster a mental framework of accepting these ideas. They tell their children, "You may be wrong, but I will defend your right to believe and say it with my life." Trust me, that is utter nonsense. But, in the post-modern existential world, how do parents tell their children that some ideas are right and some are wrong and that some are good and some are evil?

154

Understanding the Nature and Bias of All Human Beings

Not only is it naïve and dangerous to think that intellectuals and scientists are intrinsically "good", but it may be even more naïve and dangerous to think that they are less bias and more well intentioned than other groups of people.

The amount of tainted and filtered data that is now coming out publicly just in the climate warming debate shows this assertion without question to be true. If this was an isolated case, then one could "chalk" that up as an aberration. The bias of so called scientists in the global warming scandal is only a "tip of the iceberg" of the aggregate of faulty and destructive conclusions of scientific bias (Excuse the pun and analogy of the tip of the iceberg in the global warming scandal.)

This one politically motivated agenda has the potential of the most onerous usurpation of political power and taxation in recorded history. Why? Because, almost everything emits CO_2 and therefore falls under the umbrella of regulation and taxation by governments. Once science becomes motivated by political agendas and becomes funded for the same reasons, it is almost impossible to be devoid of a bias methodology and conclusions. The results tend to be motivated by the end, which is the agenda. Scientific inquiry will, as a rule rather than the exception, be controlled to accomplish the agenda so that funding and political policies can be justified fallaciously and given credence by the perceived "unsullied" and "objective" name of science. Obviously, if one takes the position that global warming is cyclic and independent of manmade emissions, then those advocates of this position are not just unlikely to get public grants, but will be ostracized publicly by their publicly funded peers and maybe even fired.

Hopefully without asserting a hyperbole, science for many is a "religious" cult. One of the primary definitions of cult is "a person, philosophy, or activity regarded with extreme or excessive admiration" When extreme or excessive admiration is present in temporal matters, then it is de facto impossible to be objective and without bias. For many scientists, science is the "end all" and their belief system does not allow them to believe anything except that which is provable by science. The problem is that many of these same scientists will lie, steal, and even use the power of the state to protect their "god" even if their hypothesis is not provable by their "god".

Michael Crichton, best known for his novels such as *Jurassic Park* was educated at Harvard University, Cambridge, Massachusetts, AB (summa cum laude) 1964 (Phi Beta Kappa). Henry Russell Shaw Travelling Fellow, 1964-65. Visiting Lecturer in Anthropology at Cambridge University, England, 1965. Graduated Harvard Medical School, MD 1969; post-doctoral fellow at the Salk

Institute for Biological Sciences, La Jolla, California 1969-1970. Visiting Writer, Massachusetts Institute of Technology, 1988. His credentials give him a very unique position of being a scientist, novelist, agnostic, and an environmentalist. His observations about those in the environmental "movement" are insightful and give some validity to at least the environmental movement as a "religious cult."

At the Commonwealth Club in San Francisco on September 15, 2003, Michael Crichton made the following remarks:

> "I studied anthropology in college, and one of the things I learned was that certain human social structures always reappear. They can't be eliminated from society. One of those structures is religion. Today it is said we live in a secular society in which many people—the best people, the most enlightened people---do not believe in any religion. But I think that you cannot eliminate religion from the psyche of mankind. If you suppress it in one form, it merely re-emerges in another form. You cannot believe in God, but you still have to believe in something that gives meaning to your life, and shapes your sense of the world. Such a belief is religious"

Today, one of the most powerful religions in the Western World is environmentalism. Environmentalism seems to be the religion of choice for urban atheists. Why do I say it's a religion? Well, just look at the beliefs. If you look carefully, you see that environmentalism is in fact a perfect twenty-first century remapping of traditional Judeo-Christian beliefs and myths.

There's an initial Eden, a paradise, a state of grace and unity with nature, there's a fall from grace into a state of pollution as a result of eating from the tree of knowledge, and as a result of our actions there is a judgment day coming for us all. We are all energy sinners, doomed to die, unless we seek salvation, which is now called sustainability. Sustainability is salvation in the church of the environment. Just as organic food is its communion, that pesticide-free wafer that the right people with the right beliefs, imbibe.

Eden, the fall of man, the loss of grace, the coming doomsday—these are deeply held mythic structures. They are profoundly conservative beliefs. They may even be hard-wired in the brain, for all I know. I certainly don't want to talk anybody out of them, as I don't want to talk anybody out of a belief that Jesus Christ is the son of God who rose from the dead. But the reason I don't want to talk anybody out of these beliefs is that I know that I can't talk anybody out of them. These are not facts that can be argued. These are issues of faith.[1]

Al Gore stated in his book entitled *Earth in the Balance*, said:

"The richness and diversity of our religious tradition throughout history is a spiritual resource long ignored by people of faith, who are often afraid to open their minds to teachings first offered outside their own systems of belief. But, the emergence of a civilization in which knowledge moves freely and almost instantaneously through the world has spurred a renewed investigation of the wisdom distilled by all faiths. This pan religious perspective may prove especially important where our global civilization's responsibility for the earth is concerned"[2]

Mark Landsbaum in an article dated 02/12/2010 and titled "What to Say to a Global Warming Alarmist" listed the following purposeful lies and malicious filtering of data to justify their presupposition position on this issue.

ClimateGate. This scandal began the latest round of revelations when thousands of leaked documents from Britain's East Anglia Climate Research Unit showed systematic suppression and discrediting of climate skeptics' views and discarding of temperature data, suggesting a bias for making the case for warming. Why do such a thing if, as global warming defenders contend, the "science is settled?"

FOIGate. The British government has since determined someone at East Anglia committed a crime by refusing to release global warming documents sought in 95 Freedom of Information Act requests. The CRU is one of three international agencies compiling global temperature data. If their stuff's so solid, why the secrecy?

ChinaGate. An investigation by the UK's left-leaning Guardian newspaper found evidence that Chinese weather station measurements not only were seriously flawed, but couldn't be located. "Where exactly are the forty-two weather monitoring stations in remote parts of rural China?" the paper asked. The paper's investigation also couldn't find corroboration of what Chinese scientists turned over to American scientists, leaving unanswered, "how much of the warming seen in recent decades is due to the local effects of spreading cities, rather than global warming?" The Guardian contends that researchers covered up the missing data for years.

HimalayaGate. An Indian climate official admitted in January that, as lead author of the IPCC's Asian report, he intentionally exaggerated when claiming Himalayan glaciers would melt away by 2035 in order to prod governments into action. This fraudulent claim was not based on scientific research or peer-reviewed. Instead it was originally advanced by a researcher, since hired by a global warming research organization, who later admitted it was "speculation" lifted from a popular magazine. This political, not scientific, motivation at least got some researcher funded.

PachauriGate. Rajendra Pachauri, the IPCC chairman who accepted with Al Gore the Nobel Prize for scaring people witless, at first defended the Himalaya melting scenario. Critics, he said, practiced "voodoo science." After the melting-scam perpetrator 'fessed up, Pachauri admitted to making a mistake. But, he insisted, we still should trust him.

PachauriGate II. Pachauri also claimed he didn't know before the 192-nation climate summit meeting in Copenhagen in December that the bogus Himalayan glacier claim was sheer speculation. But the London Times reported that a prominent science journalist said he had pointed out those errors in several e-mails and discussions to Pachauri, who "decided to overlook it." Stonewalling? Cover up? Pachauri says he was "preoccupied." Well, no sense spoiling the Copenhagen party, where countries like Pachauri's India hoped to wrench billions from countries like the United States to combat global warming's melting glaciers. Now there are calls for Pachauri's resignation.

SternGate. One excuse for imposing worldwide climate crackdown has been the UK's 2006 Stern Report, an economic doomsday prediction commissioned by the government. Now the UK Telegraph reports that quietly after publication "some of these predictions had been watered down because the scientific evidence on which they were based could not be verified." Among original claims now deleted were that northwest Australia has had stronger typhoons in recent decades, and that southern Australia lost rainfall because of rising ocean temperatures. Exaggerated claims get headlines. Later, news reporters disclose the truth. Why is that?

SternGate II. A researcher now claims the Stern Report misquoted his work to suggest a firm link between global warming and more-frequent and severe floods and hurricanes. Robert Muir-Wood said his original research showed no such link. He accused Stern of "going far beyond what was an acceptable extrapolation of the evidence." We're shocked.

AmazonGate. The London Times exposed another shocker: the IPCC claim that global warming will wipe out rain forests was fraudulent, yet advanced as 'peer-reviewed' science. The Times said the assertion actually "was based on an unsubstantiated claim by green campaigners who had little scientific expertise", "authored by two green activists" and lifted from a report from the World Wildlife Fund, an environmental pressure group. The "research" was based on a popular science magazine report that didn't bother to assess rainfall. Instead, it looked at the impact of logging and burning. The original report suggested "up to 40%" of Brazilian rain forest was extremely sensitive to small reductions in the amount of rainfall, but the IPCC expanded that to cover the entire Amazon, the Times reported.

Peer ReviewGate. The UK Sunday Telegraph has documented at least sixteen non-peer-reviewed reports (so far) from the advocacy group World Wildlife Fund that were used in the IPCC's climate change bible, which calls for capping manmade greenhouse gases.

RussiaGate. Even when global warming alarmists base claims on scientific measurements, they've often had their finger on the scale. Russian think tank investigators evaluated thousands of documents and e-mails leaked from the East Anglia research center and concluded readings from the coldest regions of their nation had been omitted, driving average temperatures up about half a degree.

RussiaGate II. Speaking of Russia, a presentation last October to the Geological Society of America showed how tree-ring data from Russia indicated cooling after 1961, but was deceptively truncated and only artfully discussed in IPCC publications. Well, at least the tree-ring data made it into the IPCC report, albeit disguised and misrepresented.

USGate. If Brits can't be trusted, are Yanks more reliable? The US National Climate Data Center has been manipulating weather data too, say computer expert E. Michael Smith and meteorologist Joesph D'Aleo. Forty years ago there were 6,000 surface-temperature measuring stations, but only 1,500 by 1990, which coincides with what global warming alarmists say was a record temperature increase. Most of the deleted stations were in colder regions, just as in the Russian case, resulting in misleading higher average temperatures.

IceGate. Hardly a continent has escaped global warming skewing. The IPCC based its findings of reductions in mountain ice in the Andes, Alps and in Africa on a feature story of climbers' anecdotes in a popular mountaineering magazine, and a dissertation by a Switzerland university student, quoting mountain guides. Peer-reviewed? Hype? Worse?

ResearchGate. The global warming camp is reeling so much lately it must have seemed like a major victory when a Penn State University inquiry into climate scientist Michael Mann found no misconduct regarding three accusations of climate research impropriety. But the university did find "further investigation is warranted" to determine whether Mann engaged in actions that "seriously deviated from accepted practices for proposing, conducting or reporting research or other scholarly activities." Being investigated for only one fraud is a global warming victory these days.

ReefGate. Let's not forget the alleged link between climate change and coral reef degradation. The IPCC cited not peer-reviewed literature, but advocacy articles by Greenpeace, the publicity-hungry advocacy group, as its sole source for this claim.

AfricaGate. The IPCC claim that rising temperatures could cut in half agricultural yields in African countries turns out to have come from a 2003 paper published by a Canadian environmental think tank not a peer-reviewed scientific journal.

DutchGate. The IPCC also claimed rising sea levels endanger the 55% of the Netherlands it says is below sea level. The portion of the Netherlands below sea level actually is 20%. The Dutch environment minister said she will no longer tolerate climate researchers' errors.

AlaskaGate. Geologists for Space Studies in Geophysics and Oceanography and their U.S. and Canadian colleagues say previous studies largely overestimated by 40% Alaskan glacier loss for forty years. This flawed data are fed into those computers to predict future warming.[3]

This scandal seems to be reminiscent of political and "religious" scandals. Many people lost faith in the important institutions in their lives because of their pervasive scandals. When pastors speak publicly and self-righteously about certain sins and then indulge in them, this taints the integrity of the entire institution. When churches protect priests that have committed the most abominable acts of pedophilia imaginable, the whole church suffers as well as its

160

adherents. When politicians use their position for personal gain and pleasure, the governmental institutions suffer as well as its citizens.

Jonanne Nova, an Australian freelance science presenter and writer, professional speaker and former TV host; author of *The Skeptics Handbook* gave this interesting quote and cartoon comparison of the skeptical and un-skeptical scientist on her website.

A transparent, competitive system where all views are welcome is the fastest way to advance humanity. The Royal Society (of London) is the oldest scientific association in the world. Its motto is essentially: *Take No One's Word For It*. In other words, assume nothing; look at the data. When results come in that don't fit the theory, a scientist *chucks out his theory*. A non-scientist has "faith", he "believes" or assumes his theory is right, and *tries to make the measurements fit*. When measurements disagree, he ignores the awkward news, and "corrects", or statistically alters, the data–always in the direction that keeps his theory alive. [4]

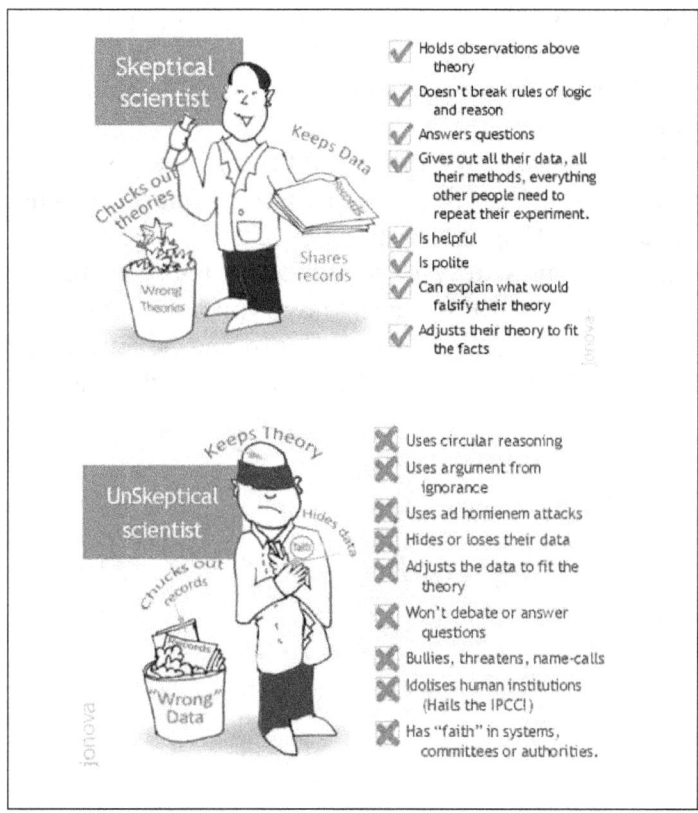

The Poverty and Consequences of Scientific Historicism

Years ago I made the statement that it is IMPOSSIBLE for macro-evolution to be true. The person I was discussing this with then said, "Prove it." Now knowing the inability of most to understand simple issues let alone complex issues, I asked, "Once I have proved it, how will you know that I have proved it? What is your standard of proof?" That ended the conversation and the person simply walked off. Every time that I have asked that question, it has ended the conversation. It shows that one of the most commonly used words in the scientific world is not generally understood to any extent by those who use it.

This concept of "Historic Science" may give some insights on why we are in such an educational mess in our sciences. In high school, I took biology as a sophomore, chemistry as a junior and physics in my senior year. I had few philosophical or theological issues in biology, no issues in chemistry, and no issues in physics. Why? Because at that time, biology was more "experimental" rather than ideological. We looked at animals, the human body, and all types of living organisms and made conclusions based on our observations through mechanism such as dissection. In chemistry, we all learned theory and then application. After that, hopefully, everyone would come to the same conclusion if their experiment was done correctly. Physics was much the same way. We learned the mathematical theory and formulas of physics and then applied that theory to practical experiments.

This was good science and proving a theory by a given methodology was required in these classes. But, because of the equating of validity of so-called "Historic Science" and "Experimental Science", the necessity to prove one's theory is not required. This same type of subjective nonsense runs rampant in our churches and the political realm where "how you feel about a proposition" is more important to many than "do you understand and believe what the scriptures or the constitution says?" A cartoon in the New Yorker illustrates this point quite dramatically.

" 'Give me liberty or give me death.' Now, what kind of person would say something like that?"

But, for those who have studied the nature of proof, it is self-evident (even axiomatic) that all proofs are NOT of the same validity or reliability. If that is the case, then there must be different methodologies for "proving" a theory or hypothesis. What are those different methodologies and what are the general principles in each method?

The most common methods of proofs currently are:

1. Mathematical and Logical Proof. This proof is "deductive" and has an extremely high reliability because you are dealing with conclusions which necessarily follow from given premises. Obviously, if one can use a mathematical or deductive logical proof, it is very difficult to refute that conclusion.

2. Scientific Proof. This proof is inductive and therefore less reliable. However, because the methodology is observable, repeatable and "testable", the ability to eliminate error is present in the method and it is very useful, albeit not perfect and subject to continual refinement and change. But, in

163

this temporal world, this has proven to provide a mechanism for repeated success in many areas of our life.

3. Legal Proof. Legal proof has its own set of rules of evidence. Obviously, since most things in a court of law deal with historic events, these events are not repeatable and therefore the conclusion of whether something has been proven is by "preponderance" of non-repeatable empirical evidence which a judge and jury determine if it is compelling beyond a "reasonable" doubt, not beyond any doubt.

4. Historic Proof. This is the least reliable of all "proofs" because what constitutes a "historic" proof is generally determined by the person's psychological, philosophical, theological, and ideological makeup. It even begs the question of, "What is history?" If I were an American Indian writing the history of America, I almost certainly would have a different perspective of history compared to an Englishman writing on the same subject.

Now, how is this pertinent to the issue of the controversy of science and to the nefarious results of ideological science? The answer is simple. Almost all of the controversy and much of the nefarious science being promoted has to do with the realm of science called "Historic Science" It is riddled with ideological bias and therefore it is needful that this bias be communicated by intellectuals and teachers; and then allows teachers and educators to question the validity of their assertions.

Again, it is naïve and even dangerous to think that intellectuals and scientists are objective. Karl Popper and David Berlinski have been adamant about making sure all know that scientists are no more objective than anyone else. But, the scientific method mitigates that bias because others have the ability to take theories and hypothesis and put them to test. However, this assumes a very important principle which is that is all those who are testing the model or hypothesis has access to unfiltered data and therefore can make conclusions based on the best set of data available.

But, what if data is systematically filtered because some scientists have become ideological activists rather than objective scientists? The result is that their conclusions are no more valid than a "philosophical" or "psychological" preference. And yet, this can actually be more harmful because the "illusion" of scientific objectivism has been fallacious applied to their ideological conjecture.

Again, what we have seen in modern science is the equating of the validity of "historic" and "experimental" science. The "historic" sciences make dogmatic assertions such as, "Macro-evolution is just as true as the theory of gravity" and few question that absurd declaration. The fact is that the fallacy IS the equating of

164

historic science with experimental science. THEY ARE NOT EQUALLY VALID AND THIS NEEDS TO BE EMPHASIZED AND UNDERSTOOD BY THOSE IN THE SCIENTIFIC REALM AND THE CULTURE AS A WHOLE.

The Results and Manifestations of Ideological Science

John Cornwell in his new book *"Hitler's Scientists. Science, War and the Devil's Pact"* asked the probing questions in the preface of this book.

> "Can we by studying the history of science in Germany in the first half of the last century draw significant conclusions about the relationship between science and the good society? Does doing science make human beings more rational, skeptical, internationalist, objective? And do we expect science to flourish better, and the discoveries of scientists to be used more responsibly and ethically, under democracies than under dictatorships?
>
> Exploring the story of the German science communities during a period that spans two world wars is an essential part of attempting to understand the nature of science and the conduct of scientists in the twentieth and the new century. Recounting the drama of ideas and discoveries, probing the behavior of Germany's researchers—towards their disciplines, towards their governments and regimes. towards their fellow human beings— cannot he separated from the task of understanding science and technology and its impact on all of us since World War II"[5]

A recounting of the human biological expermiments by brilliant Nazi scientists need not graphically be verbally depecited, but these events gives evidence of the principle of ethical nuetrality of science. Most people and scientists believe that Hitler was just an immoral thug. The fact is that Hitler promoted family values, the importance of exercise, the importance of eating right, was against abortion, and believe that homosexuality was immoral. His protrayal of his party was one of clean cut, upstanding members of society. Surely one could see the "good" of such social policies and see how they care in many respects consistent with Christian values.

But, the ugly truth is that the REASON why Hitler advocated these "virtues" was entirely different than why a Christian would advocate the same policies. But, taken independently of Hitler's other policies, it would have been difficult to see why even these policies were not "good" in a transcendent ethical sense.

What is the reason that Hitler advocated these "virtues" policies? The reason was that he and those who surrounded him were ideological Darwinists. Having experienced to some degree the reaction of well meaning Darwinists, I'm not saying that all Darwinists would act out their particular beliefs in the same ways that Hitler, Lenin, Stalin, and Mao Tse Tun did anymore than I believe that all existentialists who believe that there is no meaning or value to anything would act out that belief system by committing capricious and systemic murders. However, it is preposterous to think it is beyond the realm of intellectual and moral analysis to ask the question, "Were the greatest mass murderers (arguable of all time) vitally influenced by theories of Darwin? Did they 'justify' their worldview by these precepts and propositions?"

To illustrate this, I was once asked to debate a professor from OU on the radio on the subject of "Intelligent Design." Callers could literally ask any question except they were not allowed to state or ask questions about Adolf Hitler and the influence of Darwinists theories in his life. Why not? What would be the harm in an open format discussing this issue? We all want to simple see "where the truth leads", right? Wrong!!!! This was simply a microcosm of how "data" and "information" is filtered by ideologues to promote their agenda. Most people including scientists are "moral" and "ethical" people, albeit for different reasons.

But, let's take a "slippery slope" trip down the path of the naturalistic Darwinian world. In this debate with the OU professor, I asked him, "Did man evolve in exactly the same way as a cockroach?" He replied dogmatically, "Yes" Now, this simple question and simpler answer can have profound ethical implications. If I really believe this, then it begs the question, "Of what superior intrinsic value does man have over the cockroach?" If you answer to yourself, "None", then how does that affect my ethical and moral worldview?

I know that most people (including scientists) are not epistemologically self-conscious. In fact, most are epistemologically challenged.

Definition: Epistemology The branch of philosophy that studies the nature of knowledge, its presuppositions and foundations, and its extent and validity. Or, it can mean the theory of knowledge and how knowledge for a given subject is obtained. Self-consciousness deals with "awareness" Therefore, to be epistemologically self-conscious means you are aware of what and why you believe something. Some will call this "your Worldview"

Most people, including intellectuals and scientists, really don't know (and maybe don't want to know) why they believe what they believe and what are the implications of that belief. But some do, and if their worldview is "pure" Darwinian and "existential", then the results can be catastrophic because it will be based on an amoral or subjective ethic. I'm not speaking in just some hypothetical

166

and theoretical hyperbole. The reality is that actions proceed from ideas and ethics.

Nefarious Results of Ideological Scientists

Catastrophic nefarious results and agendas are almost always a result of theoretical ideas advocated by intellectuals and ideologue scientists, and then subsequently implemented politically by political activists. Some of the results are "unintended" and some are "intended." The extermination of the Jews by Hitler and the genocide of the Ukrainians by Stalin were "intended" and purposeful. The deaths by malaria of tens of millions in Africa after banning DDT may have been "unintended" (I'm giving the advocates of this decision the benefit of the doubt for argument sake; but, after reading a number of the works of those who advocated this, I'm now convinced it was "intended.")

It does seem that psychologically all actions by mankind need some sort of "rational" justification. I do not believe that Hitler, Lenin, Stalin, Mao Tse Tung, or Margaret Sanger were insane even though many do. Now, Nero, Kim Jung-il, Idi Amin Dada, Theodore Robert "Ted" Bundy, and Caligula were probably all insane by any standard. But, when asked, "Why did you do what you did?" many of their answers were to some degree rational. This reminds me of an existential/surrealist joke.

Question: How many existentialists does it take to change a light bulb?
Answer: Fish.
The question is, "Is this a rational or insane answer?"

Political activists, even though driven by their ideological worldview, generally need public validation of their beliefs from intellectuals and/or scientists. And, they can find such people to support their politics because they agree with the philosophy of the political activists or because they are de facto paid by those who hold the purse strings of the public treasury.

Let's review some of the more destructive historic and political "events" in recent history and see if this hypothesis holds true.

Abortion

The United States has averaged 1.3 million abortions each year since 1973. That is about the population of Maine. What would most people think if some sort of catastrophe happened each year that killed off the entire state of Maine, or New Hampshire, or Hawaii? The estimation for the catastrophe of the Katrina hurricane is about 1,836 deaths. The recent Haiti earthquake devastation resulted in about maybe 300,000 deaths.

Let's do a little math. If you take the Katrina death count and divide by the average number of abortions, the percentage is approximately 0.1412% which is about 1/7 of 1%. When you calculate the percentage of the Haiti deaths to the yearly abortion average, it is about 23%. I'm doing this only because many trivialize the whole issue of abortion.

It is obvious that only many do not see the merit of this comparison and argument because they have the ability through persuasive positions of power to define words and arguments to fit their agenda. It is not valid, in their minds, to compare the deaths of Katrina to abortions because this is an "apple" and "orange" comparison. But, why isn't it a valid comparison? Because, intellectual and scientists who are abortion ideologues have defined a pre-natal baby as a mass of tissue (like a cancer) and therefore can be discretionarily removed and destroyed.

It is a lot easier to conveniently define things as a non-human when you want to capriciously destroy it. The Nazis did that to the Jews and other undesirables. The communists did the same thing in classifying society into the proletariat and bourgeoisie. After redefining their victims, it was a "rational" decision to eliminate them. The Nazis and communists were no more insane than someone who sprays their garden to destroy destructive insects.

Why is it so easy today for women to have an abortion? The causation of anything is generally complex and compound. But, there is no doubt that the influence of Margaret Sanger was instrumental in politically changing the perception of this issue.

As was previously stated, ideas have consequences, especially when implemented by political activists. Margaret Sanger was one of the most highly esteemed activists in the "pro-choice" abortion world. What are the most prominent "ideas" of this highly esteemed activist?

Some of the better known ideas held by this "highly esteemed" intellectual "scientist" are eugenics, segregation, abortion, Darwinian, and Marxism. Here are a few quotes from Sanger who is promoted as being the "Founder of Planned Parenthood":

"The most merciful thing that a family does to one of its infant members is to kill it"*Margaret Sanger (editor). The Woman Rebel, Volume I, Number 1. Reprinted in Woman and the New Race. New York: Brentanos Publishers, 1922.*

"Birth control must lead ultimately to a cleaner race"*Margaret Sanger. Woman, Morality, and Birth Control. New York: New York Publishing Company, 1922. Page 12.*

"We should hire three or four colored ministers, preferably with social-service backgrounds, and with engaging personalities. The most

successful educational approach to the Negro is through a religious appeal. We don't want the word to go out that we want to exterminate the Negro population. And the minister is the man who can straighten out that idea if it ever occurs to any of their more rebellious members" *Margaret Sanger's December 19, 1939 letter to Dr. Clarence Gamble, 255 Adams Street, Milton, Massachusetts. Original source: Sophia Smith Collection, Smith College, North Hampton, Massachusetts. Also described in Linda Gordon's Woman's Body, Woman's Right: A Social History of Birth Control in America. New York: Grossman Publishers, 1976.*

"Eugenic sterilization is an urgent need ... We must prevent multiplication of this bad stock" *Margaret Sanger, April 1933 Birth Control Review.*

"Eugenics is ... the most adequate and thorough avenue to the solution of racial, political and social problems" *Margaret Sanger. "The Eugenic Value of Birth Control Propaganda" Birth Control Review, October 1921, page 5.*

"Birth control itself, often denounced as a violation of natural law, is nothing more or less than the facilitation of the process of weeding out the unfit, of preventing the birth of defectives or of those who will become defectives" *[no source available at this time ...]*

"As an advocate of birth control I wish ... to point out that the unbalance between the birth rate of the 'unfit' and the 'fit', admittedly the greatest present menace to civilization, can never be rectified by the inauguration of a cradle competition between these two classes. In this matter, the example of the inferior classes, the fertility of the feeble-minded, the mentally defective, the poverty-stricken classes, should not be held up for emulation ... On the contrary, the most urgent problem today is how to limit and discourage the over-fertility of the mentally and physically defective" *Margaret Sanger. "The Eugenic Value of Birth Control Propaganda" Birth Control Review, October 1921, page 5.*

"The campaign for birth control is not merely of eugenic value, but is practically identical with the final aims of eugenics" *Margaret Sanger. "The Eugenic Value of Birth Control Propaganda" Birth Control Review, October 1921, page 5.*

"Our failure to segregate morons who are increasing and multiplying ... demonstrates our foolhardy and extravagant sentimentalism ... [Philanthropists] encourage the healthier and more normal sections of the world to shoulder the burden of unthinking and indiscriminate fecundity of others; which brings with it, as I think the reader must agree, a dead weight of human waste. Instead of decreasing and aiming to eliminate the stocks that are most detrimental to the future of the race and the world, it tends to render them to a menacing degree dominant ... We are paying for, and even submitting to, the dictates of an ever-increasing, unceasingly spawning class of human beings who never should have been born at all"** *Margaret Sanger. The Pivot of Civilization, 1922. Chapter on "The Cruelty of Charity," pages 116, 122, and 189. Swarthmore College Library edition.*

"The undeniably feeble-minded should, indeed, not only be discouraged but prevented from propagating their kind"** *Margaret Sanger, quoted in Charles Valenza. "Was Margaret Sanger a Racist?" Family Planning Perspectives, January-February 1985, page 44.*

"The third group [of society] are those irresponsible and reckless ones having little regard for the consequences of their acts, or whose religious scruples prevent their exercising control over their numbers. Many of this group are diseased, feeble-minded, and are of the pauper element dependent upon the normal and fit members of society for their support. There is no doubt in the minds of all thinking people that the procreation of this group should be stopped.** *Margaret Sanger. Speech quoted in Birth Control: What It Is, How It Works, What It Will Do. The Proceedings of the First American Birth Control Conference. Held at the Hotel Plaza, New York City, November 11-12, 1921. Published by the Birth Control Review, Gothic Press, pages 172 and 174.*

"The marriage bed is the most degenerative influence in the social order ..."** *Margaret Sanger (editor). The Woman Rebel, Volume I, Number 1. Reprinted in Woman and the New Race. New York: Brentanos Publishers, 1922.*

"[Our objective is] unlimited sexual gratification without the burden of unwanted children ..."** *Margaret Sanger (editor). The Woman Rebel, Volume I, Number 1. Reprinted in Woman and the New Race. New York: Brentanos Publishers, 1922.*

"Give dysgenic groups [people with 'bad genes'] in our population their choice of segregation or [compulsory] sterilization"** *Margaret Sanger,*

170

April 1932 Birth Control Review. "As we celebrate the 100th birthday of Margaret Sanger, our outrageous and our courageous leader, we will probably find a number of areas in which we may find more about Margaret Sanger than we thought we wanted to know ..."

Faye Wattleton, Past-president of Planned Parenthood

Margaret Sanger, Founder of Planned Parenthood, *proposed the American Baby Code* that states, "No woman shall have the legal right to bear a child ... without a permit for parenthood."

Margaret Sanger, Founder of Planned Parenthood, *proposed the Population Congress* with the aim, "... to give certain dysgenic groups in our population their choice of segregation or sterilization."

Abortion was her primary means of implementing eugenics toward blacks. Today, black women have abortions at about three times the number of white numbers per capita. And, there are about 43 million abortions a year worldwide. How many people died in World War II? Good estimates seem to run around 60 million. But, that took five to six years from the period 1939 through 1945. Abortion kills that many in about 1.5 years. We have comparable deaths of World War II every eighteen months if we counted the unborn as a human being. But, all of this is simply just theoretical and philosophical hyperbole since intellectuals and ideologue scientists have redefined the unborn as a mass of tissue. But, what if they are not just a mass of tissue, but real human beings in a less mature state?

Eugenics

For thirty-two years, I developed software systems for various industries including two of the largest and most successful horse breeding and training farms in American. Even though money could be made by racing, many times the "real" money came when the stallion retired to stud. Stud fees can range from $2,500 to $500,000 per mare and have reached as high as $1 million.

Why is this important to our subject matter? As the old saying goes, "You don't get a thoroughbred by breeding two Shetland ponies" If horse breeders and all animal breeders know that the probability of getting a "superior" animal is greatly enhanced by selected breeding, why wouldn't that work for humans? In fact, it should be easy to see that physical traits are passed on to their progeny. Intellectual traits are more difficult because, even though there is a positive correlation between intelligent parents and their children, how do you prove that it is a result of genetic traits being passed on to their children and not a result of a superior environment? But, even with this question, studies of intelligence of twins who live in disparate environments show remarkable intellectual parity.

The Darwinian ideologue knows this better than almost any sector of America. Now, if he believes in Darwin's theories of natural selection and the survival of the fittest and denies a transcendent God, what would deter him to make the human race the best it could be?

Hitler simply implemented what intellectuals and ideologue scientists had verbalized in the United States. He simply was in a position to be more consistent in his policy and philosophy. Edwin Black wrote in an article, "The Horrifying American Roots of Nazi eugenics" which was published in the San Francisco Chronicle, said, "

Eugenics was born as a scientific curiosity in the Victorian age. In 1863, Sir Francis Galton, a cousin of Charles Darwin, theorized that if talented people only married other talented people, the result would be measurably better offspring. At the turn of the last century, Galton's ideas were imported into the United States just as Gregor Mendel's principles of heredity were rediscovered. American eugenic advocates believed with religious fervor that the same Mendelian concepts determining the color and size of peas, corn and cattle also governed the social and intellectual character of man.

Hitler studied American eugenics laws. He tried to legitimize his anti-Semitism by medicalizing it, and wrapping it in the more palatable pseudoscientific facade of eugenics. Hitler was able to recruit more followers among reasonable Germans by claiming that science was on his side. While Hitler's race hatred sprung from his own mind, the intellectual outlines of the eugenics Hitler adopted in 1924 were made in America.

During the '20s, Carnegie Institution eugenic scientists cultivated deep personal and professional relationships with Germany's fascist eugenicists. In *Mein Kampf*, published in 1924, Hitler quoted American eugenic ideology and openly displayed a thorough knowledge of American eugenics. "There is today one state" wrote Hitler, "in which at least weak beginnings toward a better conception [of immigration] are noticeable. Of course, it is not our model German Republic, but the United States"

Hitler proudly told his comrades just how closely he followed the progress of the American eugenics movement. "I have studied with great interest," he told a fellow Nazi, "the laws of several American states concerning prevention of reproduction by people whose progeny would, in all probability, be of no value or be injurious to the racial stock"

Hitler even wrote a fan letter to American eugenic leader Madison Grant calling his race-based eugenics book, *The Passing of the Great Race* his "bible."[6]

Did these intellectuals and ideologue scientists "CAUSE" Hitler to kill millions of Jews that he had relegated to sub-humans? This seems to be the "trump card" question of those who really believe everything that Hitler believed concerning the application of "natural selection" and "survival of the fittest" to man, they just in their pietistic verbosity were able to distance themselves from someone who could put their ideas into practice. It reminds me of the Calvin and Hobbs cartoon dealing with violence in media causing violence.

Of course, in one sense, the nefarious ideas of intellectuals and ideologue scientists did not "CAUSE" the actions of Hitler any more than the ideas of Marx

"CAUSED" the actions of Lenin, Stalin, Mao Tse Tung and Castro. But, to say that these ideas did not contribute to the actions of these despots would be foolish. When discussing causation, the ideologues who espoused the very ideas implemented by their political "children" can always assert deniability because their ideas were not "sufficient" to make something happen since that is the responsibility of the one who actually committed these heinous acts.

While nefarious ideas are not sufficient to cause onerous actions, they may very well be necessary. Many criminals had certain catastrophic events happened to them in their lives such as being molested when a child. Is that "sufficient" a reason for them becoming a hardened criminal? Even though liberal judges have moved to this conclusion, most still that the person is responsible; and, of course, there are many who went through the same type of cataclysmic experience and became model citizens. But, would the person who became a hardened criminal been different if he had not experienced that cataclysmic event? Probably not! Therefore, the event, even though it is not "sufficient" to cause someone to be criminal, it is likely that it was necessary.

Philosophical Darwinism combined with philosophical eugenics were arguable a necessary causation for the deaths of 10 to 100 million humans in the last century.

DDT

Is Rachel Carson guilty of more deaths than Adolf Hitler? Here is a prime example of what probably started out with "good" intentions from someone who wanted to preserve the "environment" and in particular preserve the hatching ability of certain birds. Sounds benign enough, doesn't it?

So, this "scientist" wrote an emotionally charged book in 1962 called *Silent Spring* which motivated probably over one million eco activists who "self-righteously" went on a Holy Crusade to ban the insecticide DDT worldwide.

Now, these "self-righteous-activists" were very aware of the fact that DDT had literally been a "God send" in eradicating malaria. Without fear of contradiction, the consequences of eliminating this insecticide have been catastrophic in many countries all over the world.

Was she really responsible for the deaths of as many as 50 to 100 million people? Obviously, these figures are just an estimate of the number of deaths caused by malaria since the ban of DDT, but it does show the enormity of the results of this action. Again, scientists and intellectuals as a rule are NOT held accountable for the results of their destructive and aberrant ideas. Did she go out and kill 50 million people? Of course not! Therefore, the nefarious ideologues in the scientific and intellectual communities give people like her a "free pass" since you can't conclusively prove causation.

What we do know is since 1962 when *Silent Sprint* was published that those who believed her "just so" story were responsible for banning DDT worldwide. And, we do know that DDT was THE most effective preventive method against malaria. And, we do know that the majority of those who have died from the new epidemic of malaria have been children. These millions of children and adults were literally condemned to death by nefarious ideological scientists and intellectuals who believed their "agenda" was more important than millions of lives of children and adults.

As a side note, if I accidentally killed someone, I would certainly show great remorse to the victim's family and would never ever forget I had inadvertently caused the death of another human. These "self-righteous-ideologues" seem to have no conscience over pronouncing a death penalty over 10s of millions of the most innocent of all humans. They not only have no conscience in this matter, they feel "morally" justified in doing it.

Concerning these types of "conscienceless/unconscionably" ideologues, C. S. Lewis once said, "Of all tyrannies a tyranny exercised for the good of its victims may be the most oppressive. It may be better to live under robber barons than under omnipotent moral busybodies. The robber baron's cruelty may sometimes sleep, his cupidity may at some point be satiated; but those who torment us for our own good will torment us without end for they do so with the approval of their own conscience"

Then, if possible, to add insult to injury, it has turned out that her "scientific justification" was "junk science" at its worse. She derived her conclusions from a misrepresentation of Dr. James DeWitt.

Problem was, few if any of the claims made by Carson were true. They were derived from a complete misrepresentation of research by Dr. James DeWitt.

Dr. J. Gordon Edwards, a well-known entomologist, wrote in the twenty-first Century Science and Technology Magazine his findings in 1992 concerning the scientific integrity of Rachel Carson.

As I neared the middle of the book, the feeling grew in my mind that Rachel Carson was really playing loose with the facts and was also deliberately wording many sentences in such a way as to make them imply certain things without actually saying them. She was carefully omitting everything that failed to support her thesis that pesticides were *bad,* that industry was *bad,* and that any scientists who did not support her views were *bad.*

According to DeWitt's work, *which Carson cited as her source,* the birds that were fed exceedingly high levels of DDT every day hatched nearly as many of their eggs (in quail) to 27% *more* of their eggs (in pheasants). The great increases in the numbers of robins were documented in the comments above, in reference to page 118. Carson's claim, therefore, that those three kinds of birds are less and less able to produce young is remarkably false—and insulting to the reader.

(On) Page 120, Carson explains the lack of young birds by saying: "... [The reproductive capacity of the birds has been so lowered by some environmental agent that there are now almost no annual additions of young to maintain the race. Exactly this sort of situation has been produced artificially in other birds by various experimenters, notably Dr. James DeWitt of the U.S. Fish and Wildlife Service. Dr. DeWitt's now classic experiments on the effects of a series of insecticides on quail and pheasants have established the fact that exposure to DDT or related chemicals, even when doing no observable harm to the parent birds, may seriously affect reproduction ... For example, *quail into whose diet DDT was introduced throughout the breeding season survived and even produced normal numbers of fertile eggs. But few of the eggs hatched"* [emphasis added]

Carson gives no indication of *how many* might be considered as "few eggs hatching." Perhaps she thought that her readers would never see the rather obscure journal in which DeWitt's results were published in 1956, the *Journal of Agriculture and Food Chemistry.* Otherwise, she surely would not have so badly misrepresented DeWitt's results! The dosage he fed the quail was 100 parts per million in all their food every day, which was roughly 3,000 times the daily DDT intake of humans during the years of the greatest DDT use!

The quail did not just hatch "a few" of their eggs, as DeWitt's data clearly reveal (Table 3). As the published data from DeWitt's experiments show, the "controls" (those quail with no DDT) hatched 83.9% of their eggs, while the

DDT-fed quail hatched 75 to 80% of theirs. I would not call an 80% hatch "few", especially when the controls hatched only 83.9% of their eggs.

Carson either did not read DeWitt's article, or she deliberately lied about the results of DeWitt's experiments on pheasants, which were published on the same page. The "controls" hatched only 57.4% of their eggs, while the DDT-fed pheasants, (dosed with 50 ppm of DDT in all of their food during the entire year) hatched 80.6% of theirs. After two weeks, the DDT chicks had 100% survival, while the control chicks only had 94.8% survival, and after 8 weeks the DDT chicks had 93.3% survival while the control chicks only had 89.7% survival. *It was false reporting such as this that caused so many leading scientists in the United States to take Rachel Carson to task.* [7]

The lengthy quote from Dr. Edwards shows the same type of "tainting" of data that we see so prevalent in the "Climate Warming" controversy. When the data does not confirm your hypothesis, then suppress the "bad" data and retain the data or misrepresent the data that does support your agenda. She claimed: "Dr. DeWitt's now classic experiments [show] that exposure to DDT, even when doing no observable harm to the birds, may seriously affect reproduction. Quail into whose diet DDT was introduced throughout the breeding season survived and even produced normal numbers of fertile eggs. But few of the eggs hatched"

Reading environmental literature leaves you with one major conclusion. That conclusion is that the modern day environmentalist movement believes that man is a great threat to the "harmony" of nature and the ecological balance. If that inference concerning this movement is correct, then that Worldview and philosophy would pollute every single "subjective" scientific agenda.

In the minds of the modern day environmentalist movement, it is very possible that this notion of man being the egregious creature on "MOTHER EARTH" is axiomatic. That means all conclusions, all proofs, all theorems, and all hypotheses are based on this conjecture. And, the consequences have been and will be catastrophic if not stopped.

Conclusion

This world has seen in the twentieth century the results of "nefarious" ideals and ideas of intellectuals. In their zeal for political utopia (the ideal), they have espoused "ideas" to justify this Worldview and have influenced political leaders to implement their agenda.

Because in their minds they know better than the "little" people, what individuals need and even want, is, therefore, of little consequence to them. They simply may have to be sacrificed for the greater good. If 30+ million have to die in Russia to setup a "proletarian" utopia, then so be it. If 50+ million have to die in China to emulate their Marxist brother country, then so be it. If 50+ million die of malaria to save a "perceived" few birds, then so be it. If well over 100 million babies are killed to justify a naturalistic Worldview, then so be it. To the leftist ideologue and scientist, it "ain't no thang" This type of cavalier attitude is not just immoral, it is amoral and that can be very, very dangerous.

We have entered an era in which ideologue politicians need ideologue intellectuals and scientists to justify their nefarious and aberrant ideas. Almost all of the twentieth century despots did this and the trend is becoming more obvious and necessary. If you will indulge me, let's take an ideologue trip. If I wanted a "NWO" (New World Order), how would I go about doing it?

Well, I could try to conquer the rest of the world and implement my "utopia." That is not easy to do. I could try to get all the countries to sit down through a United Nations conference and agree to give up their sovereignty. Again, that's not easy to do. We can't even get a consensus in our own congress on anything. I could create the perception of multiple catastrophic crises and then "entice" the people and countries of this world to "unite" to combat this perceived crises. But, how would I do this? I would need complicity of ideologue politicians, ideologue intellectuals, ideologue educators, ideologue media magnets, and ideologue SCIENTISTS. This last group may be the most important of all because of the "false" perception of objectivity. The other groups have all been historically tainted by philosophical biases; but, scientists have been "publicly" immune to this perception.

Now, as a NWO disciple, I'm a "long term" thinker and planner. I realize that this could not be done with the scientists of previous generations who generally required empirical and testable verification of a given hypothesis. I need to infiltrate the educational system to change this rigid perception of science to include more "philosophical" and "historic" science and give those sciences parity. No problem, I will only hire and tenure those ideologues who agree with my Worldview and they will insure that their intellectual progeny are placed in positions of power and authority.

Now, with the same "dogmatism" of proof as the scientific method, I can assert "philosophy" and the public will accept at the same level of veracity. Once I have the ability to substitute philosophy for empirical verification, then I can make such preposterous statements and declarations as, "Science can solve and answer all of life's problems and issues"

Now, I need a crisis. Evolution is a "problem", but not a "crises" Think about it, irrespective of what the Darwinian ideologues say, there is nothing scientific that would be affected if one does not believe in or even heard of Darwinian evolution.

I need something else. What affects every person? THE ENVIRONMENT!!!! Eureka! What I can't do through political revolution, I can

177

do through fear and intimidation. Nothing is more ubiquitous than the environment. It does affect every person. After fifty years of macro-planning, I now have the all the major components in place.

I have the ideologue media. I have the ideologue educational systems. I have the ideologue progressive politicians. I have the ideologue intellectuals. And, best of all, I have the ideologue scientists. And, they are all in unison on this one issue.

Now, I have the potential to control every country, every person, every corporation, and almost every event on the face of the earth. And, I may only have to kill off 1 or 2 billion people to implement this utopia. But, it will be for the good of all mankind and especially "Mother Earth"

I realize that my "just so" story may not be 100% valid or correct. But, the reason for the story is to jar us out of our naivety and complacency and motivate us to look carefully at the "real agenda" of all sectors of our society. To not do this is like walking across an interstate highway without looking to your left or right but just observing what is in front of you.

Frédéric Bastiat, the brilliant French economist and philosopher of the era of 1801-1850 write a work called, "Selected Essays on Political Economy" Chapter 1 is titled, "What Is Seen and What Is Not Seen" Here is what he said in his first two paragraphs.

1.1 In the economic sphere an act, a habit, an institution, a law produces not only one effect, but a series of effects. Of these effects, the first alone is immediate; it appears simultaneously with its cause; *it is seen.* The other effects emerge only subsequently; *they are not seen;* we are fortunate if we *foresee* them.

1.2 There is only one difference between a bad economist and a good one: the bad economist confines himself to the *visible* effect; the good economist takes into account both the effect that can be seen and those effects that must be *foreseen.*

Even though Bastiat was primarily speaking of economics, this principle has almost universal application. Solomon, in an almost monotonous manner, repeatedly talked about the most important thing is to "get wisdom." In Proverbs 4:7, he said, "Wisdom is the principal thing; therefore get wisdom: and with all thy getting get understanding"

And, what is wisdom? Wisdom must have, as a necessary component, knowledge. How can I be wise about something I don't know? But, wisdom must be something other than "memorizing" facts. Wisdom must to some degree be able to see the consequences of particular ideas and actions. It must understand "reality" better than sheer happenstance.

When a person enters into a "profession", there is a standard of intellectual competency as well as ethics associated with that profession. Pastors are held accountable to those standards, engineers are held accountable to those standards, politicians are held to those standards (albeit, not very stringently), and scientists should be held to those standards. Philosophers and intellectuals, however, don't seem to have any standards and therefore can't be held accountable for anything.

The intellectual competency of a scientist must be more than knowledge and the memorizing of facts. Because science has much to do with determining "reality/truth" or at least eliminating error, there must be a methodology or mechanism in place where verification of results can be independently determined. This must and will mitigate error and bias. But, science cannot retain its "integrity" without an ethical standard. I'm not talking about a Judeo/Christian ethics (even though that is certainly preferable), but an ethic of science, a system of "moral" standards or principles.

Doctors have an ethical code succinctly stated in the Hippocratic Oath. Here is one paragraph of the classic version. *"I will neither give a deadly drug to anybody if asked for it, nor will I make a suggestion to this effect. Similarly I will not give to a woman an abortive remedy. In purity and holiness I will guard my life and my art"* In the "Original" and the "Classic" version, both had as part of the oath that the doctor would not give a woman an abortive remedy. In the "Modern" version of 1964, this is left out. I believe this is analogous to what is happening to the scientific community today. Ethics are being divorced from science; all types of aberrant sciences become acceptable. If the standards are lifted, then "truth" loses its preeminence and agendas and policies determine scientific outcome. If agendas and policies determine scientific outcome, then scientific policies and agenda require a "political" philosophy. That philosophy is that only those who agree with that policy and agenda will be acceptable in that community. And because "truth" has been supplanted by "agenda/policies", the ability to question or verify or criticize must be suppressed. This is the hallmark of a despot and is antithetical to ethical science.

What does the scientific community have to fear from "scientific errors"? NOTHING! Errors are present in every hypothesis and science has a way of "culling" out error by its "ethical" standards and methodology.

What does the scientific community (and the world) have to fear from "scientific ideologues?" EVERYTHING! Errors will not be "culled" out and science will be used for the most egregious and nefarious reasons by those who control the political, educational, and financial realms. This is a very serious matter and affects every single creature on this planet.

References

Michael Crichton, Home Website – Environmentalism as Religion - Speech given to Commonwealth Club San Francisco, CA September 15, 2003 http://www.crichton-official.com/speech-environmentalismaseligion.html

Al Gore, Earth in the Balance, Rodale Books, 2006, pg 258

Mark Landsbaum, an article in the Orange County Register, 02/12/2010 http://www.ocregister.com/articles/-234092--.html

Joanne Nova, an article on her website, http://joannenova.com.au/2010/03/help-how-do-i-know/

John Cornwell, Hitler's Scientists: Science, War, and the Devil's Pact, Penguin (Non-Classics), 2004, pg 3

Edwin Black, The Horrifying American Roots of Nazi Eugenics, San Francisco Chronicle, Sunday 9 November 2003, pg D–1

Dr. J. Gordon Edwards, 21st Century Science and Technology Magazine, 1992, http://www.21stcenturysciencetech.com/articles/summ02/Carson.html

Resources

Thomas Sowell, Intellectuals and Society, Basic Books, 2009.

Richard Weikart, From Darwin to Hitler, Evolutionary Ethics, Eugenics, and Racism in Germany, Palgrave Macmillan, 2004.

Karl Popper, Poverty of Historicism, 1989.

Karl Popper, Popper Selections, Princeton University Press, 1985.

Karl Popper, Conjectures and Refutations: The Growth of Scientific Knowledge, Routledge, 1992.

Richard Weikart, Hitler's Ethics – The Nazi Pursuit of Evolutionary Progress, Palgrave Macmillan, 2009.

Ronald Bailey, Global Warming and Other Eco Myths: How the Environmental Movement Uses False Science to Scare Us to Death, Prima Lifestyles, 2002.

R.J. Rummel, Death by Government, Transaction Publishers, 1994.

Author bio

Charles W. Imes married Rema Bray in 1967 and graduated with a degree in Mathematics from the University of Oklahoma in 1969. He was a math teacher at the high school level for 9 years and taught Algebra, Trig, Logic, Analytics, and Chess. In 1980 while working on a degree in accounting, he founded a computer software company that developed software systems for a number of industries, primarily the Oil and Gas industry.

In 1993, he was hired by Chesapeake Energy to manage their IT department and watched the company grow from about twenty-five employees to over five thousand. In 2009, he retired from Chesapeake Energy. He is a past president, active member, and director of Oklahomans for Better Science Education and served as a director for the Association of Professional Oklahoma Educators. He is also co-pastor of a small conservative church. He teaches classes on logic, philosophy, and theology for various groups and churches. And, he has a personal library of at least 4,500 books with the preponderance of the library dealing with theology, scientific apologetics, philosophy, logic, sociology, math, computer technology, and history.

In 2009, he also was the recipient of a heart transplant and now spends the vast majority of his time studying, writing, developing software, creating websites, and pontificating. You may contact him at: charleswimes@cox.net.

Black Holes, Einstein and Scientific Objectivity
Stanley Robertson, Ph.D.

Editor's note

Dr. Robertson has been a good friend for over thirty years. He knew of the trouble I had in biology for my reservations about evolution, but felt because much of physics relies heavily on mathematics it was more objective. After publishing some technical papers about black holes not being as they were thought, he found a lack of objectivity in physics as well. Such prejudice seems to be part of our human nature and he discusses this point in his essay.

The reason for him getting started investigating black holes is most unusual. One of the professors he was teaching physics with was more than a little enamored with black holes and he found that amusing. Somewhat on a lark, he argued with his colleague that black holes were over rated and not as they had been perceived by many astrophysicists. He decided to pursue this and has done so now for over a decade. His discoveries led to several provocative papers and have changed the view of black holes. He also made friends with a European astrophysics that shared some raw data from a radio telescope that shed light on black holes...so to speak. Enjoy and learn.

Introduction

Although the public at large is captivated by black holes, their interest is nurtured by theoretical physicists and astronomers intoxicated by the mystique of black holes. There is almost a religious fervor in some of the articles that report observations of new black holes, or report on things that they do that ought to be impossible for a black hole. For example, in astronomy, radio emitting jet outflows are ubiquitous, occurring in young stellar objects, neutron stars, stellar mass black hole candidates and galactic nuclei. In the case of the first two of these, there are detailed computer simulations that show how the spinning magnetic field of the central star drives the jets. But when the same kinds of jets and synchrotron emissions are found in black hole candidates, it is reported as amazing and different and mysterious and they build these models that effectively use tooth fairies to do what a spinning magnet does naturally. In this essay, I will try to explain some of how this has occurred. Part of the explanation is that it is all too human to want to enjoy a special status as practitioner of a science that encompasses something as bizarre as a black hole within a bedrock theory. I have, in times past, enjoyed a bit of this smug satisfaction for myself.

The existence of black holes remains an article of faith among most physicists for several reasons. With Special Relativity theory, Einstein revolutionized physics. When his General Relativity (hereafter, GR) theory correctly predicted three (later, four) small effects that were not part of Newtonian gravity theory, it created a strong sense among both physicists and the public at large that it must also be correct. Because of this prevailing bias in its favor and the fact that it is very difficult mathematically, it has not been subject to the same degree of testing as, for example, quantum mechanics. It is also not part of every physicist's studies. Many know no more than they have read in the popular press, including the prediction of black holes. Also, the most striking features of GR only appear where gravity fields are extremely strong, but it has only been tested in controlled experiments within the solar system, where gravity is weak. It is ironic that Einstein hated black holes, but if you speak ill of them it is considered to be an attack on Einstein and you become an instant pariah.

Little did I suspect that my confidence in black holes would be shaken by a 1996 astrophysics publication that offered "proof" of the existence of event horizons. This is the same as a proof of the existence of black holes, because the very essence of a black hole is the event horizon—the boundary from which nothing can escape. The proof was conceptually simple. There are binary star systems in our galaxy that episodically flare brightly in X-rays from a compact source. Some of the compact sources are known to be neutron stars and others are believed to be black holes. The X-rays are produced when gas from the companion star swirls in toward the compact object and is compressed and heated to millions of degrees by its own viscosity. As the flares subside and the systems revert to quiescence, gas flowing at lower rates would merely pass through the event horizon of a black hole and disappear. But the flow into a neutron star surface would end with X-ray emissions from a surface impact. They should be tens of thousands of times brighter than the quiescent black holes. Unfortunately, the scholarship of the "proof" article was so poor and the treatment of observational data was so dishonest that I expected it to become the butt of crude jokes. Instead it has become the proof of what everyone wanted to believe and it is still cited several hundred times per year.

I analyzed the "proof" observations of quiescent luminosities and tabulated the minimum luminosities reported for five neutron stars and five "black holes"; (all of the measurements available at the time). The data were scattered, but with means not different by more than about 10x; and certainly not 10,000x. I re-plotted these data in a way that clearly destroyed the "proof" and submitted the plot and a note to the same journal, but it was rejected. The referee chosen was the author of the "proof" article. That made me angry. I have seen similar shenanigans in other sciences, but never in physics.

That made me determined to try to find out what are the compact objects that are believed to be stellar mass black holes. I spent a year digging through X-

184

ray astronomy data before concluding with some certitude that the objects are very similar to neutron stars, though more massive, and are strongly magnetic. Since being magnetic is not a possible attribute of a black hole, it was pretty clear that the objects were something else, but exactly what, I still don't really know. Since black holes are apparently the offspring of Einstein's general relativity theory, my first thought was that GR theory would have to be replaced by something that didn't predict such monstrosities. GR sets an upper limit of three times the sun's mass for compact (i.e. nuclear density) stars. The objects more massive than that are called black holes, though none of the distinguishing attributes of a black hole have ever been measured. There are some other candidate gravity theories, one of which I used to model the properties of compact stars. In this theory the black hole candidates were essentially fat neutron stars. In support of this calculation, I offered a table of observations and analysis that strongly implied that the observed objects were magnetic and submitted a paper for publication.

It was eventually published in *Astrophysical Journal*, (which means that the conclusions cannot be easily rejected by the establishment, or they surely would have been) but for the response it got, I might as well have mailed it off to a black hole. Only Darryl Leiter, a theorist who began an e-mail correspondence, showed interest. He took the evidence seriously, and we worked together via e-mail exchanges for a couple of years trying to put some rigor into the gravity theory that I was using. We finally both gave up. That suited him fine. He is a General Relativity theorist from a long ways back; a Ph.D. student of Nathan Rosen, one of Einstein's most important co-authors. Eventually, we found that although GR, as presently used, seems to allow black holes as solutions of the field equations, this is not a physically realizable result. This conclusion has been published by others in several forms over the past twenty years, but it has not made a dent in the establishment that controls grants, grad students, jobs and journals.

Only after several years of study have I learned that the very concept of a black hole is ill founded and not really an essential part of GR. The first solution of Einstein's GR field equations was obtained one year after Einstein published the equations. It is the solution for the gravity field of a point mass and was obtained by Karl Schwarzschild, a soldier in the German army in World War I. Schwarzschild died at the Russian front in World War I, but Einstein saw to it that his solution was published—in German. Ironically, the solution known as Schwarzschild's solution is believed to predict the existence of black holes, though it did not. In a paper published in 1979 in *The Physical Review*, America's most prestigious physics journal, Leonard Abrams showed that David Hilbert, the most famous mathematician of the first half of the last century, had reworked

Schwarzschild's solution and made a mathematical error in the process. To give Schwarzschild due credit for the first solution, Hilbert called this erroneous solution the "Schwarzschild Solution." Unfortunately for Schwarzschild, his name has been attached to Hilbert's error. In the erroneous solution there is a distance from the point mass such that time stands still and space becomes infinitely stretched out. This is known as the "event horizon." Einstein commented that the event horizon "just didn't smell right" and he rejected the notion of black holes.

Although Hilbert's erroneous mathematical solution of the Einstein field equations would lead to a black hole condition, there is more to the GR theory than that. Also encompassed by the theory is that nothing can travel faster than the speed of light. But the "black hole" solution would require this part of the theory to be violated by anything falling into the black hole. So what we really have is an inconsistency between an apparent solution of the equations and other requirements of the theory.

In Hilbert's erroneous solution, any mass that becomes compact enough to reside entirely inside its event horizon would be a black hole. So the practice that has become the norm in astrophysics is that any object that has such a compact mass is called a black hole without verification of any further properties of a black hole. Only about one in ten authors now a use the more appropriate term, "black hole candidate", and Hilbert's theoretical error is the basis for calling astronomical objects black holes.

There are two classes of objects that appear to have this compactness and even neutron stars are within about a factor of two of being compact enough. Neutron star masses appear to all be less than about 1.5 solar mass. But there are similar compact objects over 5 solar mass that, at neutron densities, are too massive to be neutron stars according to Hilbert's solution. These are, of course, called black holes. There are also galactic nuclei that have million to billion solar mass that are also variable sources of light from X-rays to infrared. To account for their rapid and large variations of brightness, the object sizes must be not much larger than the distance light would have to travel to get from one side to another. Thus these objects are compact enough to also qualify as black holes, if such exist.

There are several reasons why I do not like the practice of uncritically calling these compact objects black holes. First, it is simply sloppy science; a biological equivalent might consist of identifying every organism with a cylindrical body as a snake, or calling every creature that produces eggs a chicken. Secondly, General Relativity has been tested only under weak gravity conditions. There is a dimensionless parameter, a gravitational potential energy that would have the value 1/2 at an event horizon. It has the value 6.9×10^{-10} at the earth surface, 1.7×10^{-6} at the surface of the sun, and about 0.2 at a neutron star surface. To get from the solar system, where the theory has been tested, to the realm of neutron stars and black holes require an extrapolation by about a factor of a million. Such a long extrapolation ending in a bizarre state of stretched space and time standing still should not be trusted. Thirdly, every other stellar mass and larger object is known to possess a magnetic field. But this is something that a black hole cannot possess. In my work, I have shown that the spinning magnetic field of the neutron stars drives a great deal of their radiation. I can account quantitatively for their X-ray emissions and jets. The same methodology applied to the stellar mass black hole candidates also accurately accounts for their similar emissions. Fourth, the accepted theory of how black holes produce these emissions relies on matter passing through an event horizon. This theory cannot apply to the neutron stars. They have surfaces that would light up from the surface impact of matter falling in. So to account for what we see, somehow, a bit of matter falling in would have to know what lay at the bottom of the well and know not to radiate if it reached a neutron star surface at the bottom. This is absurd. Fifth, there are other profound similarities between the neutron star systems and the stellar mass black hole candidates. I can show multiple instances of published statements that essentially say that "… we can rule out the involvement of surfaces and magnetic fields as explanations of these phenomena, because we see the same things from neutron stars and black holes …"; i.e. we cannot see what the theory says that we cannot see, even when what we are seeing is synchrotron (magnetic) or surface radiation. This is simply appalling.

I think that I have demonstrated that it is possible to publish contrary views in top-flight journals if there are no holes in your logic or facts, but you may wind up being ignored anyway. Especially if you make the mistake that I made early on of suggesting that GR might be less than perfect. This is unhealthy, but it is reality. Things changed a little with Darryl Leiter on board, and this is where the story begins with three of my recent papers, which are listed below. The 2002 *Astrophysical Journal* article analyzes the same data, plus extensions, as my first published paper on the subject, showing magnetic effects, but not trying to explain them with any theory at all. In the second one (2003) we found new solutions of the Einstein equations and showed that GR actually allows some magnetic beasts that are about as compact and even more bizarre than black holes. They are composed primarily of an electron-positron pair plasma. We could not take the entire power output of all nations on earth and produce a cubic micron of a pair plasma! So we have an object compatible with GR and observations that is not a black hole. This is likely the most hated hypothetical object in all of physics at the present time, but it is being taken seriously. Lastly, the MNRAS article shows how these objects can produce the jets and correlated radio and X-ray emissions of all of the black hole candidates, whether star-like (GBHC) or active galactic nuclei (AGN can be millions to billions of times more massive than the sun). The last of the last sentence of the abstract ends with "… GBHC and AGN have observable intrinsic [magnetism] and hence do not have event horizons." This is a deliberate thumb in the eye of the establishment. I doubt that it would have been possible to get such a sentence into print four years ago. I may not live to see this widely accepted, but I think that eventually it will be.

I am not sure what will happen in the near future. I can get my heretical ideas into the "best" journals, but I am often ignored anyway, just as Abrams was and several others have been. The black hole machine keeps rolling along. They find new "black holes" every day. At present it is merely rude to question their status as black holes. In time it may become impossible. If remembered at all, I may merely be a "flat-earth" kind of physicist. There is an orthodoxy that controls grants, jobs, grad students and journals. Who knows how long it will continue? So, is it possible to get good science published in good journals? Maybe, but maybe not. Questioning the existence of black holes makes me a flat-earth sort of physicist and I would have problems obtaining grants or working in a graduate department. Now imagine what questioning aspects of evolution must be like in Biology. One really can't work as a normal biologist with grants, grad students and a job while questioning what are regarded as the foundations of biology. Never mind that a lot of the foundations have distinctly ad hoc qualities. We invoke evolution to explain changes that took place over time, but no one really knows if the appropriate and necessary biochemistry changes took place. The rapidity with which some species changes occurred certainly suggests mechanisms beyond random point mutations. With respect to knowledge of the biochemical basis of life, we are in our infancy.

The one thing that can be published, in any field, is reproducible observations and these are also freely shared. Interpretations are a different matter. There are prevailing orthodoxies in most sciences that control interpretations in the absence of compelling data. For example, if it can be reproducibly and reliably shown that Carbon-14 is found in pre-Cambrian diamond, that result can be published in the best of geology or physics journals. You can bet that it would engender discussion and testing until the interpretation issue it raises was settled! But if you need grant money and facilities for making the measurements, then you might be out of luck. Objectivity may not stretch that far.

Recent references

Evidence for Intrinsic Magnetic Moments in Galactic Black Hole Candidates. Stanley L. Robertson & Darryl J. Leiter, 2002. *The Astrophysical Journal* V565, p. 447. In this paper, we showed that the similar quiescent x-ray spectra of neutron stars and black hole candidates in X-ray nova systems can be quantitatively explained as driven by spinning magnetized objects.

On Intrinsic Magnetic Moments in Black Hole Candidates. Stanley L. Robertson & Darryl J. Leiter, 2003 ***The Astrophysical Journal Letters***, V596, pL203.
Black holes cannot possess intrinsic magnetic fields, consequently if the black hole candidates are magnetized, they must be exotic objects of some other kind. Here we propose that they are hot, collapsing and slowly radiating away their mass. We call them magnetic eternally collapsing objects (MECO) and show that they are permitted by General Relativity.

On the Origin of the Universal Radio-X-Ray Luminosity Correlation in Black Hole Candidates. Stanley L. Robertson & Darryl J. Leiter, 2004 ***Monthly Notices of the Royal Astronomical Society***, V350, 1391. We show that the MECO model provides a quantitative explanation of x-ray and radio emissions from neutron stars, galactic black hole candidates and active galactic nuclei (quasars and Seyfert galaxy nuclei).

The Magnetospheric Eternally Collapsing Object (MECO) Model of Galactic Black Hole Candidates and Active Galactic Nuclei. Stanley L. Robertson & Darryl J. Leiter, Nova Science Publishers, chapter of book, ***Black Holes Research***, in press. This is a full exposition of the MECO model, including its consistency with relativistic gravity theory and quantitative aspects of broadband spectra, temperatures, jet emissions, interactions with accretion disks, etc. We provide scaling laws from neutron stars to stellar mass black holes to galactic nuclei. The MECO model is thus the first unified, consistent model of gravitationally compact objects.

Observations Supporting the Existence of an Intrinsic Magnetic Moment Inside the Central Compact Object Within the Quasar Q0957+561.
Rudolph E. Schild, Darryl J. Leiter and Stanley L. Robertson, submitted to ***Astronomical Journal***, May 2005 The image of quasar Q0957+561 is split into four by gravitational lensing by an intervening galaxy. Within the multiple images are light fluctuations caused by microlensing by objects of planetary size within the quasar's own galaxy. The analysis of these fluctuations permits a determination of small scale structures, such as the accretion disk and compact jets of the quasar. These features match the predictions of a MECO model that is scaled up a billion fold from the realm of stellar mass black hole candidates.

Co-authors

Darryl J. Leiter is a General Relativity theorist, the last Ph.D. student of Nathan Rosen, who was one of Albert Einstein's frequent and important coauthors. In a continuation of Rosen's work, Leiter is author of several papers

189

dealing with the role of gravitational potentials in General Relativity. These functions determine both the geometry of space and time and gravitational forces that affect matter.

Rudolph Schild is a Professor of Astrophysics at Harvard University's Center for Astrophysics (CFA). He is a recognized expert in the analysis of microlensing within the multiple gravitationally lensed images of distant quasars. He has found that the MECO model (above) of Robertson and Leiter provides a consistent description and explanation of his observations.

Author Bio

After graduation from Maud (Okla.) High School in 1957, Robertson worked as an oil field roughneck for several years. Eventually it occurred to him that wrestling pig iron might have a limited future so he enrolled at East Central State College in Ada, Okla. but still supported himself by working in the grease trees. He chose to major in Math in college, because it was easy to spell, and then found to his dismay that Physics, which is harder to spell, was a required minor. He responded to this challenge by earning pathetic grades for the first three courses. As fate would have it, enrollments in physics were so small that he was pressed into service as a paper grader anyway and thus was forced by irate fellow students to learn some physics. This was reminiscent of an analogous circumstance in high school, by which he had become a running back on the football team. Actually, he was slow enough to be a lineman, but he wasn't big enough. Always the lucky devil, Robertson married Lois Ward, the prettiest and smartest (excepting this inexplicable lapse) girl on campus in 1962. In his last semester at East Central, Robertson took an independent study course and, with the help of a friend, borrowed some materials from the Physics Dept. Library at Oklahoma University. While returning them, he accidentally stepped on the foot of Prof. Richard G. Fowler, who, after apologies, by happenstance, had a research assistant's position open at the beginning of the next semester. At this point, the unwitting Robertson had become hopelessly ensnared in physics, where he remains to this day.

After completion of a doctorate in physics in 1969, Robertson worked for eight years as a physics professor at Fort Hays State University in Hays, Kansas. He left the windy high plains for Tahlequah, Oklahoma and Northeastern State University after his new minnow bucket, complete with minnows and water, blew into Cedar Bluff Lake. (It is rumored that only full professors have a chance against the Kansas winds, not to mention the dreadful lack of woodpeckers in western Kansas). Three years of Northeastern, Lake Tenkiller and the Cookson Hills were great fun for a young family with five children and a father who was an SPS sponsor and T-Ball coach and awash in new research funds. But when an offer from Oklahoma University arrived in 1980, it was back to Norman along with more little league coaching until the kids were all demonstrably better baseball players than he. This was not to last, however. Oklahoma in 1982 was in tall cotton from a drilling boom and a large pay increase enticed Robertson into a petroleum reservoir engineering job which he held for the next eight years (in the manner of a rat clinging to the flotsam of the bust for the last few). In 1990 he was hired to teach physics at Southwestern Oklahoma State University and eventually retired there in 2004. While there he enjoyed serving as the sponsor of the Physics and Engineering Club for many years. He also taught many engineering oriented courses as well as electronics, physical geology, and from time to time, astronomy. A Procrustean approach to research at Southwestern eventually forced him to publish some quixotic papers dealing with astrophysics, black holes and general relativity. At this point, at least on paper, he is qualified to withstand the winds, but he does not plan to give Kansas another chance. You may email Dr. Robertson at: stanrobertson@itlnet.net.

Global Catastrophe and Historicity of Noah

Jay Hall, MS

Editor's note

As a public school mathematics teacher, Jay provides fresh insight into some of the problems our high school students face in today's increasingly anti-Christian society. He and I have known each other since 1980 when I owned an airplane. We flew to Ann Arbor, MI for a board of directors meeting for the Creation Research Society. That meeting had a powerful influence on his life and he has remained active in the creation science arena. This chapter provides and interesting look at some of the arguments for and against the catastrophe detailed in the Genesis account of the global flood. Sadly, discussion of this important topic is forbidden in many of our public schools and is deemed "religious" and not "science" This should not be surprising as over 3,000 professors have been fired for rejecting evolution dogma in favor of the creation science. Our religious freedom is rapidly eroding, yet few people seem aware of this growing tragedy in our society today.

Abstract

Global catastrophe is a potential threat to our planet. Comets, asteroids and super-volcanoes pose a real danger to the healthy ecology of the earth. Consideration of episodic events on a world-wide scale is often discarded when it comes to historical geology. There are a number of indicators pointing to a global event in the past. This connects well the global traditions of a great flood that covered the earth. Noah's story is the one version that provides an actual historical account.

Introduction

Many geologists accept the impact theory for the extinction of the dinosaurs at the end of the Cretaceous period. The possibility of future disasters caused by cosmic collisions is also readily admitted. However, the evidence for a global catastrophe just thousands of years ago is often ignored. Taken as a basis, this historical event naturally brings to mind the historical narrative of Noah in Genesis. There are a number of lines of argument that validate a global flood and support the version of the Deluge given in Genesis as an accurate reportage of what happened in the past.

The K-T Boundary Disaster

According to the impact theory for the demise of the dinosaurs, a comet (or asteroid) hit the earth near the Yucatan peninsula of Mexico and left a crater 100 km wide and 30 km deep. This allegedly happened at the Cretaceous-Tertiary (K-T) boundary 65 million years ago. Forest fires erupted on a global scale. Massive dust in the atmosphere produced a runaway greenhouse effect that lasted for centuries. The impact may have triggered flood basalts which may form deposits with more than two million cubic kilometers of basalt. The great amount of sulfur dioxide added to the atmosphere from this volcanism would have created acid rain. About seventy percent of all species died, including the dinosaurs (Desonie, 1996).

What about similar events in earth's future? If Comet Shoemaker-Levy 9 had hit the earth instead of Jupiter (as it did in 1994), then "our fate might have been similar to that of the dinosaurs" (Levy, Shoemaker and Shoemaker, 1996). Isaac Asimov (d. 1992) estimated that meteorite-induced tsunamis may occur every 71,000 years (Asimov, 1979).

Mainstream Response to Catastrophism

At one time the Global Flood view was widely accepted. Consider this passage from the 1797 *Encyclopedia Britannica*, "we may reasonably suppose the deluge to have been the cause of all or most of the fossil appearance of shell, bones, & c., we meet with …" (Repcheck, 2003). Even James Hall, a close friend of uniformitarian James Hutton, postulated that the Flood may have been caused by a sudden uplift of the ocean floor (Baxter, 2003).

Writing in the *Journal of Geological Education*, Edgar Heylmun made this startling confession: "Many instructors dismiss the possibilities of global catastrophes altogether, whereas others ridicule and scoff at the early ideas. These same instructors will implore their students to think scientifically and to develop the principles of multiple-working hypotheses. The fact is the doctrine of uniformitarianism is no more 'proved' than some the early ideas of worldwide cataclysms have been disproved" (Comninellis, 2001). In contrast, William Ryan and Walter Pitman claim that, "… the mythical flood has not been substantiated in the record of earth history" (Ryan and Pitman, 1998).

Episodic (rare and rapid) sedimentation is now accepted as the rule for many formations in the rock record. But why not use the traditional term "catastrophic" instead of "episodic"? Professor Robert Dott gave the answer in his 1982 Presidential Address at the meeting of the American Society of Economic Paleontologists and Mineralogists: "Episodic was chosen carefully over other possible terms. "Catastrophic" has become popular recently because of its dramatic effect, but it should be purged from our vocabulary because it feeds the neo-catastrophic-creation cause" (Morris and Morris, 1989). Is this an attitude that fosters objectivity?

The book *Grand Canyon: A Different View*, which adopts Flood Geology, was attacked by the presidents of seven leading scientific organizations when they attempted to ban the book from the Grand Canyon National Park bookstore (Looy, 2004). Is this any way to promote openness in scientific research?

Writing in their textbook *Introduction to Geology*, William Stokes and Sheldon Judson provide a more balanced approach: "A catastrophist might contend that the twisting and breaking of strata, the transportation of huge blocks of rock, the violent cutting of canyons, and the wholesale destruction of life is within the power of a great universal flood—and he would be right (Stokes and Judson, 1968).

Evidence for a Global Catastrophe
Many layers in the geologic record are acknowledged to be the result of high-energy events. Consider the following scenario: One bed was deposited rapidly by a storm surge. Then a long period of time elapsed with no deposition or erosion. On top of the storm surge deposit, turbidity currents laid down graded beds which formed rapidly. Did the "missing" time really exist?

A vast amount of time is claimed to have elapsed between beds in the rock record. Ripple marks and animal tracks are two features that must be covered quickly in order to be preserved. Bioturbation, where plants and burrowing animals (like worms or clams) disturb the sediment, is rare in the rock record. This indicates that no great time lapse occurred between the layers. Similarly, soil layers are also rare. Yet, exposed rocks normally form soils through weathering and chemical breakdown given sufficient time (Morris, 1994).

The extremely flat bedding plane between the Hermit shale and the overlying Coconino sandstone in Grand Canyon, without evidence of erosion, points to continuous deposition despite the ten million year gap that mainstream geology assumes (Morris, 1994). There are sand-filled intrusions at the base of the Coconino which penetrate into the Hermit shale below. Some of these classic dikes are U-shaped and get smaller going up which defies any long time gap between these beds (Whitmore, 2005). The Coconino sandstone, and its equivalent formations, can be found in Arizona, New Mexico, Texas, Oklahoma, Colorado and Kansas and cover more than 200,000 square miles. Such continental flooding implies currents of tens of meters per second (Vale, 2003). These observations correspond nicely with Derek Ager's comment that, "… sedimentation was at times very rapid indeed and that at other times there were long breaks in sedimentation, though it looks both *uniform and continuous"* (Ager, 1995, emphasis added).

Billions of cone-shaped nautiloids are found in a six foot thick layer of the Redwall Limestone of Arizona and Nevada. The orientations of the nautiloids indicate rapid sedimentation. The high density of fossils (one nautiloid per square yard), in such vast numbers, over a thin layer are evidence of a catastrophic flood (Vale, 2003).

In 1975, John Morris predicted that the trees in the Yellowstone Petrified Forest would show from their tree rings that they existed at the same time. Mike Arct researched trees from a number of layers and found that the tree ring "fingerprints" matched and so lived contemporaneously. Thus, the Fossil Forest of Yellowstone Park did not take many thousands of years to form (Morris, 1994).

Another way to look at tree rings is to consider the tree in the fossil record with the most number of tree rings. Some petrified trees may have as many as 1000 rings (Morris, 1995). Today's Redwoods may live for thousands of years. Where are the trees in the fossil record with 2000, 3000 or 4000 tree rings? If we take our chronology from Genesis, with the Flood occurring less than two millennia after Creation Week, we would not expect to find any.

Arthur Chadwick has shown that Paleozoic sediments generally moved from east and northeast to west and southwest across North America. This continent-wide trend can best be explained by the Global Flood (Coffin, Brown and Gibson, 2005).

Radiocarbon dating actually supports the Flood Geology position. The RATE team (Radioisotopes and the Age of The Earth) had ten coal samples analyzed. These coal samples were from the Paleozoic, Mesozoic and Cenozoic eras and thus were supposedly many millions of years old. In addition, several diamonds which are highly resistant to contamination were also studied. Based on standard assumptions, these coal and diamond samples indicated carbon-14 ages between 44,000 and 57,000 years. Another RATE study of deep zircons crystals, allegedly over a billion years old, yielded a date of 6000 ± 2000 years, based on measured helium diffusion rates (DeYoung, 2005). These results support Young Earth Science (YES).

Traditional Flood Stories

Fred Warshofsky, the science editor of the CBS series *"The 21st Century"*, made this bold claim, "Worldwide flood myths arise when catastrophic floods wash over a good part of the earth. And there is a great deal of evidence that this is precisely what happened" (Warshofsky, 1977).

Stephen Baxter claims that "Chinese written history stretched back centuries *before* this date [of the Flood, 2300 BC], and made no mention of a disastrous global deluge" (Baxter, 2003, emphasis in original). Baxter is mistaken on both counts. The standard beginning of Chinese civilization is pegged at 2200 BC. (King, 2000). Mark Lewis, the Kwoh-ting Li Professor of Chinese Culture at Stanford University, wrote a whole book on *The Flood Myths of Early China*. The Lolos in northern Myanmar (Burma) and southern China tell of the hero Du-mu and his four sons who survived the Flood. The Kammu of northern Thailand speak of a brother and sister who survived the Deluge sealed in a drum and got married to each other because a bird told them to do it (Lewis, 2006). Yao is said to be the first sage-king of China and in *The Canon of Yao* (Yao dian) in *The Book of History* (Shang shu) the Flood is described: "Destructive in their spread are the waters of the deluge. In their vastness they embrace the mountains and submerge the hills, rising to the heavens with their swell" (China Heritage, 2007). One Chinese tradition has the flood due to a crab being offended by a bird (Bailey, 1989).

William Ryan and Walter Pitman, who promote the Black Sea flood as the source of flood legends, speak of a "... puzzling absence of a flood mythology in Europe" (Ryan and Pitman, 1998). In contrast to this assertion, we find an Irish account that tells of Queen Ceseair and her court boarding a ship and surviving a flood that lasted seven and a half years. The Welsh version of the Flood story involves Dwyfan and Dwyfach and is recorded in *The Third Catastrophe of Britain* (Berlitz, 1987). There exist flood legends among the Gypsies of Transylvania and the Voguls of eastern Russia (Frazer, 1923). In the Lithuanian account, the survivors of the Flood travel in a nutshell which a god dropped from the sky (Bailey, 1989).

The Babylonian account has the Ark in the shape of a cube 197 feet on each side which would be very unseaworthy (Bailey, 1989). Berossus, the Greek historian, gives a version of the flood story where the ship of rescue has a length-to-width ratio of 5-to-2 which is not as stable as the 6-to-1 ratio for the dimensions of Noah's Ark as given in Genesis. The Greek flood legend of Deucalion has the Deluge lasting only nine days. Manu, the hero of the flood story from India, was warned of the deluge by a talking fish (Frazer, 1923). The tale of Atrahasis (c. 1700 BC), has the Deluge caused by people being too noisy and only lasted seven days (Bailey, 1989).

Event	Number	Reference
warning period	120 years	Gen. 6:3
dimensions of Ark	450 x 75 x 45 ft.	Gen. 6:15
position of roof	18 ins.	Gen. 6:16
number of decks	3	Gen. 6:16
unclean animal count	by 2's	Gen. 7:2
clean animal count	by 7's	Gen. 7:2
rain alert	7 days	Gen. 7:4
continuous rain	40 days	Gen. 7:4
Noah's age at start	600 years	Gen. 7:6
start of Flood	2/17	Gen. 7:11
number of survivors	8	Gen. 7:13
min. draft of Ark	22.5 ft.	Gen. 7:20
duration of increase	150 days	Gen. 7:24
receding begins	150 days	Gen. 8:3
Ark on Ararat	7/17	Gen. 8:4
mountains visible	10/1	Gen. 8:5
raven sent out	10/1 + 40 days	Gen. 8:6
1st dove sent out	10/1 + 47 days	Gen. 8:10
2nd dove sent out	10/1 + 54 days	Gen. 8:12
Noah's age at end	601 years	Gen. 8:13
ground becomes dry	1/1	Gen. 8:13
earth completely dry	2/27	Gen. 8:14

The chronicle of the Flood in Genesis is highly superior to these strange legends as shown by the careful numerical detail. In the chart below the dates are given in Month/Day format and the 18 inch cubit is used.

Noah's Flood was a Historical Event

The exact geographic center of the earth (latitude 39° N, longitude 34° E) is near Ankara, Turkey and not very far from Mount Ararat (Morris, 1973). The fact that the Ark landed near the center of the earth facilitated the dispersal of people and animals across the globe.

Physical evidence supports the historicity of Noah. A Roman coin over 1700 years old, on display at the Israel Museum in Jerusalem, depicts Noah sending out the dove and the side of the Ark is labeled "Noah" in Greek (ΝΩΕ). This coin was minted in Apameia Kibotos in Turkey (Sellier and Balsiger, 1995).

There is also a Sumerian map which shows where Utnapishtim, the survivor of the Flood, resided (Ryan and Pitman, 1998).

We can even find Noah in the stars. The constellation Arago the Ship (Argo Navis) is often depicted as resting on a mountain. A centaur from the ship sacrifices a beast on an altar. The Milky Way seems to be smoke rising from the altar. Columba the Dove is also associated with Arago (Henry, 2008).

John Chrysostom (d. 407 AD) boldly asked, "Have you heard of the Flood—of that universal destruction? ... Do not the mountains of Armenia testify to it, where the Ark rested? And are not the remains of the Ark preserved there to this very day for our admonition?" Berossus (c. 275 BC) claimed that some part of the Ark still remained on a mountain in Armenia. Marco Polo spoke of the "Mountain of Noah's Ark" in Greater Armenia (Montgomery, 1974).

In 1670, Domingo Alessandro, a hermit who lived on Mount Ararat, gave Jans Janszoon Struys a cross and testified, "I myself entered that Ark and with my own hands cut from the wood of one of its compartments the fragment from which that cross is made. I informed the same Jan Janszoons [Struys] in considerable detail as to the actual construction of the Ark ..." (Montgomery, 1974).

Since dinosaurs were on the Ark, we should expect to find evidence of their existence in post-Flood times. There are many historical accounts of dragons which match our knowledge of dinosaurs. Marco Polo tells of a giant snake with legs and said it was "so large that it would well swallow a man" in the Yunnan Province of China. Shi Lizhuo, a magazine editor in that region, confirms the legend of a huge reptile with legs that ate people (Edwards, 2001). Job 40:15-24 clearly speaks of a dinosaur when Behemoth is described. Isaac Asimov claimed that, "... *legend makers expanded [Behemoth] to enormous dimensions no animal could truly have*" (Asimov, 1979). In 1994, geologist John Whitmore, Buddy Davis and others went to the Liscomb Bone Bed on the North Slope of Alaska and found frozen dinosaur bones. The tentative identification is of the duckbilled dinosaur *Edmontosaurus* (Davis, Liston and Whitmore, 1998). Surely, these frozen bones are not 65 million years old.

The oldest living organism on the planet is the Methuselah tree, a bristlecone pine, in the White Mountains of California and is considered to be 4800 years old (Stokes and Judson, 1968). This age fits in well with the traditional date of the Flood just a few thousand years ago.

John Woodmorappe has given reasonable explanations for common objections regarding the logistics relating to Noah's Ark. Woodmorappe deals with such issues as the construction of the Ark, number of animals on the Ark, waste management, fresh water and the feeding of the animals (Woodmorappe, 1996).

In conclusion, the evidence of a global catastrophe is consistent with the results of the RATE team indicating a shortened chronology and favors Young Earth Science (YES). This version of earth history, involving only thousands of years, naturally brings to mind Noah's Flood as the primary candidate for this cataclysm. The worldwide distribution of Flood sagas and other facts provide vindication of the historicity of Noah and the Global Deluge.

References

Ager, Derek 1995. *The New Catastrophism*. Cambridge University Press.

Asimov, Isaac 1979. *A Choice of Catastrophes*. Fawcett Columbine Books, New York.

Bailey, Lloyd 1989. *Noah: The Person and the Story in History and Tradition*. University of South Carolina Press, Columbia, South Carolina.

Baxter, Stephen 2003. *Revolutions in the Earth: James Hutton and the True Age of the World*. Weidenfeld & Nicolson, London.

Berlitz, Charles 1987. *The Lost Ship of Noah*. Ballantine Books, New York.
China Heritage. 2007. *Chinese Myths of the Deluge*. **China Heritage Quarterly No. 9**, March 2007, http://www.chinaheritagenewsletter.org/articles.php?searchterm=009_de-luge.inc&issue=009 Accessed 3/25/09.

Coffin, Harold, Robert Brown and James Gibson 2005. *Origin by Design*. Review and Herald Publishing.

Comninellis, Nicholas 2001. *Creative Defense*. Master Books, Green Forest, AR.

Davis, Buddy, Mike Liston and John Whitmore 1998. *The Great Alaskan Dinosaur Adventure*. Master Books, Green Forest, AR.

Desonie, Dana 1996. *Cosmic Collisions*. Henry Holt and Company, New York.

DeYoung, Don 2005. *Thousands ... Not Billions*. Master Books, Green Forest, AR.

Edwards, Mike. 2001. *Marco Polo in China (Part II)*. **National Geographic. 199(6)**:20-45.

Frazer, Sir James 1923. *Folk-lore in the Old Testament (abridged ed.)*. Tudor Publishing, New York.

Henry, Jonathan 2008. *Constellations: legacy of the dispersion from Babel*. **Journal of Creation. 22(3)**:93-100.

King, Margaret 2000. *Western Civilization(combined ed.)*. Prentice Hall, Upper Saddle River, NJ.

Levy, David and Carolyn & Eugene Shoemaker, forward to Desonie, Dana 1996. *Cosmic Collisions*. Henry Holt and Company, New York.

Looy, Mark 2004. Shenanigans recently uncovered in Grand Canyon book controversy.
http://www.answersingenesis.org/docs2004/1214shenanigans.asp Accessed 3/20/09.

Montgomery, John W. 1974 (2nd ed.). *The Quest for Noah's Ark*. Bethany Fellowship, Minneapolis, MN.

Morris, Henry 1973. The Center of the Earth. http://www.icr.org/article/50/ Accessed 3/20/09.

Morris, Henry and John Morris 1989. *Science, Scripture, and the Young Earth*. Institute for Creation Research, El Cajon, CA.

Morris, John 1994. *The Young Earth*. Master Books, Green Forest, AR.

Morris, John 1995. The Yellowstone Petrified Forests. http://www.icr.org/article/-yellowstone-petrified-forests/ Accessed 3/21/09.

Repcheck, Jack 2003. *The Man who Found Time*. Perseus Publishing , Cambridge, MA.

Ryan, William and Walter Pitman 1998. *Noah's Flood*. Simon & Schuster, New York.

Sellier, Charles and David Balsiger 1995. *The Incredible Discovery of Noah's Ark*. Dell Publishing, New York.

Stokes, William and Sheldon Judson 1968. *Introduction to Geology*. Prentice-Hall, Englewood Cliffs, NJ.

Vale, Tom (ed.) 2003. *Grand Canyon: A Different View*. Master Books, Green Forest, AR.

Warshofsky, Fred 1977. *Doomsday: The Science of Catastrophe*. Reader's Digest Press, New York.

Whitmore, John. 2005. *Origin and significance of sand-filled cracks and other features near the base of the Coconino Sandstone, Grand Canyon, USA*. Creation Research Society Quarterly 42(3):163-180.

Woodmorappe, John 1996. *NOAH'S ARK: A Feasibility Study*. Institute for Creation Research, El Cajon, CA.

Author bio

Jay Hall has both a Master's and Bachelor's degree in Mathematics from the University of Oklahoma. He also has 43 credit hours in natural science courses. He has been an origins activist since 1976 and has been published in the *Creation Research Society Quarterly* and *Origins Research* (later named *Origins and Design*). Hall also has a piece on the MathWorld website. In 2002 he gave a talk on "Math and Creation" to the Inland Empire Creation Science Association of Riverside, California. Both the beauty and applicability of mathematics point to a wise Creator. Hall was the co-founder of the First Coast Creation Club of Jacksonville, Florida and gave a presentation on the Global Flood to the group. He maintains the AdamsLostDream blog (www.adamslostdream.blogspot.com). You may email him at hallofmath@yahoo.com.

Religion of Global Warming
Fritz Ward, Ph.D.

Editor's note
In a former life, both Fritz and I were truck drivers. We met while refueling our big rigs side by side at a truck stop and became instant friends. I met his wife as well as his father while I was trucking. He has a doctorate from the University of California at Riverside. There are actually a surprisingly large number of educated truck drivers; because the pay is excellent...over twice what one can make teaching at a university. He and his wife enjoy backpacking and camping in the scenic mountains of California. Such times communing with God's creation provides time to think in an environment most can only dream about.

Science is unique in that it does base itself on evidence rather than superstition, upon authority, upon holy books, or upon revelation. From Richard Dawkins, Interview with Fox News host Bill O'Reilly, 10/9/2009

Science, Religion and Public Policy
During a May 26, 2010 press conference at Solyndra, a solar company, President Obama announced, "Climate change poses a threat to our way of life."[1] This dramatic comment came about in the context of discussing the environmental disaster wrought by the British Petroleum (BP) oil spill in the Gulf of Mexico. Following the lead of his Chief of Staff Rahm Emanuel who early in the administration announced that no good crisis should go to waste, President Obama was clearly using the oil spill disaster as an excuse to pass an energy bill that would tax carbon emissions, though under the guise of a "Cap and Trade" policy. More than a few political commentators recognized the president's call for a new comprehensive energy bill as a political ploy, even among those who generally approved the policy.[2] Almost no one, however, questioned the president's claim that climate change is a serious threat. This is in part due to the perception that climate change is a matter of science.

Science plays a fairly unique role in modern political (often liberal) rhetoric. Because science is widely perceived to be a neutral means of establishing truth, it is considered beyond question. Indeed, many liberals, who have otherwise embraced post modernism with its rejection of absolute values and beliefs still nonetheless, insist that science is somehow different from other forms of knowledge. And there is some basis for this belief. To the extent that science is recognized as a method for deriving limited conclusions to specific questions, it does offer some reasonably sound answers to certain problems. But "science," as it is used in public policy debates is often considered to be more than simply a method of addressing limited problems. It has taken on the aura of a worldview that, as Richard Dawkins expressed in the quote above, is in opposition to traditional worldviews, particularly religion.[3]

This perception of science is doubly odd when it comes to the global warming hypothesis because the latter resembles nothing so much as it does a religion. It has an origins story, an "Eden," a "Fall" due to the sins of mankind, principalities and powers which conspire against goodness (notably "greedy" corporations), an apocalypse, and a priesthood who interprets holy texts for believers. In this essay we shall examine how global warming resembles a religion and consider what type of religion it is. Undoubtedly this perspective will offend some and confuse others. After all, there are "scientific" papers that claim to analyze the effects of global warming, and some will think that the very existence of such papers is sufficient to "prove" that global warming is not a religion. But this view is naïve on two accounts. In the first instance, it is based on the presupposition that science and religion are opposing, rather than complementary proposals. Aside from the assertion of various humanist writers, most recently Richard Dawkins, there is no reason to suppose this is true. Science may use different methods, but this does not necessarily produce different results. In fact, numerous scientists from the early modern period to the present have actually been inspired by and used theology as a complement to their scientific research.[4] On the other hand, it is entirely possible for religious commitments to skew scientific results.[5] Science should therefore be viewed as part of a continuum with religion. Some theories are either more or less influenced by faith commitments. In the case of global warming, it is clearly more.

Aside from the obvious point that there is no clear delineation between science and religion (both are, after all, human enterprises), there are other reasons for considering the global warming movement mostly as a non-scientific enterprise. In the first instance, many of its most vocal proponents are not scientists. Former Vice President Al Gore, for example, took only two science classes during his undergraduate years and received a D and C- in them respectively.[6] And he is hardly alone. Many of the loudest defenders of the "science" of global warming share Gore's lack of credentials, including (economic) noble laureate Paul Krugman and the various protestors who busy themselves making snowmen to show the dangers of global warming.[7] However much melting snowmen might illustrate the "crisis" of global warming in the minds of some, one cannot say that this is an example of a scientific representation. In any event, the "science" behind global warming has been effectively answered in a series of articles and books. Notable among these are the recent Douglass and Christy paper, "Limits on CO2 Climate Forcing from Recent Temperature Data of Earth" (2009), the Science and Environmental Policy Project's *NIPCC* report of 2008, the research of Henrik Svensmark popularized in his book *The Chilling Stars* with Nigel Calder, and of course the essays by Edward F. Blick and Rod Martin in this volume.[8] Indeed, there have been so many cogent criticisms leveled against the global warming hypothesis from the very start of this movement that the relevant question is not, what does the research say, but rather, why do people continue to believe it at all? And it is on this point that a study of how the belief in global warming is like a religion is valuable.

It is also worth noting that this essay is hardly the first time that a supposedly scientific movement is considered in terms of its religious implications. Darwinian thought, widely viewed as a "scientific" paradigm, is actually dependent upon a series of theological developments in Anglican theodicy.[9] Environmentalism itself can be seen as a religious movement.[10] Even global warming has been explicitly compared to a religion before in Roy Spencer's modern broadside, *The Bad Science and Bad Policy of Obama's Global Warming Agenda.*[11] But if environmentalism is a new religion, then global warming is its most fundamentalist form, albeit with an ancient heretical slant. Humans have a built in belief in religion, and if they reject traditional forms of it, they will find a replacement. The culture that has arisen around global warming is an almost perfect illustration.

The Fall

Nearly all religious traditions postulate a mythical past when harmony prevailed. This past may be identified as an Eden or paradise where all of creation is at peace. As the late student of world mythology, Joseph Campbell, noted this theology is shattered by the discovery of knowledge by man. Knowledge includes perceptions of dichotomy, good and evil, for example, and it is man's knowledge that leads to the "fall" of creation, in Christian terms and the end of harmony in other religious traditions.[12] Like all other religious traditions, environmentalism and global warming offer a similar narrative.

One of the most central tenants in environmentalist thought is the concept of "balance" From a very early age, children are taught, as part of biological science, that ecosystems are complex structures that exist in harmony. Often ecosystems are pictured as a fragile building of boxes. Remove as little as one piece, and the whole can tumble down.[13] Yet what is so interesting about this vision is that, for the most part, neither mainstream science nor certain religious traditions seem to accept it, at least not wholeheartedly. Biological science is committed to Darwinian thought which presupposes that most species are now extinct. Some biologists, notably the late Stephen Jay Gould and his successors, believe that evolution occasionally caused dramatic changes in a very short period of time.[14] This is hardly the expected outcome of a system which is in perpetual harmony. One might think that creation "scientists" would have a different view. After all, balance and harmony might be expected to come from a designed world and this vision is consistent with their reading of Genesis. But they too, it seems, assume catastrophic events which wipe out whole ecosystems.[15] So where does this idea of harmony and balance come from?

The answer, which should be apparent to any student of mythology and religion, is that ecosystems represent a modern version of the garden of Eden, where all is in harmony, the lamb lays with the lion (or at least, the lions don't eat too many lambs in this modern version) and all is well except that, alas, man is tending the garden and his sins are destroying the harmony that once prevailed. Is this explanation of ecological science overwrought? Consider how often those who claim to represent "science" actually appeal to a much older language choice in their discourse. Julia Butterfly Hill climbed a redwood to save it from the lumber company, who, incidentally, owned the land the redwood grew on. Did she write a scientific paper about the unique species she found in the canopy? No. She painted a large sign demanding that people respect their elders. (In this case, the tree is the "elder") We hardly need note that this is pantheism primarily and whatever science may exist is merely along for the ride.[16] And Ms. Hill is hardly an isolated voice. Consider the comments of Bruce Babbitt, former secretary of the Interior under the Clinton administration. We have an obligation, he insists, to "protect the whole of creation."[17] From whence, one might ask, does this obligation derive?

These are very odd views for those who claim to believe in modern science. Many people would like to think science is beyond the creation metaphor. From a strictly materialist perspective, humans are part of various ecosystems themselves. So why do we have a special obligation to preserve "wilderness" areas ("places where man is an infrequent visitor") and the like? The answer of course is that just as the religious view of ecosystems is that they represent an "Eden," man himself is considered something of a second creation with stewardship responsibilities.[18] Of course, this perspective makes perfect sense in light of western religious thought, most especially Christianity, but the people who claim to see in ecosystems a last Eden do not generally espouse this view. And there is, it should be noted, one crucial difference. In the traditional religious view, man is fallen. In the new religion, man is an active destroyer.

If much of ecology can be written as a search for a harmonious past before the corrupting influence of man, then perhaps the best known paper of the global warming crusade represents the ultimate vision of the fall of nature due to man.

206

That paper, of course, is the famous hockey stick graph of Michael Mann. First published in *Nature* and then appearing on the cover of several Intergovernmental Panel on Climate Change (IPCC) reports, Mann's graph had an immediate visual impact that, for many, summarized all the dangers associated with global warming.[19] Purportedly based on tree ring data and various other proxies for climate study, this graph showed virtually no change in temperature for a thousand years, followed by a dramatic rise in temperature in the twentieth century as humans (those greedy sinful consumers) pumped large amounts of carbon dioxide into the atmosphere.

Mann's graph was immediately seized upon as the smoking gun proving the global warming case. As a visual, it certainly provided a powerful image. It was also close to a complete fantasy. Researchers immediately noticed that the graph did not include such well-known temperature fluctuations as the Medieval Warming Period (MWP) or the subsequent Little Ice Age that lasted from roughly 1500 to 1800. Indeed, critics discovered that the algorithms used by Mann actually produced hockey stick shaped graphs, even when fed random data. One mathematician, Edward Wegman, even suggested that qualified statisticians regularly check the work of climate scientists before they publish. Undoubtedly, such an innovation would be a death blow to the whole global warming alarmist industry.[20]

For our purposes, however, the salient point is not that Mann's famous graph is fundamentally flawed. That was obvious from the moment it appeared. Rather the question should be, why was it ever accepted for publication by *Nature*, much less given such prominence by the IPCC? Here again, the answer is that the graph matches the religious sensibilities of the global warming proponents. Like much of ecology, this graph effectively postulates a world in harmony and balance but then mankind sins against the planet and destroys the Garden of Eden. It is a powerful story (and indeed it has been for thousands of years) but the new storyline does differ from the older one in one key aspect. In Genesis, man sins against God leading to the fall of creation. In the new story, man sins against creation itself.

The Magisterium and the Heretics

A fundamental problem for all religions is deciding upon which texts and perspectives are to be considered authoritative. Students of early Christian church history quickly become aware of many movements that identified themselves as "Christian" Not surprisingly, not all of these groups came to the same conclusions about what it means to be a Christian, and debates about everything from the nature of Christ to the nature of the gospel to the very texts that would be regarded as authoritative scripture at one time or another threatened to tear the early church apart. The church rather quickly developed a structure to settle these disputes. Those within the church were defined by their participation in communion and communion, in turn, required one to assent to the leadership of bishops, who, within less than a century following the crucifixion, were widely regarded as the successors to the apostles. The bishops in turn resolved disputes through councils of the whole church and eventually certain bishops came to be regarded as more important than others.

Naturally, not everyone found themselves at home within this hierarchy. Apologists and philosophers who were in the church, but whose writings were not regarded as scripture came to be included in what was called "the magisterium" of the church. In other words, their writings became part of the church's sacred tradition and could be cited as authoritative, though not as authoritative as scripture. Those who, on the other hand, wrote outside the church, were considered heretics. Since, as we know, the church (which would eventually become the Catholic and Orthodox churches respectively) was the winner, we often do not have much in the way of the writings from the heretics. Scholars are sometimes reduced to trying to discover what heretics thought from reading their critics, a dangerous proposition.[21]

Interestingly enough, the Christian experience in this regard is not unique. All organized religions must find some way of distinguishing what is part of the "canon" of religious thought and what is heretical. Indeed, Christianity itself was later split asunder over the question of whether authority lay in the church hierarchy and tradition or in the scriptures. Global warming, considered as a religion, is no different. It has its holy text, the IPCC reports, a recognized priesthood who interprets the texts for the masses, and most importantly, a means of distinguishing the true writings or magisterium from dissenting heresy. That means, incidentally, is peer review.

Peer review is simply a process by which papers submitted to an academic journal are vetted, supposedly anonymously, by (traditionally) three scholars to determine if a paper is suitable for publication. It is a relatively new phenomena. As recently as the early twentieth century a journal's editor was also the sole reviewer who determined whether a paper was worthy of publication. On the whole, peer review might appear relatively innocuous, though it does have its critics. Among other problems, peer review tends to support the status quo and actually limits the publication of truly ground-breaking research.[22] Still, one would think that peer review is simply an academic convention of little interest to the general public. And yet, this is precisely the opposite of the truth, at least insofar as the general public also happens to include advocates of the climate change religion.

Consider, for example, Ed Begley, Jr.'s comment to Fox News reporter Stuart Varney. Begley is a television actor of modest talents and a committed environmental activist. He is clearly not a scientist. But when asked about the damning comments made by various climate scientists in some released emails, he reiterated that really we should only be looking at "peer-reviewed studies."[23] And Begley is hardly unique. Anyone who debates the results of climate science will sooner or later be hit with the charge that we cannot believe what the critics say because after all, they don't publish in peer-reviewed journals. (In a debate on Amazon under my review of the book, *Green Hell*, I challenged the defenders of climate change orthodoxy to refute the Douglas et al. paper that actually was published in a peer reviewed journal. A critic promptly responded that the journal was not as good as other peer-reviewed journals, an instance of raising the bar that he felt was a sufficient enough rebuttal as to not even have to read the article in question.[24])

Even professional historians are not above falling for the line about peer review establishing what constitutes legitimate discussion of the topic. Naomi Oreskes, for example, produced an article for *Science* in which she claimed to have found 928 papers on global climate change in peer reviewed literature, not one of which contradicted the "consensus" view that man was causing global warming.[25] Needless to add, Oreskes's methodology and even her use of search engines has been seriously and effectively criticized.[26] But Oreskes has also received a lot of praise from activists for her "famous" paper.[27] For our purposes, however, the important point of note is that Oreskes's work was the real beginning of the consensus claim that all scholars agree on climate change. Of course, such a claim is not true in the literal sense, but it is true in the religious sense. What Oreskes actually did was create a magisterium for the new global warming religion. Anyone who has any desire at all can find lots of dissenting scholarly pieces, many very well written, all over the Internet (the page, "Watt's Up with That?" maintained by Anthony Watts is a good place to start) and even within the peer-reviewed literature, but these are not part of the magisterium and so, they are not part of the consensus.

The irony in all this is that we now know just how rigged the "peer review" system is. In November of 2009 an anonymous hacker released thousands of emails from the Climate Research Unit (CRU) at East Anglia University in England. These emails include some truly damning comments such as threats of violence against dissenting researchers.[28] But one of the most conclusive things that we have discovered in reading these emails is the extent to which peer review is manipulated. A fairly small body of scholars can use their influence to prevent publication of articles that dissent from the thesis that man is causing climate change. When they cannot monopolize review of the articles proper, they attempt to manipulate editors and in one case even got an editor removed from his position. So, effectively critics of climate change are forced to publish outside the peer review system and, conveniently enough, that puts them outside the magisterium for the faithful. Indeed, the latter never even have to read, much less address, the arguments of the critics because they are not "really" science.[29]

Unfortunately for this view, science proper does not require consensus. Indeed, it functions best when questions and skepticism abound. Religion, on the other hand, generally does require some "consensus" And in the religion of global warming, the magisterium of texts available to true believers is controlled by peer review, and peer review, to a greater and greater degree, is controlled by the priests and bishops of the new faith. It is not quite an "Index" of forbidden and heretical writings, but it's clearly close enough for government work.

The End of the World

Many religious traditions, particularly in the West, postulate an end to the world as we know it. Of course, in the broadest sense, such a conclusion is consistent with what we know from physics. The universe had a dramatic beginning and it will eventually burn out.[30] But religious traditions beginning with Manichaeism (founded by Zoroaster) and including both Christianity and Islam postulate something a bit more dramatic: an apocalypse during which time man will be judged. Not surprisingly, end of the world scenarios are also popular with the global warming crowd. They predict heat waves, droughts, wildfires, and a host of other perils that will ultimately destroy our world.

The interesting thing about the various end of the world predictions is that these, more than just about any other area of global warming "research," are routinely dismissed by both experts and the course of events. Consider for example, the claim that global warming would lead to more powerful hurricanes. Such comments were rampant during the aftermath of Hurricane Katrina in 2005. But they were also unfounded. Christopher Landsea, an expert on hurricanes rejected the link and withdrew from participation in the IPCC reports when his research was ignored. In the end, after several mild hurricane seasons, even the mainstream media began to downplay the connection between supposed global warming and hurricanes. As Dr. William Gray noted, "The degree to which you believe global warming is causing major hurricanes to increase is inversely proportional to your knowledge about those storms."[31]

But if you thought the true believers would be humbled by how dramatically wrong they were about hurricanes, you had best think again. After all, Christian fundamentalists did not abandon their fascination with Armageddon stories just because every substantial prediction found in Hal Lindsey's *Late Great Planet Earth* (1973) turned out to be wrong. One need look no further than the success of the *Left Behind* series to see that. But whereas Christians are now not quite so specific about just when the end will occur, the global warming proselytizers are as alarmist as ever. The polar ice caps will melt! Polar bears are dying out! Sea levels are rising! And so it goes. Every one of these claims are demonstrably false, but that hardly matters.[32] In a dire warning against those congressmen so degenerate as to vote against the cap and trade energy bill, economist Paul Krugman declared, "we are facing a clear and present danger to our way of life, perhaps even to civilization itself."[33] Some protesters avoid making predictions at all and instead just share images of the world afire. Thus ends the world with remarkable biblical imagery [34]. And there you have it. The end is upon us and only enlightened regulation forcing a reorganization of our way of life can save the planet. Man has sinned, but there is still time and hope for the future, if only we give up our SINs (or SUVs, as the case may be).

Conclusions

Global warming is still widely portrayed as a scientific proposition. Although the movement is not totally bereft of scientific claims (e.g., Aspen trees grow better in warmer climates), the fact remains it is much more of a religious proposition than a scientific one.[35] That the two go hand in hand may come as something of a shock to the humanist/atheist set, but there is no reason to take their presumptions as authoritative. Why, after all, would a strictly scientific perspective need to be defended by titles such as James Hoggan's *Climate Cover-Up: The Crusade to Deny Global Warming?* Pure science would hardly need to fight or defend against a crusade. (One of the supreme ironies of James Hoggan's fantasy title above is that it came out just before the CRU emails were released. Interested reviewers were thus treated to evidence for an actual conspiracy at the same time that Hoggan's fictional one came out.) Only religious zealots think in these terms and global warming is first and foremost, a religion.

But what type of religion is it? I suggested at the beginning of this little essay that global warming is a fundamentalist version of a broader religion of environmentalism. And indeed, global warming shares some of the dogmatic traits found among some forms of fundamentalism: it focuses on the beginning and end of its gospel story and not so much on the middle. But faith in global warming is "fundamentally" different from traditional faiths as well. Belief in a broad salvation of mankind is lacking. Indeed, many of the global warmers openly wish for the destruction of much of humanity.[36] Our planet, after all, can only "sustain" so many people. Salvation, therefore, is really only available for a precious enlightened few.

In this regard, global warming is more of a Gnostic religion than a broad faith tradition like Judaism, Christianity, Buddhism, and Islam. Believers know the deeper truths that are simply unavailable to the less enlightened. Thus while the masses are told that we must adopt some treaty similar to the failed Kyoto protocols, privately the enlightened ones realize that will not be enough. Ultimately the masses will simply have to be led and controlled, for their own good of course. Some would argue that at various times and places Christians have acted in the same way. But that simply reinforces the point. Global warming is a religion. But, like other religions that promise heaven on earth, it will invariably promote misery and suffering. The same cannot be said for more traditional faiths.

References

1. The full transcript of the Obama speech can be found at www.whitehouse.gov/the-press-office/remarks-president-economy-0. Obama reiterated his belief that we needed to pass the bill in his press conference the following day about the BP spill.

2. Praise for Obama came from Climateprogress.org (http://climate-progress.org/2010/05/27/Obama-climate-change-poses-a-threat-to-our-way-of-life) and the same announcement was, naturally enough, received less favorably by Fox News (http://www.foxnews.com/politics/2010/06/15/-obamas-pitch-energy-speech-gulf-crisis-infuriates-republicans/?loomia_ow=t0:s0:a16:g2:r2:c0.0- 98586:b34919344:z6)

3. Although Dawkins likes to believe that he is something of a leader in pointing to conflicts between "science" and religion, the truth of the matter is that he offers little that is new. Readers who want to read an incisive commentary on the same topic written some 80 years before Dawkins was making a name for himself should examine Harry Elmer Barnes's little broadside, *Science v. Religion as a Guide for Life* (Girard, KS: Haldemann-Julius, 1927). Barnes essentially covered most of the themes that are, in the early 2000s considered "new atheist."

4. A full survey of scientists whose research was inspired and guided by Christianity can be found in the book, *Evolution or Creation: Consider the Evidence before Deciding* by E. Norbert Smith.

5. Usually creationists are faulted on this account. And indeed, the argument for a "young" creation involving literal 24 hour days is difficult to defend. It is worth noting ancient writers rarely did so. The debate on how Christian scholars should read Genesis continues to this day. See for example David G. Hagopian, ed., *The Genesis Debate: Three Views on the Days of Creation* (Mission Viejo, CA: Crux Press, 2001). But the influence of religious commitments is much more common than the example of the creationists would suggest. A recent biography of Darwin has noted that the great biologist was committed first to a naturalistic theory and sought evidence for it later. See in this

regard Benjamin Wiker, *The Darwin Myth: The Life and Lies of Charles Darwin* (Washington, DC: Regnery, 2009).

6. Al Gore's academic record was first exposed by David Maraniss and Ellen Nakashima in their March 19, 2000 Washington Post article "Gore's Grades Belie Image of Studiousness" The article was in response to a constant portrayal of Gore as a more scholarly candidate than then candidate George W. Bush. As it happened, their academic record was essentially identical: both were mediocre.

7. The snowman image can be viewed at http://blogs.glam.com/glamspirit/2009/02/20/eco-friday-hundreds-attend-global-warming-protest/.

8. Douglass, David H, and Christy, John R., "Limits on CO2 Climate Forcing from Recent Temperature Data of Earth" Energy and Environment 20, 1-2, 2009. The NIPCC report is can be found as a pdf file on the SEPP website at http://www.sepp.org/ under publications. Henrik Svensmark and Nigel Calder, *The Chilling Stars* (Cambridge, UK: Icon Books, 2007).

9. The religious basis of Darwinian thought is well established in a series of books by Cornelius G. Hunter. See in particular his *Darwin's God: Evolution and the Problem of Evil* (Ada, MI: Brazos Press, 2001) and *Darwin's Proof: The Triumph of Religion over Science* (Ada, MI: Brazos Press, 2003).

10. See in this regard Robert H. Nelson's recent paper, "Ecological Science as Creation Story" *The Independent Review* 14,4 (Spring, 2010) 513-34, and his popular editorial, "Environmentalism is the New Secular Religion" which can be accessed at http://www.independent.org/newsroom/article.asp?id=2773. Both of these essays follow the lead of the late author Michael Crichton whose 2003 speech to the Commonwealth Club of San Francisco, "Environmentalism as Religion" which can be accessed at http://www.oism.org/news/s49p1521.htm.

11. Roy Spencer, *The Bad Science and Bad Policy of Obama's Global Warming Agenda* (New York: Encounter, 2010)

12. This theme is explored in many of Campbell's works but most fully in his collection of essays *Myths to Live By* (New York: Penguin, 1972).

13. This particular illustration is shown to great effect in the popular children's science series, Bill Nye: The Science Guy, episode 9, Biodiversity.

14. Eldredge, Niles, and Gould, S.J., "Punctuated Equilibria: An Alternative to Phyletic Gradualism" in TJM Schopf, ed., *Models in Paleobiology* (San Francisco: Freeman Cooper, 1972): 82-115.

15. Generally speaking, creationists emphasize catastrophic events by giving a literal reading to the flood story found in Genesis. The classic portrayal of this perspective is found in Henry Morris and John C. Whitcomb, *The Genesis Flood* (Philadelphia: Presbyterian and Reformed Publishing, 1961.)

16. Hill's quasi mystical experience with her tree, "Luna" can be found in her book, *The Legacy of Luna: The story of a Tree, a Woman and the Struggle to Save the Redwoods* (San Francisco: Harper, 2000). While religious sentiments inform virtually all "ecological" science, one can find the science a little more heavily represented on spectrum of redwood preservation in the book *Wild Trees: A Story of Passion and Daring* by popular science writer Richard Preston (New York: Anchor, 2007). The book is largely about Steven Stillet's study of the old growth redwoods, but the secrecy he and his colleagues hold to "protect" the trees goes far beyond what one would reasonably expect from pure science.

17. Cited in Nelson, "Ecological Science as Creation Story," p.513.

18. Readers might think I am being rhetorical in noting that supposedly wild ecosystems are considered remnants of "Eden" by the new religion. Actually, the term is in such common useage that a Google search of the term "last Eden" generates 17,700,000 hits. Some of these, of course, reference a Slider's television episode, or assorted other memorabilia from pop culture, but the vast bulk of the hits cover remote regions of the globe like Logando National Park in Africa or the Makay Mountains in Madagascar. All these places are routinely described as Eden simply based on their (apparently) untouched ecosystems. Indeed, there are so many "last" Edens that a cynic might almost conclude Eden is in the eye of the beholder. Or, considered another way, the Kingdom of Heaven is within you.

19. The image of the graph is still widely circulated and can be viewed here: http://www.planetthoughts.org/?pg=pt/Whole&qid=2392

20. The whole story of the hockey stick controversy can be found in Christopher C. Horner's *The Politically Incorrect Guide to Global Warming* (Washington, DC: Regnery, 2007), 120-131. Wegman's report, titled "Ad Hoc Committee Report on the 'Hockey Stick' Global Climate Reconstructions" is found at http://www.uoguel-ph.ca/~reckitri/research/WegmanReport.pdf.

21. The most important of the ante-Nicean texts detailing heretical thought is St. Irenaeus, *Adversus Haereses* (Against the Heresies), a refutation of the Gnostic writers. Since the discovery of the Gnostic writings at Nag Hammadi in 1945, however, some scholars have suggested Irenaeus's portrayal is a little overwrought. Rodney Stark, however, in his book *Discovering God* (Newy York: Harper Collins, 2007) pp 325ff suggests that for the most part Irenaeus was correct.

22. The strengths and flaws of peer review are discussed in Frank J. Tipler's thoughtful essay, "Refereed Journals: Do they Insure Quality or Enforce Orthodoxy?" in William A. Dembski, ed., *Uncommon Dissent: Intellectuals who Find Darwinism Unconvincing* (Wilmington, DE: ISI Books, 2004): 115-30. That peer review tends towards enforcing orthodoxy, even with regard to topics

without public policy implications, is apparent from the treatment of Lofti Zadeh, founder of fuzzy set theory. His work was widely rejected and criticized. Indeed, he was unable to publish his seminal paper, "Fuzzy Sets," under traditional peer review. Ultimately the paper appeared in *Information and Control* 8 (1965): 338-53 as a favor on the part of the editor. Of course, fuzzy sets are now widely used in engineering and have numerous applications. They just would not make the grade if peer review were the only academic standard of validity. Zadeh's long attempt to gain a respectful hearing for fuzzy sets is found in Bart Kosko's *Fuzzy Thinking: The New Science of Fuzzy Logic* (New York: Hyperion, 1993): 145ff.

23. Quoted in Mark Steyn, "CRU's Tree Ring Circus" Nov. 28, 2009, *National Review Online*. The article can be found at http://article.nationalreview.com/416113/crus-tree-ring-circus/mark-steyn.

24. The critic in question was one WHM and the whole increasingly caustic debate can be found under my Amazon review of Stephen J. Milloy's *Green Hell: How Environmentalists Plan to Control your Life and What You can Do to Stop Them* (Washington, DC: Regnery, 2009). The review is found at http://www.amazon.-com/Green-Hell-Environmentalists-Plan-Control/dp/1596985852/ref=cm_cr-mr-title#noop. Amazon readers and reviewers are some of the more educated members of the American public, but lack the financial ties to academe. As a result, one finds a much more wide ranging discussion there than is common in other settings. Judging by "helpful" votes, it is clear that a significant percentage of Amazon readers are not convinced by the "science" of climate change. A small minority of true believers, however, try to monopolize discussion on this topic. Their discourse is often quite revealing about their actual motivations.

25. Oreskes, Naomi, "Beyond the Ivory Tower: The Scientific Consensus on Climate Change" *Science* 306 5702 (2004): 1686. Oreskes's paper is the most famous of the attempts to create a consensus in the minds of climate scientists and has spawned a miniature industry of its own. The most recent contribution to the genre is Anderegg, William R.L., et. al., "Expert Credibility in Climate Change" forthcoming in *Proceedings of the National Academy of Sciences*. The article can be accessed here: http://www.pnas.org/content/early/2010/06/04/1003187107.-abstract. The authors examine "an extensive dataset of 1,372 climate researchers" and conclude that 97-98% support the consensus. Of course, to be included in the data set, a researcher had to publish 20 peer-reviewed papers which, on the face of it, means most of the prominent critics like physicist Freeman Dyson, are excluded. More troubling, however, is the active intent of the authors to influence public policy by directing journalists to only contact the scientists they list for information on climate change. Roy Spencer argues in his blog that this is little more than a blacklist. I think an index might be a better description, but in either

case Spencer also notes that much of the "research" actually described is not so much proving as simply assuming human activity causes global warming. His informed commentary is found at http://www.drroyspencer.com.

26. The most thorough criticism of Oreskes remains Benny Peiser's thorough review of all 928 abstracts she found. His conclusion is that only 13 actually supported the global change consensus she hoped to establish. See Peiser's list of abstracts at http://www.staff.livjm.ac.uk/spsbpeis/Oreskes-abstracts.htm. Two of the papers Oreskes listed as supporting the consensus actually opposed it.

27. See for example popular science writer Chris Mooney's blog entry, "Who 'Framed' Naomi Oreskes" (found at http://www.desmogblog.com/who-framed-naomi-oreskes). Mooney never bothers to actually defend Oreskes from any of Peiser's criticisms ("Peiser and all that" is his throw away line) but he does imagine a conspiracy of skeptics who are busy trying to undermine the consensus which, he assures us, "is alive and well" He finally stresses the importance of defending Oreskes, but insists we not address actual arguments in so doing, lest we allow skeptics to "fire up their base" None of this makes any sense in terms of ordinary scientific discourse, but read as an attempt to protect the faithful from dangerous ideas, Mooney's arguments suddenly appear quite reasonable.

28. The climategate emails are available for all to view at http://eastangliaemails.com/.
As one writer (otherwise sympathetic to the religion of global warming) put it, "the charge that climate skeptics 'are not published in peer reviewed journals' just lost most of its power..." See Megan McArdle's column, "Climategate" in The November 23, 2009 issue of *The Atlantic* accessible here:
http://www.theatlantic.com/business/archive/2009/11/climate gate/30702

29. Similarly, during the religious conflicts of the reformation, Catholics were not even expected to read Protestant tracts that were listed on the Index of forbidden boxes. Faithful Protestants, for their part, were told that Catholic apologists worked for the Anti-Christ and their works were full of deceit. The attempt to protect the faithful from seeing another perspective is common to religious thought. In science proper alternative perspectives are valued, except of course when the "science" in question is actually a religion.

30. For a discussion of how discoveries in modern physics are, in the broadest sense, confirming traditional Judeo-Christian views of the universe, see Robert Jastrow, *God and the Astronomers* 2cd edition (New York: WW Norton, 2000). Jastrow began his work as something of an agnostic, but he correctly notes that the claim, found so often from the new atheists that "science" is working on (insert unresolved problem here) is a poor justification for a materialist world view. As he effectively demonstrates, there simply are places we cannot go with

the methods of science, and those areas have been the domain of religion for generations.

31. Quoted in Horner, *Politically Incorrect Guide to Global Warming*, 162.

32. On the melting of the ice caps see the wonderfully alarmist article, "North Pole May be Ice-Free for the First Time this Summer" published by *National Geographic News* on June 20, 2008. Of course, in the winter of 2007-8, the arctic ice cap had dramatically expanded over the previous year, and the fear was that the "new ice" would melt rapidly. Certainly, the article assured us, we would see the ice cap free of ice within 5 years. This gem of an article is, amazingly enough, still accessible at:

> http://news.nationalgeographic.com/news/2008/06/080620- north
> pole.html.

Of course, nothing of the sort happened. In fact, the arctic ice cap expanded dramatically again the following winter to the point that it reached the same extent as 1979, the base line for all studies of the decline in arctic ice. On this note, see the Daily Tech Science blog which summarized the arctic ice tracking reported by the University of Illinois Climate Research Center. This blog is found at http://www.dailytech.com/Article.aspx?newsid=13834. The perceived threat to the polar bears (from declining ice sheets!) led the US government to list the polar bear as threatened. This may mark the first time in the history of the Endangered Species Act that a creature with a dramatically increasing population has been so listed. In the 1960s, for example, the polar bear population was only 5,000. It is now closer to 25,000. This has not prevented the faithful from proclaiming that the polar bear population will cease reproduction in 2012 (another of those remarkably testable claims that you can bet will not come true) and die off within another decade. A good discussion of the polar bear situation can be found in Bjorn Lomborg's book, *Cool It: The Skeptical Environmentalist's Guide to Global Warming* (New York: Knopf, 2007) 4-8. One of the faithful tried to rebut Lomborg's sobering look at population counts by noting "global warming can temporarily boost polar bear populations" In other words, there is absolutely no evidence that polar bears are threatened, but they will be. Such logic requires great faith. See the article, "On the myth that Polar Bear Populations are Flourishing" found at http://grist.org/article/will-polar-bears-go-extinct-by-2030-part-i/ Of course, even if the polar ice were to completely melt, it does not follow polar bears would become extinct. After all, we have fossils of polar bears from the last interglacial period when summer temperatures were significantly warmer than now (approximately 7-9°F). Bottom line, there is no reason to fear for the polar bears. On the other hand, polar bear researchers who do not tow the line on global warming are endangered. Dr. Mitchell Taylor was not invited to the

Copenhagen talks in 2009 because, as he was told, his expertise on polar bears "was extremely unhelpful" See in this regard Christopher Booker's report, "Polar Bear Expert Barred by Global Warmists" from the June 27, 2009 Guardian and accessible here:

http://www.telegraph.co.uk/comment/columnists/christopherbooker/56640 69/Polar-bear-expert-barred-by-global-warmists.html.

On the matter of wildfires, unnamed experts were regularly cited by news agencies in 2003 and 2007 as fires raged across southern California. These experts uniformly claimed that "global warming" would result in more such fires in the future. That may be, but the fires in the present were due to forest mismanagement and particularly, the abandonment of selective cutting. The story on how this particular environmental catastrophe was due to the very measures environmentalists use to "preserve" the forest is found in Iain Murray, *The Really Inconvenient Truths: Seven Environmental Catastrophes Liberals Don't Want You to Know About—Because they Helped Cause Them* (Washington, DC: Regnery, 2008), pp. 147-185.

33. Paul Krugman, "Betraying the Planet" found in the June 28, 2009 *New York Times*. The full op-ed can be accessed at: http://www.nytimes.com/2009/06/29/opinion/29krugman.html?_r=1.

34. One such image can be viewed here: http://www.123rf.com/photo_3089963_photoshopped-image-based-on-a-nasa-public-domain-image-of-a-red-hot-glowing-burning-earth-on-a-fire-.html

35. Climate change proper is not causing aspen growth rates to accelerate, but increasing levels of CO_2 are. See in this regard, http://sciencedaily.com/releases/-2009/12/09/091204092445.htm.

36. The most extreme version of this vision is found in the group Voluntary Human Extinction Movement which argues that "Phasing out the human race by voluntarily ceasing to breed will allow Earth's biosphere to return to good health" (See their website, http://www.vhemt.org) Of course, not everyone thinks we need to kill off all humans to attain a better planet. We really only need to "rethink" reproductive rights. This altogether more scary vision (after all, the human extinction movement is voluntary) can be found in Carter J. Dillard's article, "Rethinking the Procreative Right" *Yale Human Rights and Development Law Journal* 10 (2007) cited in Milloy, *Green Hell*, 29-31. Milloy notes that virtually all the "Greens" (his term for the new secular religion) would like to see at the very least limits on population growth and many would like to see a reduction in population. Who gets to decide which "populations" to reduce is, of course, another issue.

Author bio:

Fritz R. Ward received his BA cum laude from University of Idaho in History. He received his MA and Ph.D. in English and early American history from University of California, Riverside, in 1990 and 1995. He has published articles and reviews in *John Locke Newsletter*, *Anglican and Episcopal History*, and *The Eighteenth Century: A Current Review*. He has also written the preface to E. Norbert Smith and Joanna Jones satire, *Battleground University*. Dr. Ward also has education credentials in math and science and is pursuing a second master's degree in mathematics education. He is currently a teacher at a prominent public school academy in California and is a top 500 reviewer on Amazon.com. You may email him at: Fritz7ntd@aol.com.

Bible Perspective on Global Warming
Rod J. Martin

Editor's note

Rod and I met through a mutual friend George Howe. Like some others that have contributed chapters, he is an engineer with an analytical mindset and training. He provides a fresh approach to topic of global warming from a biblical perspective. I have long felt human caused global warming or climate change is nonsense for it would mean our Creator made an error when he told humans to multiply and fill the earth, not once but twice. He did so after the original creation and again after the global flood. Our Creator-God did not make such an error in judgment. As will be abundantly clear from the following essay, carbon dioxide is NOT a pollutant, but instead acts as a plant fertilizer. An increase in the amount of atmospheric carbon dioxide would be a good thing and certainly not something to be avoided. Industries releasing carbon dioxide into the atmosphere should be praised, not reprimanded or fined. It is certainly time to put this sacred cow of science out of its misery.

Abstract

Media coverage of global warming has been increasing for over twenty years. Major proponents include the United Nations, politicians, environmentalists, and celebrities. Oddly, the church has had little to say on the issue and has made scant use of Scripture to evaluate the alleged problem. This paper will identify the major goals of global warming advocates, propose a biblical (young-earth creationist) framework for evaluating the issue, and highlight basic scientific data related to the alleged claims. It will be shown that the Bible provides sufficient counsel to enable Christians to evaluate the claims of global warming and arrive at a confident position that is in accord with real science. The contention that man's activities are causing global warming, as described in the media and by its advocates, is a myth. There is no reason either biblically or scientifically to fear the exaggerated and misguided claims of catastrophe as a result of increasing levels of manmade carbon dioxide (CO_2).

Introduction

Al Gore contends that the greatest moral issue of our times is global warming. In addition, he and others characterize global warming, which he considers to be predominately caused by man, as a moral, ethical and spiritual challenge. These claims are in his slide show presentations, his book and his film, *An Inconvenient Truth* (Gore 2006, introduction). If he is right, then Christians should examine this issue and take a strong biblical position. Moral, ethical and spiritual issues are the domain of the church. At the very least, global warming should be evaluated to see if indeed it is a moral issue. Few Christian groups have publicly taken a side regarding global warming. Two associations of well-known evangelicals, however, made statements on global warming during 2006. In mid-February, 2006, the Evangelical Climate Initiative (ECI) came out in support of legislation to control CO_2. They issued a four-page statement called *Climate Change: An Evangelical Call to Action* (Evangelical Climate Initiative, 2006). Later in 2006 the Interfaith Stewardship Alliance (ISA) issued a 22-page statement called *A Call to Truth, Prudence, and Protection of the Poor: An Evangelical Response to Global Warming* urging caution (Cornwall Alliance, 2006, website, formerly Interfaith Stewardship Alliance). Also in 2006, several members of ECI were featured in an hour-long television program describing growing support for the global warming agenda among evangelical Christians. When ECI leaders were asked what the Bible had to say on this issue, they merely referred to general "creation care" concepts such as: be a good steward, and do not hurt the poor. A more detailed understanding of "creation care" can be gained by reading the above cited reports. Claim #3 in the ECI paper is a good summary of the concepts. Both evangelical groups resorted to these concepts, yet they both also stated that they wanted to bring a decidedly Christian perspective to the debate on global warming. Are very general "creation care" concepts all the guidance the Scriptures provide? Are concern for the poor and a desire to wisely steward the earth exclusively Christian positions? Many non- Christians also share these concerns. This paper affirms these concerns while searching Scripture for additional counsel and a uniquely Christian perspective. It will be shown that the Bible provides a clear framework for evaluating the claims of humanly produced global warming and coming to a credible decision. This paper is not intended to answer all the questions on global warming. The primary objective is to offer a biblical framework for evaluating the major claims of global warming advocates and demonstrate that this framework is consistent with basic science. Obviously, not even all creationists will agree with every assertion in this paper. Hopefully, however, interested creationists will be encouraged to expand the biblical and scientific framework for understanding this issue. The spiritual implications of accepting evolution have been eloquently and comprehensively argued by many creationist organizations. Yet, for far too long the creation-evolution debate has been viewed by many, even in the church, as an abstract, academic topic with little relevance to real life. Man-made global warming is a direct product of evolutionary thinking, and the potential impacts are very applicable to real life.

Proposed secular solutions to the alleged claims of global warming will directly impact everyone who depends on fossil fuels for their current life style. The issue of global warming presents biblical creationists with an opportunity to demonstrate not only the efficacy of Scripture in addressing life's issues, but also to show how ignoring Scripture leads to unnecessary, expensive, and harmful actions. Global warming is an arena where the battle between biblical truth and evolutionary untruths is currently raging, and it will affect everyone in very practical ways. Contrary to what advocates say, a consensus does not exist on global warming, the debate is not over, and a biblical (young-earth creationist) perspective has not yet been widely discussed.

God is the creator of the universe. In His Word, the Bible, God has addressed every area of life (family, state, church, science, man, sin, etc.). God's Word is truth. The revelation given to us in Scripture is sufficient to enable man to understand the world around him and make decisions that will honor God and benefit mankind. When faced with a challenge, a follower of Christ should first ask, "What has God said that will help me understand this issue and respond in a manner that honors Him?" This paper is an effort to answer that question regarding the alleged issue of global warming.

Definition of Terms

Before proceeding any further it will be helpful to present the following definitions. These definitions are simply stated in order to make them clear, easy to understand, and easy to apply.

Weather

Weather refers to atmospheric conditions at a particular time, for example: temperature, humidity, wind, barometric pressure, precipitation, and so forth.

Climate

Climate comprises the average weather conditions present in a particular location at a particular time of year. Climates are measured over several decades.

Climate change

Climate change, obviously, refers to long-term changes in average weather conditions.

Global warming

Global warming is an assertion that the entire earth's surface is warming. Unfortunately, many individuals, and the popular media, often use the terms "climate change" and "global warming" interchangeably. As shown above, they are not synonymous terms. Both climate change and global warming are commonly attributed to human activities like burning fossil fuels and harvesting forests.

Primary Issues

Media news on global warming tends to be confusing. Dissimilar terms are used interchangeably (global warming and climate change), the scientific issues are unfamiliar to the general public (chemical analysis of ice cores, reef bleaching, ocean current stagnation, etc.), and an unusual mix of scientific experts, politicians, and celebrities claim that devastating consequences will occur if we ignore their advice (massive floods, epidemics, drowning polar bears, etc.). A means must be found to cut through the confusion and emotional rhetoric in order to grasp the core issues and concepts. Identifying what global warming advocates want to control helps bring the issue into clearer focus. From this perspective, two issues are of primary concern to global warming advocates: CO_2 emissions and the harvesting of forests. They want to control both CO_2 emissions and the harvesting of trees. Global warming advocates are concerned that certain "greenhouse gases" (GHG), principally CO_2, are being generated by mankind in quantities sufficient to adversely affect the long-term climate of the earth. They claim that since the start of the industrial revolution, the burning of fossil fuels has unnaturally increased the atmospheric content of CO_2. This in turn is retarding the earth's emission of long-wave radiation and artificially increasing the earth's surface temperature. This is called the GHG effect. Many adverse and catastrophic conditions are alleged to arise from this temperature rise, namely: melting ice caps, rising sea level, expanding deserts, more storms, more severe storms, accelerating species extinction, growing threats from pestilence, and others. If these claims are true, we should certainly be concerned.

The proposed solutions for the alleged problem are to control CO_2 emissions, reduce the cutting and burning of forests, and plant more trees. The United Nation's Kyoto Protocol is ostensibly designed to reduce CO_2 emissions. Although the United States has not joined the agreement, some states have adopted legislation to reduce CO_2. California, for example, has committed to reduce greenhouse gas emissions to the 1990 level by 2020 (AB 32, *Global Warming Solutions Act*, 9-27-06). Some legislators are now wondering how this can be accomplished. Before our country commits to spending billions (probably trillions) of dollars on CO_2 reduction, we need to consider what light the Bible can shed on this issue.

Exactly why are global warming advocates so concerned about burning fossil fuels and the harvesting of forests? It must be kept in mind that global warming advocates are predominantly evolutionists. Al Gore readily admits that he is an evolutionist (Gore 2006, p. 160). Accordingly, they believe that there was a time in the distant past when earth's atmosphere contained a much higher percentage of CO_2 (over 21%) and no oxygen (O_2). They believe the earth's atmosphere developed O_2 only as a result of photosynthesis by plants or bacteria (Bergman and Renwick 2003, p. 137). Advocates believe that forests, especially tropical rain forests, are the largest reservoir for storing carbon and generating oxygen on land. This helps explain their strong desire to protect rain forests. From an evolutionary perspective it is easy to see why preserving forests and reducing CO_2 is important, even if the projected catastrophes are unfounded or exaggerated.

Development of a Biblical Framework

Most Christians rightly believe the Bible to be their foundation for faith and practice. It determines what they believe and consequently how they behave. The Bible provides frank and absolutely reliable direction for every moral issue experienced by mankind. The biblical position on moral issues like abortion and homosexuality are clear to those who accept the inspiration of Scripture and who understand the straightforward implications of Scripture on these issues, but other issues require thoughtful study of Scripture. With respect to global warming, the Bible provides much more guidance than "creation care" concepts. The following is a proposed biblical framework for evaluating the claims of global warming.

Foundation for a Biblical Interpretation

This paper accepts the verbal plenary inspiration of the Bible (all of the words in the original manuscripts are inspired), and follows a literary interpretation protocol. Passages dealing with the Creation, the Flood and the tower of Babel are treated as narrative in keeping with the historical-grammatical approach to Scripture. The Bible-science movement is keenly interested in determining the original intent of biblical passages. A joint study by the Creation Research Society and the Institute for Creation Research called *Radioisotopes and the Age of the Earth* (RATE) illustrates this point. The study team included a Hebrew scholar, Dr. Steven Boyd, whose task was to determine if the Genesis creation verses are narrative or poetry, a critical question. If the passages are poetry then they merely illustrate a spiritual truth, but if they are narrative then they describe real events and real people. Dr. Boyd determined that Genesis 1:1 to 2:3 is narrative with a 99.996% probability at a 99.5% confidence level (Vardiman et al. 2005, p. 690).

Relevant Biblical Data

The Bible does not speak directly about what we call global warming. It does, however, provide a framework for evaluating the merits of global warming claims. To reiterate, the global warming discussion centers on CO_2 (the atmosphere) and trees (plants). The Bible, of course, addresses the atmosphere and plants. The biblical framework for evaluating global warming is primarily found in Genesis. The RATE study mentioned above established that Genesis 1:1 to 2:3 (the Creation account) was narrative. The study also determined that the Flood account (Genesis chapters 7 to 9) is also narrative (Vardiman et al. 2005, pp. 661 and 667). This paper will also briefly reference the dispersion of the nations at the tower of Babel in the summary. Although the RATE study did not evaluate the Tower of Babel, I believe that if the creation and the Flood passages are narrative, then the tower of Babel passage is narrative also. These passages describe real events and real people. The following sections briefly discuss passages related to the atmosphere and plants.

Creation of the Atmosphere

In Genesis 1:1 we are told that "God created the heavens and the earth" Creation obviously includes the atmosphere. In fact, if the atmosphere was not created on Day One, it certainly was in place by Day Two when God "separated the waters which were below the expanse from the waters which were above the expanse" (Genesis 1:6–8). This "expanse" was the atmosphere in which the birds flew on Day Five. Regardless of the exact day, the central biblical point is that the atmosphere was created, it did not evolve. The atmosphere was intentionally designed and created by God to support life, including plants, animals, and mankind, which He subsequently created. Contrary to evolution theory, the atmosphere is not a constantly changing mixture of gases, which billions of years ago were poisonous to life but now has evolved to the point where it can support a precarious array of life. The original created atmosphere contained the right amount of CO_2 for the plants that would be created on day three and sufficient O_2 for the soon to be created animals and mankind. This is a far different atmospheric history than the evolution story. A created atmosphere has purpose, stability, and is more robust than a randomly evolved atmosphere.

Creation of Plants

Aside from all the other reasons for which God may have created plants, the Bible specifically states that He made them for human and animal food, and this is largely being ignored by global warming advocates (Genesis 1:29–30). Since all animals and mankind were vegetarians originally, plants were created as R. J. Martin a reliable and sustainable source of food. As people began eating meat, they became even more dependent on vegetation as a source of food because the animals we eat all must consume multiple pounds of vegetation for each pound of meat produced. As an example, the grain conversion ratio for poultry is about four while beef is 15 (Bergman and Renwick 2003, p. 320). This means that on average a cow would need to consume 15 pounds of feed (vegetable matter) to generate one pound of meat. Consequently, as the human population grows, and as proportionately more people become meat eaters, substantially more land must be allocated for agriculture. By the way, the areas most useful to man in producing edible plants and animals are not the forests, but the plains. The useful carrying capacity of grasslands far exceeds the useful carrying capacity of forests. Consider, for example, the millions of bison, antelope, elk and bear that once inhabited the western Great Plains. Today these plains are producing record amounts of grains such as corn and wheat, along with other edible crops. Most forests, including tropical rain forests, are climax communities. This means that new growth is nearly offset by decaying vegetation, yielding little if any net gain (Oberlander and Muller 1987, p. 240).

While it is true that harvesting of forest products should be done in line with intelligent use of that ecosystem, unless forests are periodically harvested, allowing new growth and providing a useful product, they have little direct economic benefit for mankind. As the human population increases then it is reasonable to convert forests to the production of food and building material. From a creation perspective there is nothing sacred about preserving forests. They are to be efficiently and effectively managed for the benefit of mankind. Nonetheless, there is little justification for the wanton destruction of forests for short-term economic benefit. As stewards accountable to God we should manage all earth resources with a long-term, biblical, perspective. It should also be noted that as plants began growing and covering the earth following Creation week, they were removing CO_2 from the environment. Land plants removed CO_2 from

227

the atmosphere while marine plants removed CO_2 from the ocean. In addition, marine animals that developed carbonate shells also removed CO_2 from the ocean.

Dominion Mandate

God purposely created mankind to rule over the earth, including both the plants and the animals. According to Genesis 1:26-29, man was told to: fill the earth, subdue the earth, and rule over all of the earthly creation. This mandate was repeated to Noah and his family after the flood (Genesis 9:1–3). Genesis 2:15–16 further indicates that man was initially also commanded to cultivate and keep the Garden of Eden. These commands indicate that man is God's appointed representative on the earth. Having been created in the image of God, man is uniquely separated from, and elevated above, the rest of creation. The earth was created for the benefit of man, but he is ultimately accountable to God in his exercise of this commission. Genesis 2:11–12 identifies the location of gold, resin, and onyx. According to Genesis chapter 4, later generations raised livestock, developed musical instruments and worked with bronze and iron. God never rebuked mankind for mining, farming, ranching, or cutting trees for building projects. All of these activities are part of man's God-given rule over the earth. Throughout Scripture, however, God has repeatedly rebuked man for disobedience to His moral commands. Eating the forbidden fruit resulted in God's curse on both man and creation. Man's wickedness in the days of Noah resulted in God destroying all air breathing creatures and men, except for the few saved on the ark. The Flood also entirely reworked the surface of the earth. Following the Flood, God confused man's languages because, among other things, mankind lingered in Mesopotamia rather than filling the earth as commanded. This resulted in various language groups slowly migrating around the earth. When Israel disobeyed God's moral commands he sent them into exile and allowed their land to grow over with thorns. Using earth resources for the benefit of mankind has never been a moral issue. Ignoring God and disobeying His commands is a moral issue.

Noah's Flood (Destruction of the earth)

The year-long Genesis Flood (Genesis 7:17–8:9) buried great volumes of plants and animals. During the Flood there were 40 days and nights of heavy rain, and the fountains of the deep were open for 150 days. These flows added significant volumes of water to the existing ocean. It is reasonable to assume that more water was added to the ocean from the fountains of the deep (150 days) than from rain (40 days). Water from the earth is warm. The average geothermal gradient is 1° F (0.6°C) for each 60 ft (18.3 m) of depth (Landes 1959, p. 169). The deeper this water originated, the warmer it would be. The Flood likely increased the temperature of the ocean. As we will see later, a warm sea following the Flood helps explain another important post-Flood phenomena, the ice age.

According to the Genesis account, the Flood waters increased for 150 days until all the high mountains everywhere on earth were covered to a depth of 15 cubits (about 22.5 ft [6.9 m]). The waters then receded for another 220 days as the present continents and mountains rose out of the ocean (Psalm 104:6–9). During this time valleys and plains were eroded and the major drainage systems were established. Noah, his family, and the animals stayed on the ark during the five months the waters were increasing, as well as the seven months while the waters were receding. God did not allow anyone to leave the ark until the earth had dried and a sufficient number of plants were growing to provide food for all life on the ark. It is important to recall that during the Flood all land plants were destroyed, yet there was sufficient oxygen in the atmosphere for all life on the ark to breath. After the Flood plants again began growing and covering the earth, just as they did at Creation. At Creation and immediately after the Flood, plants were just beginning to cover the earth yet there was no shortage of oxygen in the atmosphere. God established enough oxygen in the original atmosphere to sustain life throughout the duration of the earth. This highlights the fact that plants are not necessary for generating oxygen. More will be said on this topic in the section on CO_2.

Plants, however, are essential as food for man and animals. In addition, plants stabilize the soil, provide habitat for various animals, and are a source not only of medically useful drugs but also inspiration and beauty. As plants again covered the earth, both on land and in the sea, they once more removed CO_2 from the environment. In a like manner, shelled animals in the sea removed CO_2. As an aside, during the Flood every man and animal on the ark would have been classified as an "endangered species" according to current definition. All animal life today is descended from one or a few pairs of animals that were carried on the ark. Plants were not endangered. Plants buried in sedimentary rocks during the Flood now exist as fossil fuels (Groombridge and Jenkins 2002, p. 10). Coal, oil, gas, tarsand, and oilshale are all partially decomposed plant material. When fossil fuels are used today in furnaces and engines we are burning plants that lived and grew prior to the Flood. The CO_2 released during burning was taken from the pre-Flood atmosphere and ocean. Even secular scientists acknowledge that fossil fuels are remains of past plants and burning them releases energy stored long ago (Northen 1968, p. 71).

The argument over burning fossil fuels versus ethanol can be reduced to a question of whether it is best to burn old plants or new plants. Burning old plants (fossil fuels) is much more efficient, and therefore "green" The massive fossil carbonate formations seen across the earth contain remains of pre-Flood shelled animals. Approximately 15–20% of the sedimentary rocks worldwide are carbonate (Ehlers and Blatt 1982, p. 251). Considering the total volume of fossil fuel captured in rocks, and the volume of carbonate rocks, it can be seen that a significant amount of CO_2 has been removed from the pre-Flood environment (atmosphere and ocean) and locked up in sedimentary formations. Another significant volume of CO_2 has been removed since the Flood and is tied up in plants and animals that have subsequently developed. As a result of burying a major proportion of earth's plant and animal life, the Flood likely caused far greater changes to atmospheric gases than any current global warming scenario.

Following the Flood, God assured Noah that there would be no other worldwide water catastrophe as long as the earth remains (Genesis 8:22). According to this promise, "seedtime and harvest, and cold and heat, and summer and winter, and day and night shall not cease" Along the same line, Peter mentions that in the last days people will say that "all continues just as it was from the beginning" (2 Peter 3:3–7). Christ also mentioned that in the days prior to His second coming all would continue routinely, "just like in the days of Noah" (Matthew 24:37–39).

From these verses it appears that until the tribulation occurs no worldwide catastrophe will affect the earth. Global warming is described as a worldwide catastrophe by the radical environmentalists and the media. The tribulation of Revelation certainly contains events that sound like some of the dire predictions associated with global warming. Unlike global warming, the tribulation is initiated directly by God, as judgment on sinful mankind, and is a sudden, not a gradual change. People undergoing the tribulation realize that it is from God, as a result of their sinful behavior, but they intentionally refuse to repent. We should not confuse the claims of global warming with tribulation events.

God's Control of Creation

God is in absolute control of His creation. He is the Creator (Genesis 1 and 2). God destroyed His creation in the days of Noah with a worldwide flood (Genesis 7–9). God sets the boundary for the seas (Job 38: 8–11, Psalm 104:9, Jeremiah 5:22) and controls the weather: lightning (Job 28:26, 37:3), hail (Job 38:22, Psalm 147:17, Haggai 2:17), rain (Job 28:26, 37:6, Psalm 147:8), and snow (Job 37:6, 38:22, Psalm 147:16). Someday God will destroy this earth and establish a new heaven and a new earth (Revelation 21:1). Man is not in control of the weather and this present earth is temporary.

Summary of Biblical Framework

Keeping in mind that the Genesis accounts of creation and the Flood are narrative (they describe real historical events), the above discussed biblical framework can be summarized as follows:

Creation week

1. CO_2 and O_2 were created early in the Creation week. Neither of these gases evolved. 2. Plants were created primarily for food. 3. Man was given dominion over the earth. 4. The earth was created for man's use, enjoyment, and occupation as he honors God. 5. Man is neither an animal nor a random accident of evolution.

Between Creation and the Flood

1. Following creation, plants, both marine and land, reproduced around the world. 2. Animals with carbonate shells also multiplied worldwide. 3. The above growth removed CO2 from the environment and replaced it with O2.

The Flood

1. The Flood added large quantities of water (likely warm) to the ocean. 2. The Flood buried substantial volumes of plants. 3. Plants began growing during the last half of the Flood, as the waters receded.

Post Flood

1. Plants and animals, both marine and land, again begin multiplying worldwide. 2. Growing plants and shelled animals removed CO_2 from the environment and added O_2. 3. Under the influence of temperature, pressure, and an O_2 free environment, the buried plants become fossil fuels. Note: God is in absolute control of the earth and all it contains, not man. As you can see, the Bible has quite a bit to say regarding atmospheric gases and plants. This biblical framework relates directly to our understanding of global warming and climate change. The atmospheric gases were created, they did not evolve. We should not expect the types of atmospheric gases to have been substantially different at creation than now. It is unlikely that the creation atmosphere contained any gases not present in the current atmosphere. Oxygen has obviously been present since creation and likely has increased as CO_2 decreased. The contribution to atmospheric gases by volcanoes from creation to the present is unknown. What is known, of course, is that CO_2 stored in plants and shelled animals that existed prior to the Flood is now stored in sedimentary rocks worldwide. The pre-Flood plants currently exist as fossil fuels and the shelled animals are contained in

carbonate deposits. We also know that currently living plants and shelled animals have taken additional CO_2 from the environment.

Consequently, in view of the massive volume of fossil fuels and carbonate rocks, it is highly probable that today's atmosphere contains measurably less CO_2 than the Creation atmosphere and a correspondingly higher O_2 concentration. Polar seas are quite cold today. In the years since the Flood, the warm worldwide ocean has gradually cooled at the poles. God created the earth for man's use. Man received a commission from God to manage the earth, including the plants and animals. This includes old plants and new plants. Man has the right to use earth resources for the benefit of mankind, but in a reasonable manner that honors God. In Table 1 we compare creation and evolution on several issues relevant to our discussion.

Issue	Creation	Evolution
O_2 in original atmosphere	<21%	0%
Source of current O_2	Created	Product of photosynthesis
CO_2 in original atmosphere	>0.03%	>21%
Source of plants	Created	Evolved
Value of plants	Food	Generate O_2
Purpose of earth	Man's home	Purposeless
Source of man	Created	Evolved animal
Purpose of man	Steward	Purposeless
Man's relation to animals/plants	Ruler	Co-equal
Source of fossil fuels	Plants	Plants

Table 1. Issues relevant to earth's climate within the creation evolution models

Creation and evolution agree on only one point. Fossil fuels were once plants. The above framework provides significantly more depth than "creation care" concepts. The value of this framework will become clearer as we consider some basic science related to global warming.

Science Related to Global Warming

Let us now consider some basic science related to global warming issues. Four topics will be discussed: glaciers, CO_2, climate, and temperature. These topics have been chosen as they are crucial to the global warming argument. We will review these sections with a Bible-science perspective. Following this we will fit the scientific data into the biblical framework previously discussed in an effort to develop a comprehensive perspective on global warming.

Glaciers

As evidence that the earth is experiencing global warming, advocates point to melting glaciers around the world. Since this is the first and strongest argument offered by Al Gore, the United Nations, and other global warming advocates, it is appropriate to spend some time discussing glaciers. The glaciers remaining around the world are remnants of the once extensive ice age. Ice ages are poorly understood (Oberlander and Muller 1987, p. 479). Although numerous evolution-based theories have been advanced to explain how an ice age is initiated, none of them are satisfactory. The most popular theory at the moment is the astronomical theory. According to this theory, small changes in the earth's orbit, tilt and wobble combine approximately every 100 thousand years to create a colder winter, especially at the poles (Dott and Batten 1988, p. 596). Proponents of this theory believe that if winters are colder then glaciers will grow and advance. Such conditions, if they actually occurred, would not start an ice age, but merely a cold-age. An ice age is characterized by thick, extensive, ice-sheet glaciers and advancing mountain glaciers. The indispensable ingredient for a glacier is lots of snow. Massive precipitation of snow requires massive evaporation of sea water. Massive evaporation only occurs from warm water. Water evaporation increases exponentially with temperature (Oard 1990, p. 5) (see Fig. 1).

So, an ice age requires warm seas in close proximity to the poles. The only viable explanation for an ice age has been clearly and thoroughly explained by Oard (1990). Warm seas worldwide following the Flood would provide optimum conditions for initiating the ice age. It would have been like lake-effect storms greatly enhanced. Oard, a meteorologist, estimated that 500 to 700 years would have been required for the ice age to reach its maximum (1990, p. 97). During this time, more snow would have been precipitated in the winter than would have melted in the summer. Consequently, the snow cover would have increased in thickness and lateral extent. As the polar seas cooled, less evaporation would have translated into less snow and eventually snowfall would equal melting, stabilizing the extent of glaciation. Further cooling of the seas would have resulted in more melting than snow accumulation and the glaciers would have retreated. Naturally, the glacial advance and retreat would have been somewhat erratic as yearly storm events varied in intensity.

Today, the polar areas are deserts due to the cold seas. The high ice plateau of Antarctica receives only about one inch (2.5 cm) of precipitation each year. Even Gore acknowledged this fact (2006, p. 176). Today's precipitation rate does not allow sufficient time to accumulate the nearly two mile (3.2 km) glacier thickness from a biblical time frame. There is overwhelming evidence that glacial ice sheets once covered most of Canada, extending as far south as northern Washington, Illinois, Ohio, New York, and New England. Glaciers also covered much of Siberia and northern Europe. The massive ice sheets covering these areas melted prior to historical times. In fact, the majority of the glacial ice melted in the distant past. As corroboration that huge volumes of glacial ice melted in the past, there is strong geologic evidence that ocean levels have risen several hundred feet (61+ m) (Groombridge and Jenkins 2002, p. 35). Obviously, all this melting occurred long before mankind began burning fossil fuels on a large scale. In other words, glacial melting has been going on for thousands of years and mankind was not the cause. Most of the melting, and subsequent sea level rise, occurred long before the recent increase in atmospheric CO_2.

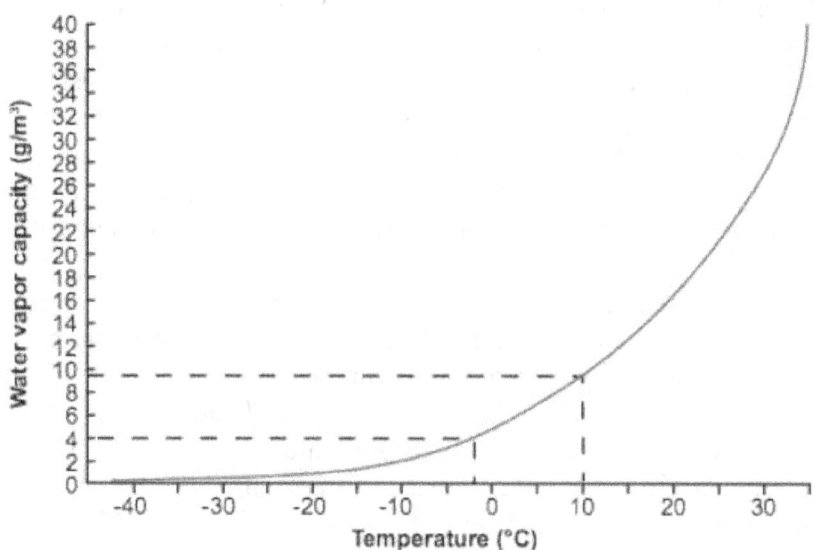

Figure 1. Graph of water vapor capacity at saturation (11% relative humidity) versus temperature. Note the 60% drop in capacity as temperature cools from 10°C to 2°C

234

Incidentally, sunken Mediterranean cities also provide historical evidence for rising sea levels. For example, ancient Alexandria disappeared from history about 1,600 years ago. It was subsequently discovered in 1999 by Franck Goddio directly offshore from present Alexandria in about 15 ft (4.6 m) of water. The fortified island of ancient Tyre was destroyed by Alexander the Great in 322 BC. The ruins of ancient Tyre now lie offshore in about 20 ft (6.1 m) of water. Other Mediterranean cities could also be cited. From this evidence, it appears that over the last 2,000 years the Mediterranean Sea has risen about 1 ft (0.3 m) per 100 years. This average sea level rise is greater than estimates of the rise over the last 100 years (4–10 inches [10–25 cm]). Apparently, sea level rise is diminishing with time. The level of the Mediterranean Sea rose because melting glaciers added water to the oceans.

In summary, Bible-science provides the only viable explanation for an ice age: warm polar seas following Noah's Flood. It also provides a reasonable explanation for the end of the ice age and the subsequently experienced large-scale glacial melt: cooling seas. Contrary to what global warming advocates are saying, the glaciers melted because the seas cooled, not because they warmed. Since seas account for nearly 71% of earth's surface area, and contain 1,000 times more heat than the atmosphere, they are obviously a major variable in determining the earth's temperature and its various climates (Solomon et al. 2007, p. 389). At first glance this may sound incredible, but it is in agreement with the biblical record and science. Biblically, there has only been one ice age, and it was a direct and inevitable result of the Flood. Melting glaciers are nothing new. The impressive glacial melt experienced since the peak of the ice age was not due to increased CO_2, warming oceans, or anything man had done. Why should we now think that man is responsible for melting glaciers? Clearly, melting glaciers are not proof of global warming.

Carbon dioxide

Carbon dioxide is being described as a pollutant by global warming advocates. In 2007 the Supreme Court ruled that the U.S. Environmental Protection Agency (EPA) has the authority to regulate vehicular green house gases (Massachusetts versus EPA, Case #05-1120, decided 4-2-07 by a 5/4 margin). This was the conclusion of a suit filed by several states, including California, that were concerned that the federal government was not doing enough to avert a global warming disaster. After reviewing this issue, the EPA proposed regulating CO_2 as a pollutant under the Clean Air Act. What exactly is an air pollutant? In the past, an air pollutant was defined as contamination of the air by noxious gases and minute particles of solid and liquid matter (particulates) in concentrations that endanger health. Does CO_2 fit the description of an air pollutant? The following discussion will demonstrate that CO_2 is not a pollutant.

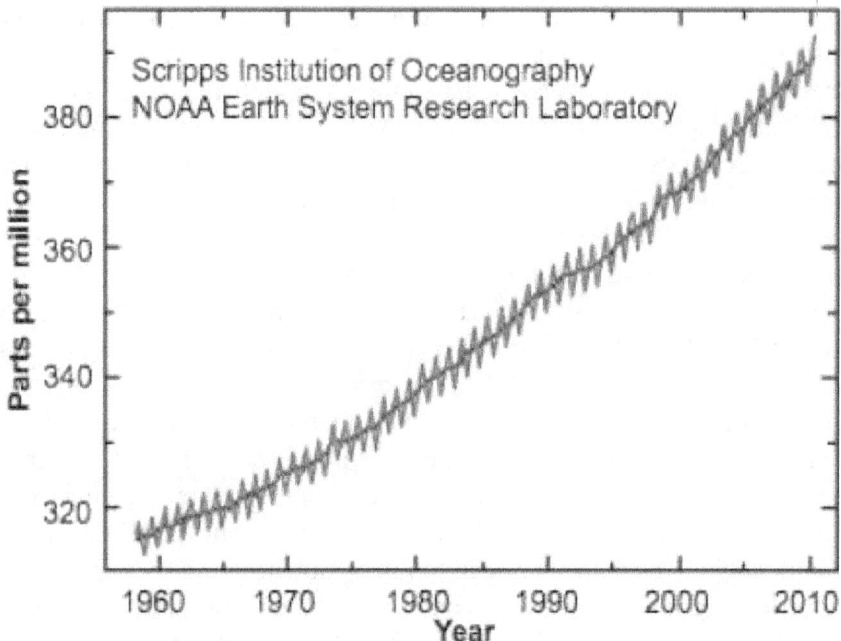

Figure 2. History of measured atmospheric CO₂

At what concentration can CO_2 be considered a health hazard (the point where it would be an air pollutant)? This is a question of critical interest to underground miners. Underground mines closely monitor the buildup of several gases which could prove hazardous to miners. Accordingly, mine safety thresholds have been established for numerous gases, including CO_2. The U. S. federal threshold level for CO_2 in underground mines is currently 5,000 ppm (30 CFR 75.321 [a]) (US Dept. of Labor). During the 1940s and earlier, the threshold level was over 12,000 ppm (Peele 1941, sec. 23, p. 20). This is not a hazardous level. It is the concentration at which miners can be safely removed from the mine and the passageways ventilated. The level of CO_2 in our atmosphere could increase over 1,300% before reaching the current mine safety limit, and this level has been reduced to only 42% of the prior safe limit. Today's atmospheric concentration of CO_2 is clearly safe for humans, and will be for over a thousand years at today's rate of increase. It is doubtful, however, if fossil fuels will last for another thousand years. Are there any benefits to CO_2? Carbon dioxide is naturally occurring and, rather than endangering life, it is necessary for life. Plants cannot live without CO_2 and man cannot live without plants. In addition to this indispensable benefit, there are other major benefits. Without an atmosphere containing GHGs the earth could not support life. Carbon dioxide is one of the

236

atmospheric gases that help moderate earth's temperature. Furthermore, for over 100 years the agricultural industry has known that CO_2 is a plant fertilizer (Northen 1968, p. 74). Some growers intentionally increase CO_2 up to ten times its normal concentration to encourage plant growth in greenhouses. This is termed "carbon dioxide enrichment" Elevated levels of CO_2 encourage faster growth, larger and more productive fruit bearing, and increased tolerance to both heat and cold. As a result of increasing levels of CO_2, plants can extend both their growing season and the extent of their habitat. Plants need CO_2 to exist. If CO_2 levels drop to about 220 ppm plants grow very slowly, and if the concentration falls to 150 ppm growth stops entirely. There is far greater danger in lowering the CO_2 level, than in increasing the level. Agricultural schools, and farmers, around the world have noted increased crop yields and enhanced forest growth as CO_2 has increased in the atmosphere. One hundred years ago the atmosphere contained approximately 280 ppm CO_2. Today the concentration has increased to around 380 ppm (Solomon et al. 2007, p. 137). See Fig. 2 for the concentration of atmospheric CO_2 as measured at the Mauna Loa Observatory and Fig. 3 for a comparison of measured atmospheric CO_2 with the current safe limit for CO_2 in underground mines. Fig. 3 shows that CO_2 is far from being a pollutant that endangers the health of humans. This increased concentration is helping farmers worldwide to feed a hungry world.

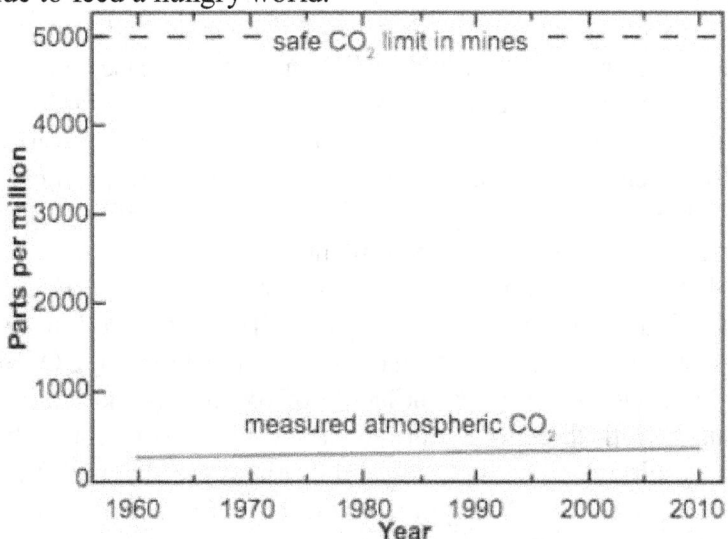

Figure 3. Atmospheric CO_2 compared with the safe limit in underground mines

All plants respond favorably to more CO_2. Is CO_2 only found in the atmosphere? The atmosphere is also in contact with the oceans. Gases are continually being exchanged between these two environments. At present the ocean contains at least 60 times more CO2 than the air (Barry and Chorley 1987, p. 5). The ocean is a large buffer for atmospheric gases. As the ocean cools, more CO_2 goes into solution, and, as the ocean warms, it gives off CO_2. Since the post-flood ocean was much warmer than now, a large volume of atmospheric CO_2 has been absorbed by the ocean as it has cooled. Is CO_2 the most important GHG? According to climatologists, water vapor and clouds account for about 60 to 95% of the GHG effect, while CO_2 has a much smaller effect. More importantly, many global heat budget parameters and the relationships between them are not adequately measured or understood (Kiehl and Trenbert 1997, pp. 197–208; NOAA, website FAQs). Global climate models are too imprecise, and the key input data too limited, to justify initiating major changes in world economics.

Climatologists who are pushing the global warming agenda are focusing on a minor GHG component and ignoring the major contributors to the GHG effect, water vapor, and clouds. What is the estimated contribution of CO_2 to global warming? Global warming is blamed on CO_2 increasing in the atmosphere. Is this reasonable? Over the past 100 years atmospheric CO2 has increased 36% (from 280 ppm to 380 ppm). Over this same time interval global temperature is alleged to have increased 1° F (0.06°C). This is an increase of 0.2% (510° F [283.2°C] to 511° F [283.7°C] on an absolute scale). Even if all the increase in CO_2 is attributed to burning fossil fuels (which it is not) and the increase in temperature is due entirely to CO_2 (which is likewise not the case) the correlation between CO_2 and temperature is quite weak. At best, advocates are claiming that a 36% increase in CO_2 is responsible for a 0.2% increase in temperature. What is the relationship between CO_2 and plants? The photosynthesis/respiration equation is as follows: $6CO_2 + 6H_2O + energy (sunlight) \rightarrow C_6H_{12}O_6 + 6O_2$. This equation shows a simplified relationship between plants and the atmosphere. During daylight hours plant cells containing chlorophyll remove CO_2 and H_2O from the atmosphere, generate a simple sugar, and give off O_2 (Northen 1968, p. 68).

As you can see, if the plant removes one molecule of CO_2 from the atmosphere it will replace it with one molecule of oxygen. Although in daylight this equation runs in both directions, during the night this equation only runs in reverse. During respiration O_2 is combined with sugar to fuel the plant's metabolism and CO_2 and H_2O are emitted (Northen 1968, p. 83). When a plant dies the equation runs in reverse (respiration) during the entire decay process until all the O_2 previously emitted is recaptured and all the CO_2 is returned to the atmosphere (Northen 1968, p. 435). Over their life-cycle plants generate neither excess O_2 nor excess CO_2. This is a zero-sum game, but with a lag-time measured in years. The implications of a plant's life-cycle are noteworthy. If over their life-cycle plants generate neither excess CO_2 nor excess O_2 then two conclusions follow: 1) plants did not generate the large volume of O_2 in our atmosphere, and 2) planting trees will not provide permanent carbon offsets. This, of course, agrees with Scripture. The atmosphere was created (it did not evolve) and plants were created as food (not a source of O_2). What is the mix of gases in the atmosphere? Approximate concentrations are shown in Table 2.

Gas	Symbol	%
Nitrogen	N_2	78.07%
Oxygen	O^2	20.94%
Argon	Ar	0.93%
Carbon dioxide	CO_2	0.04%
Miscellaneous		0.02%

Table 2. Composition of the atmosphere

Currently there is about 550 times more O_2 than CO_2 in the atmosphere. One hundred years ago, when CO_2 was 280 ppm, there was 750 times more O_2 than CO_2. Obviously, converting all the CO_2 into O_2 would have a minimal impact on total O_2 concentration, but would be devastating for plant life. As you will recall, according to evolution earth's atmosphere once contained no O_2. We are told that O_2 only exists in our atmosphere as a result of photosynthesis. As discussed above, photosynthesis converts one molecule of CO_2 into one molecule of O_2. If evolution is correct then earth's atmosphere once contained over 21% CO_2. Biblically, CO_2 is good. It is needed for life to exist. God created CO_2. It is a plant fertilizer, not a pollutant. The hazardous level for humans is far above concentrations attainable by burning all our fossil fuel reserves. In addition, the correlation between CO_2 and an alleged global temperature increase is weak at best and most likely spurious. Carbon dioxide is a minor GHG. It should also be remembered that the CO_2 released by burning fossil fuels was taken from the atmosphere that existed in the pre-Flood world. If the CO_2 wasn't a problem in the lush pre-Flood earth, it shouldn't be a problem now. Increasing levels of CO_2 are not proof of global warming.

Climate

Some global warming advocate's claim that climates were relatively fixed over the last 10,000 years until man started burning fossil fuels and affecting the world's climate. Is this really true? Climatologists realize that climates vary over time (Groombridge and Jenkins 2002, p. 33). World climate maps are based on averages collected over a few decades in the mid-twentieth century (Bergman and Renwick 2003, p. 85). Clearly, climates have steadily been changing with time, requiring plants and animals to adjust accordingly.

Think of all the climatic change initiated by the Flood. The initial uniformly-warm world ocean generated greatly enhanced evaporation and precipitation worldwide. The results of this precipitation were not only an ice age, but also lush rainforests. As the ice age glaciers grew they encroached on vegetated land, forcing plants and animals to migrate. Interestingly, it appears that during the ice age there was a highly productive grassland community along the fringe of the warm Arctic Ocean (Oard 2004, pp. 29–31). Ice sheets eventually covered a large percentage of the Northern Hemisphere; most of Canada and the northern states, much of Siberia and northern Europe, along with all the high mountain ranges worldwide. As these great ice sheets retreated, plant communities followed their migration. The plants were subsequently followed by animals. The western US, between the Rocky Mountains and the Sierra Nevada, once contained numerous large lakes. Salt Lake is the remnant of one of these lakes. Archaeological finds indicate sizeable and diverse populations of people living in this region in the past.

As the lakes evaporated, plant communities changed, forcing men and animals to migrate elsewhere. In a like manner, North Africa was once much wetter, supporting more cities and extensive agriculture. As the desert expanded the cities and agricultural lands were abandoned. Similar scenarios occurred on every continent following the ice age. Climate has been dynamic since the ice age requiring plants, animals, and man to adapt. From a biblical time frame (Ussher 2003), the flood occurred about 2349 BC and the glaciers began retreating approximately 1850 BC (earliest estimate of ice age peak according to Oard, and also the time of the patriarchs).

Many geologists believe that past ages were much warmer than historical times. Geological textbooks estimate some ages were as much as 25° F (13.9°C) warmer (Dott and Batten 1988, p. 593). This is evident when viewing museum dioramas, park displays and National Geographic shows. Past ages are shown as tropical or subtropical. This is because most fossil plants are tropical or subtropical. It should also be noted that the divisions between the geologic periods were initially based on mass extinctions. Many evolutionist geologists still support this theory. Creationists realize that most of the sedimentary rocks, and their included fossils, were deposited during the Flood, not over millions of years. Consequently, there really was only one mass extinction, cause by the Flood. The tropical and subtropical plants assigned to the evolutionary geological ages were all living at the time of the Flood. We are told that global warming will increase both the frequency and severity of storms. Storms, however, are driven by the temperature difference between a warm equator and cold poles. This temperature difference sends cold fronts down from the north and warm fronts up from the south. Since northern and polar regions are the areas expected to warm the most from global warming, the temperature difference will decrease. Thus warming, if it actually occurred, would result in fewer and less severe storms.

Biblical history provides the only viable explanation for the ice-age (warm polar seas following the Flood), the melting of glaciers, and the development of deserts (cooling seas since the Flood). From a biblical perspective the past 4,350 years since the Flood have witnessed a vast change in climates around the world, none of which can be attributed to man-made causes. The mere presence of climate change is not evidence of man-made global warming. Climate change is normal, and was initiated by the Flood.

Temperature

We are warned by Al Gore, and on the news, that global surface temperatures have warmed 1° F (0.6°C) over the past 100 years, and that it is now warmer than it has ever been in the history of the earth. As a result of this "huge" temperature increase we must take immediate and extreme action to avert sudden and imminent global disaster. As mentioned in the prior section, most geologists would dispute this claim. It is also stated that temperature records are being broken all over the world, thus verifying that we are on the brink of this global disaster. These are bold statements, but are they accurate? In addressing these claims we will consider the temperature history record in three parts: temperature data collection, data handling, and data interpretation. It will be shown that the margin of error in each of these areas significantly exceeds the global temperature increase reported for the past 100 years.

Collection methodology

The earth is huge. We simply do not have a sufficient number of collection points (weather stations) to accurately determine earth's average surface temperature. The problem is complicated by the seas. When approximately 71% of the earth is covered by ocean, but most of the weather stations are on land how can we truly know the temperature of the entire earth? The National Weather Service (N. W. S.) establishes standards for official weather stations (Leffler and Redmond 2004, p. 11). According to these standards, if a weather station is moved five miles (8 km), or 100 ft (30.5 m) in elevation, then it must be designated as a new station. In other words, the N. W. S. believes that an accurate determination of temperature over a large area requires a temperature measuring station at least every five miles (8 km).

There are approximately 1,221 climate-monitoring stations overseen by the N. W. S. in the continental United States (Watts 2009, p. 1). Following the N. W. S. five-mile guideline there should be at least 124,800 stations in the continental United States. The actual number of stations is less than 1% of the recommended number if an accurate average temperature for the U. S. is desired. In other words, 99% of the U. S. is unrepresented by temperature monitoring stations. The average spacing of weather stations over the entire globe is much sparser than even in the United States. How can these stations be representative of the entire globe? A recent survey of 70% of the N. W. S. stations revealed that 89% did not even meet the NWS siting requirements (Watts 2009, p. 1). Over half of these stations were expected to experience an error of over 2° F (1.1°C) just due to siting deficiencies (Watts 2009, p. 16). In view of a 2° F (1.1°C) temperature error due to siting, what is the significance of a 1° F (0.6 C) temperature change in 100 years? No significance.

In a typical U. S. city, temperature measurements can easily differ by more than 3° F. (1.7°C) between various parts of town. Consequently, the official temperature reported may have a margin of error of several degrees. If the temperature reported for a single town is not truly representative of that town, then how is the global average of such temperatures representative of the entire world? Large variations in temperature also exist in the countryside, depending on land cover, elevation, slope, and aspect. It is well known that weather stations near large cities are impacted by what is called the heat island affect (Oberlander and Muller, 1987 p. 71). Weather stations that once were in the country have been encroached by asphalt and concrete, thus raising the average temperature in the vicinity of the station. Cities become anomalously warm and are not representative of the larger surrounding area. Temperatures in cities can be 6–14° F (3.3–7.8°C) warmer than the surrounding countryside (Barry and Chorley 1987, pp. 358–360).

Temperature proxies (tree rings, glacial-ice cores, and ocean-sediment cores) are sometimes used in an attempt to reconstruct earth's temperature history far into the past. Proxies are extremely imprecise and obviously not representative of the entire earth. How can tree-ring thickness be accurately correlated to a specific temperature? Is the ring wider due to higher temperature, greater moisture, both of the above, or some other factors? Ice cores and ocean cores are even more difficult to interpret, especially since temperature is estimated from O_2 measurements and age is interpreted from an evolutionary time scale. As an added point, in order to accurately measure the temperature history of the entire earth it would be necessary to measure temperatures simultaneously. A 24-hour day in New York is not the same time interval as a 24-hour day in Los Angeles or Honolulu. Unless simultaneous time intervals are captured, and averaged, the calculation introduces an additional error. As you can see, temperature data collection is not very accurate, even in the United States. Weather stations in most of the world don't attain to the U. S. standards. A majority of the stations experience an error of at least 2° F (1.1°C). Can we believe a long term warming trend of only 1° F (0.6°C) poses a significant risk?

Data Handling

The global surface temperature history is basically a weighted average of numerous temperatures from weather stations around the world. Unfortunately, the number of weather stations is constantly changing. Between 1950 and 2000 the number of weather stations in the Global Historical Climatology Network has varied from over 15,000 to slightly over 5,000 at present, most of which are on land (McKitrick, 2003 p. 6). How can a consistent and accurate global temperature be calculated when the number and location of stations is changing drastically? This procedure places the significance of an alleged 1° F (0.6°C) temperature increase over 100 years into question. What is the impact of eliminating approximately two-thirds of the stations within a 50 year interval?

Of course, when looking back 100 years it is impossible to maintain a constant number of weather stations and also pretend to be measuring global temperature. One hundred years ago there were far fewer stations (about 10% of current) and most of them were in the United States and Europe. The idea of a reliable 100-year history for the earth's temperature is an oxymoron. It fails by definition. Temperature wasn't even measured over much of the earth 100 years ago. The average temperature used by climatologists in the 100-year history is merely the average of the high and low readings (maximum and minimum temperature) at each weather station for each day of the year (Bergman and Renwick 2003, p. 70). This procedure introduces a significant and unpredictable error for each station and is clearly not accurate if you want to capture the true temperature of the surface of the earth. This method would only be representative if temperature varied uniformly and symmetrically between the high and low temperature reading each day. This is an atypical event. The true average temperature can vary by several degrees from a simple average of the high and low temperatures, especially if partial cloudiness is experienced.

If individual stations can experience a daily temperature error of several degrees why should we be alarmed by a 1° F (0.6°C) change in 100 years? The alleged temperature increase is well within the margin of error for each station's daily reading. Even if highly accurate daily average temperatures were available from all of the stations, and the number of stations was constant, the global averaging technique would introduce an error which must be considered. How do you calculate an accurate, and representative, global average temperature from the approximately 5,000 stations? Do you calculate a simple mean, are the stations weighted by area, or is an isotherm map developed? If you weight by area how do you determine the area represented by each station? Do you consider topographic boundaries like mountains? What happens to the large areas unrepresented by stations? If you develop an isotherm map you must select from an assortment of methods for using the area of each isotherm to determine an average temperature. The average temperature calculated using these mathematically acceptable techniques can easily vary by more than 1° F (0.6°C) between themselves. The global averaging technique selected introduces a margin of error which must be considered and reported. Data handling techniques also introduce an error greater than the alleged temperature increase due to global warming. When considered in perspective, there is no cause for alarm over a stated 1° F (0.6°C) temperature increase.

Interpretation

A wide range of surface temperatures exist on the earth simultaneously. At the same time it may be -100° F (-73.3°C) in Antarctica and +130° F (54.4°C) in Death Valley. There will be places on the earth experiencing every temperature between these two extremes. A number of areas on earth will have temperatures between -9° F (-22.2°C) and +9° F (-12.8°C), single digits. According to the rule of significant digits, the end product of a calculation cannot have more significant digits than the component with the least number of significant digits. If we are measuring the earth's average surface temperature then the end product can have no more than one significant digit, in this case 1° F (0.6°C). We can only know the temperature of the entire earth within 1°F (0.6°C). Therefore, a 1° F (0.6°C) change in 100 years is within the margin of error of the calculation. It is common in scientific disciplines to indicate a margin of error when reporting summary calculations. As we have seen, there are significant errors in both temperature collection and data handling procedures. These errors can range from a few degrees to as much as 14° F (7.8°C). In view of this, the margin of error for the global temperature history is well over 2° F (1.1°C). If a true margin of error is reported, then a 1° F (0.6°C) temperature rise in 100 years fades into insignificance and the alarm over increasing temperatures evaporates. In the western U. S., and many other parts of the world, reliable temperature records simply do not exist farther back than 100 years.

If evolution is correct and the earth is 4.6 billion years old then we have no reliable temperature records for most of earth history. The last 100 years represents only 0.0000002% of earth's history. Even with a 6,000-year old earth (from the biblical account), the temperature record covers merely 1.7% of earth history. Is this a sufficient history to contend that earth's temperature is zooming out of control? Do we have a reliable temperature base from which to confidently predict the earth's future temperature and commit to spending trillions of taxpayer dollars? Given that reliable temperature records are a relatively recent event, and climates are constantly changing, we should expect temperature extremes to be regularly broken. Broken temperature records do not prove either global warming or climate change. They merely indicate that we have a small sampling of earth's temperature history. It is safe to say that even if we had accurate temperature records for the past 100 years it is impossible to know with confidence either the historical range of earth's surface temperature or if we have exceeded a safe level and are heading towards a disaster. What really is the significance of average surface temperature? People, plants, and animals live in areas where the average surface temperature is very cold and also very hot. Even if the surface temperature was accurately known, it would have little real significance for global warming, since the atmospheric layer in which heat is constantly being transported around the earth is six to ten miles thick (troposphere). Isn't the temperature of the rest of the troposphere important?

Based on the above discussion, it can be concluded that global warming advocates are attributing an accuracy to earth's current temperature measurements that is not justified by the raw data. We have a short temperature history acquired from a small number of widely-spaced, constantly changing, poorly sited, land-based, inaccurately averaged, and unrepresentative weather stations. A 1° F (0.6°C) temperature increase in 100 years is well within the acceptable margin of error of the measurement system and certainly does not justify any alarm. In truth, we have no idea what the average surface temperature of the earth is. Because the temperature change is well within the margin of error, the only conclusion we can make is that earth's average temperature is steady! There is no reliable scientific data to prove a worldwide global warming problem today. The predictions of disaster are all based on a questionable temperature history and an even more suspect array of highly biased computer projections. As we all know, computer output is only as good as the input data and the calculation components. Both are highly suspect.

Finally, as mentioned in the climate section, reputable scientists recognize that much of earth's past was notably warmer than at present (Groombridge and Jenkins 2002, p. 34). The earth is obviously not warmer than it has ever been, and the current surface global temperature measurement system is too imprecise to identify a reliable trend. Obviously, the reported surface temperature history does not prove global warming.

Scripture and Science Summary

Combining the previously discussed biblical framework with the basic scientific data just reviewed, allows construction of a brief, yet very useful, history for CO_2 and plants. This history will help put the global warming issue in proper perspective.

Creation: • God created the atmosphere. The atmosphere contained adequate CO_2 and O_2 initially. It did not evolve. • God created plants and animals. Plants and animals did not evolve, they were created, and the atmosphere contained all that they needed to live (CO_2 and O_2 in suitable concentrations). • Plants were created as food for the animals and man. Plants were not needed to provide oxygen for life. • Plants (both land and marine) and animals with carbonate shells removed CO_2 from their environment (atmosphere and ocean) as they reproduced and covered the earth. • God created man and gave him dominion over the earth. Man was commanded to fill, subdue, and rule over the earth, plants, and animals. • The earth was created for man's sustenance, use, and enjoyment.

Flood: • God judged the world with a Flood. Large volumes of plants and shelled animals were buried in the year-long, global, Flood of Noah. • During the Flood a significant volume of warm water was added to the original ocean. Most of the warm water flowed out of the earth from the fountains of the deep. The ocean was likely well mixed from the Flood which resulted in warm oceans surrounding the poles.

Post Flood: • After the Flood, plants (land and marine) and shelled animals again began removing CO_2 from their environment as they once more inhabited the earth. • The buried plants became fossil fuels (coal, oil, gas, tarsand, oilshale), and shelled marine animals became carbonate deposits. • The warm ocean surrounding the poles triggered an ice age. Massive volumes of water were evaporated from the warm polar seas and precipitated as snow. This rapidly generated large sheet glaciers inland from the ocean. Land immediately adjacent to the ocean produced lush vegetation which supported large and diverse communities of animals (for example, woolly mammoth, horse, bison, musk ox, moose, antelope, bear, etc.). • As the glaciers grew the ocean level dropped and numerous land connections were developed between the continents. • About 100 years after the Flood, God stopped people from working on the Tower of Babel by creating different languages among the people. Language groups congregated together and many began migrating around the earth. The migration was facilitated by warm seas which provided abundant freshwater and lush vegetation, as well as the land bridges created by falling sea level due to glaciation. • With

246

time the polar seas cooled, which decreased precipitation of both snow and rain. Eventually, the glaciers began to retreat as melting exceeded snowfall. As the glaciers melted, sea level rose and the land bridges were slowly covered. Inland lakes evaporated and deserts developed. Some deserts are still expanding. • Plants and animals migrated to accommodate to the changing climates. In addition, the cooling ocean absorbed more CO_2 from the atmosphere. As the poles became much colder the once lush grasslands and thriving animal herds along their margin became extinct. • Mankind converted wilderness land to agricultural use as their population increased. • Sea level continued to rise as the glaciers continued to melt. • Climates continued to change, setting new temperature records all around the earth.

Why There Is No Reason for Alarm • O_2 and CO_2 in the atmosphere were created, they did not evolve. • Today's atmosphere likely contains significantly less CO_2 than before the Flood. • CO_2 is necessary for life, and was created prior to plants and animals. • CO_2 is not a pollutant. • Increasing levels of CO_2 are beneficial for plants. • Decreasing levels of CO_2 could be a serious problem. • Burning fossil fuels simply returns CO_2 to the air, from which it originated, in the pre-Flood atmosphere. Increasing CO_2 in the atmosphere does not reverse a billion year old evolutionary trend and upset the delicate balance of nature. • The present levels of oxygen in the air are adequate without any unusual efforts to plant trees or to further limit the forestry industry. • Plants were created as food for humans and animals. They are not necessary for storing carbon or for generating O_2. • Glaciers have been retreating for thousands of years since the Flood. Most of the glacial melt occurred before man began burning fossil fuels. • Ice age glaciers melted due to cooling seas, not warming seas. • Climates have been constantly changing since the Flood. Consider all the major climate changes since the Flood and initiated by the Flood. • Plants, animals and mankind have been adapting to climate for thousands of years. • Recent global temperature histories are insufficient for developing reliable conclusions about trends or impending catastrophes. • Increasing the concentration of CO_2 in the atmosphere will continue to improve crop production around the world, benefiting mankind. • Neither melting glaciers, increasing CO_2, changing climates, nor earth's surface temperature history are proof of global warming. • God is in control of history and the earth's climates, not man.

Conclusion

The biblical history of the earth, contained in the first 11 chapters of the book of Genesis, provides a useful and sufficient framework for evaluating the current global warming issue. As we have seen, CO_2 is a natural atmospheric gas that is essential for man's existence. It is not a pollutant. The atmosphere is likely deficient in CO_2 compared with the original created atmosphere. Reducing CO_2 would definitely create problems, but increasing it will not. Burning fossil fuels merely returns CO_2 to its place of origin. Forests are to be used for man's benefit. They are not needed to produce O_2 and they have no intrinsic rights, but should be managed responsibly and effectively.

Basic science is consistent with the biblical history and argues strongly against the global warming hypothesis. Melting glaciers and changing climates are not an indication of man-made global warming. These natural phenomena have been operating for thousands of years. Temperature histories are imprecise and unreliable. Global warming is built on an evolutionary earth history and an evolutionary time scale. Anything built on a faulty foundation cannot stand. Global warming is an offshoot of evolutionary thinking and is needlessly creating mass hysteria. God is in control of the earth, not man. It can be expected that several trends evident since the Flood, however, will continue: sea level will rise as polar glaciers continue to melt, and deserts will expand. These trends, as we have shown, have little to do with CO_2, they are a consequence of a God-ordained event, the Flood. Governments with either ocean boundaries or deserts should consider how to efficiently and economically address these trends. There is no viable justification either biblically or scientifically for limiting the generation of CO_2 or restricting logging of forests. In view of the great benefit of CO_2 it is absolutely unnecessary to consider spending billions of dollars to restrict something that is extremely good for mankind and the earth. We cannot properly understand creation apart from God's Word. Viewing global warming within a Bible science perspective brings much needed clarity to this issue. As stated in Psalm 119:105, "Your word is a lamp to my feet and a light to my path" Those interested in reviewing scientific arguments not treated in this paper are referred to the skeptics reading list included at the end of this paper.

Postscript

Two questions remain to be answered: what must global warming advocates do to prove there is a real problem, and what should the church do regarding the global warming allegations? A proposed answer to each of these questions is outlined below.

What must global warming advocates prove? • Global warming actually exists • Global warming is causing climate change • Global warming is caused mainly by CO2 • Burning fossil fuels is the primary cause of CO2 increasing •

Global warming will absolutely cause serious harm • Proposed solutions are effective, fair, and economic

What should the church do regarding global warming? • Commit to viewing the world from God's perspective • Understand and rely on Scripture as a foundation for life • Use the Bible to understand the world and evaluate all problems • Help inform other believers • Promote the truth and oppose false beliefs with gentleness and respect It is imperative that the church disciple believers so that they know God's Word, think biblically, act biblically (grow in sanctification as disciples), and share God's Word. Let's honor God by being influenced and led by His truth (the Bible), and not by man's error.

Acknowledgments
The author thanks Dr. George Howe and Dr. Dennis Englin for their very helpful comments on the original draft of this paper and also a Creation Research Society peer review committee for their critique.

References
Barry, R. G., and R. J. Chorley. *Atmosphere, Weather and Climate*. New York, New York: Methuen & Co. Bergman, 1987.

Dott, R. H., Jr., and R. L. Batten. *Evolution of the Earth*, 4th ed. New York, New York: McGraw-Hill Book Co., 1988.

E. F., and W. H. Renwick. *Introduction to Geography: People, Places and environment*. Upper Saddle River, New Jersey: Prentice Hall, 2003. Cornwall Alliance. Retrieved from www.cornwallalliance.org. 2006.

Ehlers, E. G., and H. Blatt. *Petrology: Igneous, Sedimentary and Metamorphic*. New York, New York: W. H. Freeman and Co., 1982. Evangelical Climate Initiative. *Climate Change: An Evangelical Call to Action*. Retrieved from www. christiansandclimate.org, 2006.

Gore, A. *An Inconvenient Truth: The Planetary Emergency of Global Warming and What We Can do About It*. Emmaus, Pennsylvania: Rodale, 2006.

Groombridge, B., and M. D. Jenkins. *World Atlas of Biodiversity: Earth's Living Resources in the 21st Century*. Berkeley, California: University of California Press, 2002.

Interfaith Stewardship Alliance. *A Call to Truth, Prudence, and Protection of the Poor: An Evangelical Response to Global Warming*. Retrieved from www.cornwallalliance.org, 2006.

Kiehl, J. T., and K. E. Trenbert. Earth's annual global mean energy   budget. *Bulletin of the American Meteorological Society* 78, no. 2:197– 208, 1997.

Landes, K. K. *Petroleum Geology*, 2nd ed. New York, New York: John   Wiley & Sons, 1959.

Leffler, R., and K. Redmond. *Factors Affecting the Accuracy and Continuity of Climate Observations.* NWS online training. PCU6-Unit No. 2,   2004.

McKitrick, R. *An Economist's Perspective on Climate Change and the Kyoto Protocol.* Presentation to the Department of Economics Annual   Fall Workshop. The University of Manitoba, Canada, July 11,   2003.

NOAA (National Oceanic and Atmospheric Administration) 2010. Greenhouse gases frequently asked questions. Retrieved from www.ncdc.noaa.gov/faqs/climfaq13.html.

Northen, H. T. *Introductory Plant Science*, 3rd ed. New York, New York: The Ronald Press Co, 1968.

Oard, M. J. *An Ice Age Caused by the Genesis Flood.* El Cajon, California: Institute for Creation Research, 1990.

Oard, M. J. *Frozen In Time: The Woolly Mammoths, the Ice Age and the Bible.* Green Forest, Arkansas: Master Books, 2004.

Oberlander, T. M., and R. A. Muller. *Essentials of Physical Geography Today*, 2nd ed. New York, New York: Random House, 1987.

Peele, R. (ed.) *Mining Engineers' Handbook.* New York, New York: John Wiley & Sons, 1941.

Solomon, S., et. al. *IPCC Assessment Report #4: The Physical Science Basis of Climate Change.* New York, New York: United Nations, 2007.

U. S. Department of Labor. Mine Safety and Health Administration. Retrieved from www.msha.gov.

Ussher, J. *The Annals of the World.* Trans. and ed. L. Pierce. Green Forest, Arkansas: Master Books, 2003.

Vardiman, L., A. A. Snelling, and E. F. Chaffin (eds). *Radioisotopes and the Age of the Earth: Results of a Youngearth Creationist Research Initiative.* El Cajon, California: Institute for Creation Research, 2005.

Watts, A. *Is the U. S. Surface Temperature Record Reliable?* Chicago, Illinois: Heartland Institute, 2009.

Skeptics Reading List

R. J. Martin Although most books published on global warming are written by advocates, a few have been published by skeptics. Listed below are several books written by evolutionists who question the scientific arguments used by advocates. The authors are not young-earth creationists but they offer critical scientific arguments.

Bethell, T. *The Politically Incorrect Guide to Science*. Washington DC: Regnery Publishing, 2005.

Dears, D. *Carbon folly: CO2 Emission Sources and Options*. Reston, Virginia: TSAgust, 2008.

Hayden, H. C. *A Primer on CO2 and Climate*, 2nd ed. Pueblo West, Colorado: Vales Lake Publishing, 2008.

Horner, C. C. *The Politically Incorrect Guide to Global Warming and Environmentalism*. Washington DC: Regnery Publishing, 2007.

Lawson, N. *An Appeal to Reason: A Cool Look at Global Warming*. New York, New York: Overlook Duckworth, 2008.

Lomborg, B. *The Skeptical Environmentalist: Measuring the Real State of the World*. New York, New York: Cambridge University Press, 2001.

Lomborg, B. *Cool It: The Skeptical Environmentalist's Guide to Global Warming*. New York, New York: Alfred A. Knopf, 2007.

Michaels, P. J., and R. C. Balling Jr. *The Satanic Gases: Clearing the Air about Global Warming*. Washington DC: Cato Institute, 2000.

Michaels, P. J. *Meltdown: The Predictable Distortion of Global Warming by Scientists, Politicians, and the Media*. Washington DC: Cato Institute, 2004.

Michaels, P. J., and R. C. Balling Jr. Climate of Extremes: Global Warming Science They Don't Want You to Know. Washington DC: Cato Institute, 2009.

Nordhaus, T., and M. Shellenberger. *Break Through: From the Death of Environmentalism to the Politics of Possibility*. New York, New York: Houghton Mifflin Companym 2007.

Singer, S. F., and D. T. Avery. *Unstoppable Global Warming: Every 1,500 Years*. Lanham, Maryland: Rowman & Littlefield Publishers, 2007.

Global Warming Skeptics Websites

www.answersingenesis.org

Note: Answers in Genesis produced *Global warming: A scientific and biblical expose of climate change*.

Available from http://www.answersingenesis.org/publicstore/product/Global-Warming-A-Scientific-and-Biblical-Expose-of-Climate-Change-DVD,5733,229.aspx www.channel4.com/science/microsites/G/great_global_warming_swindle .

Note: BBC produced *The great global warming swindle.* www.heartland.org www.lomborg.com/ www.oism.org/pproject/ www.petitionproject.org/ www.sepp.org/ www.SurfaceStations.org

Author bio

Rod J. Martin has held engineering and management positions in oil & gas, mining, and the electric power industries. His career has taken him from installing offshore oil platforms in the Santa Barbara Channel, to helping develop a pioneering tarsand mine in Northern Alberta, Canada, as well as expanding heavy oil production in California, operating oilfields in Texas, drilling oil wells in Wyoming, surveying for the U.S. Forest Service in Colorado, teaching college earth-science and physical-science classes part-time, and managing a coal-fired steam-electric cogeneration plant in the Czech Republic. He holds an M.S. degree in geology (Cal. State, Northridge), a P.E. degree in petroleum engineering (Colorado School of Mines), is currently working on an M.A. degree in biblical counseling (The Master's College), and has taken additional graduate study in business, engineering, and theology.

Anthropogenic Global Warming
Edward F. Blick, Ph.D.

Editor's note:

There is indeed an "Inconvenient Truth" about Global warming, but it is not as Al Gore imagined. In fact it is the opposite … the climate is cooling NOT warming. Dr. Blick supports this profound and unexpected reality with copious facts and references. Once again an engineering background provides fresh insight into this important contemporary topic. Once again it is clear any industry that releases carbon dioxide into the atmosphere should be rewarded, not punished. More carbon dioxide would be a good thing and would stimulate plant growth all over the world and provide additional food for man and beast. Many today simply have this totally wrong.

Introduction

Carbon dioxide induced global warming is a huge sacred cow of science. Our government is using it as a "presumed evil" to frighten the public and promote legislation to limit the use of fossil fuels. They are pushing for massive taxes on all forms of fossil fuel because when burned, they emit carbon dioxide (CO_2). Our government tells us the "mean molecule" CO_2 is causing a climate disaster by warming the Earth, which melts our polar ice caps, raises the sea level, and inundates our coastal cities. There is no credible scientific evidence to support any of this nonsense. Anthropogenic Global Warming (AGW) and its new name, "Climate Change", are politicized science. In August, 2009, the UN Secretary General Ban Ki-moon warned the world that we have just four months to save our planet before the global warming induced sea level rise of 20 ft. will drown our coastal cities. The UN stated a similar warning back in 1998, when they said the planet would be finished by the year 2000 unless we solved the global warming problem. The UN was wrong then and they are wrong now. This politicized junk science scheme is right out of H. L. Mencken's *Aim of Practical Politics*: "Keep the populace alarmed and hence clamorous to be led to safety by menacing them with an endless series of hobgoblins, all of them imaginary." The government's radical solution is always more taxes and more power over the people. Over 31 thousand American Scientists, including the author, have signed the following petition rejecting global warming:

We urge the United States government to reject the global warming agreement that was written in Kyoto, Japan in December 1997, and any other similar proposals. The proposed limits on greenhouse gases would harm the environment, hinder the advance of science and technology, and damage the health and welfare of mankind. There is no convincing scientific evidence that human release of carbon dioxide, methane, or other greenhouse gases is causing or will, in the foreseeable future, cause catastrophic heating of the Earth's atmosphere and disruption of the Earth's climate. Moreover, there is substantial scientific evidence that increases in atmospheric carbon dioxide produce many beneficial effects upon the natural plant and animal environments of the Earth.

Evidence against Global Warming

United Nations politicians, while admitting their lack of evidence, gave birth and nurtured the fraud of Anthropogenic Global Warming (AGW). This myth proposes that man's burning of fossil fuels since the inception of the industrial age a hundred and fifty years ago has increased the amount of atmospheric CO_2 by 20%. This supposedly caused the earth to warm and threatens to melt the polar ice caps, increase the sea level and flood our coastal cities. Their Malthusian purpose is to frighten people into accepting the UN as the "centerpiece of democratic global governance" and let the UN, direct sovereign nations in rationing fossil fuels.

During the twentieth century the Earth warmed up about 0.6° C due to increased solar activity, not from man's burning of fossil fuels. World temperature records, (Fig. 1) show no evidence of AGW. From Fig. 1 it is seen that only Antarctica had a record high temperature since 1942. Where is the footprint of man-made global warming? Addition evidence refuting man-made global warming is shown in Fig. 2, which shows the distribution of all-time high records for our 50 states. It is obvious that again, there is no evidence of manmade global warming. Twenty-four, or almost half the states had their all-time record high temperatures in one ten year period the 1930s (seventy years ago). Thirty-three, two-thirds of all states had their record high temperatures in the 1880s-1930s! During this period of time there was far less CO_2 put into the atmosphere by the burning of coal, and other fossil fuels than during the period of 1940-2008. Only sixteen had their record temperatures set in last fifty years! Where is the fingerprint of "AGW"? How can any reasonable person with half a brain, look at Figures 1 and 2 and say that man-made global warming is true? There is absolutely no evidence for global warming.

The global warming myth assumes, with zero evidence, that the increase in atmospheric CO_2 is due is due to man, and this increases the Earth's temperature. Figure 3 is a plot of atmospheric CO_2 concentration from 1880–2008 using data from Beck's (2007) data and the Mauna Loa observatory. Notice in Figure 3 the peak concentration in atmospheric CO_2, occurred in 1940, just after the peak temperatures of the US in the 1930s (Figure 2). The warming and peak

Continent	Record High °F	Year
Europe	122	1881
Australia	128	1889
South America	120	1905
Oceania	108	1912
North America	134	1913
Africa	136	1922
Asia	129	1942
Antarctica	59	1974

Figure 1. All-time record high temperatures for continents

temperatures occurred before the peak in atmospheric CO_2. This is the opposite of what the global warming theory predicts! Ernst-Georg Beck's CO_2 data from the 1800s until 1960 is the gold standard for CO_2 measurements since it is a compilation of 90000 accurate chemical analysis of CO_2 in the air. These were published in 380 scientific papers. The accuracy was better than 3%. Several scientists who won the Nobel Prize made some of these measurements. Omitted from Fig. 3 was the fraudulent UN-IPCC CO_2 data taken from ice cores. See Jaworowski's (2007) discussion in the later section "CO_2 Not Harmful to Man or Earth", for an explanation of why the ice core data used by the UN is flawed.

The twentieth century warming correlated with the great increase in solar activity (Fig.4). Astronomers have noted that Jupiter, Mars, Saturn, Neptune and Pluto all warmed up in the twentieth century (Archibald, 2008). Man driving his SUVs had nothing to do with the warming of Mars, etc. Since 1998, global warming has taken a vacation. We've had global cooling from a lazy Sun with its reduced sun spot activity (Archibald, 2008).

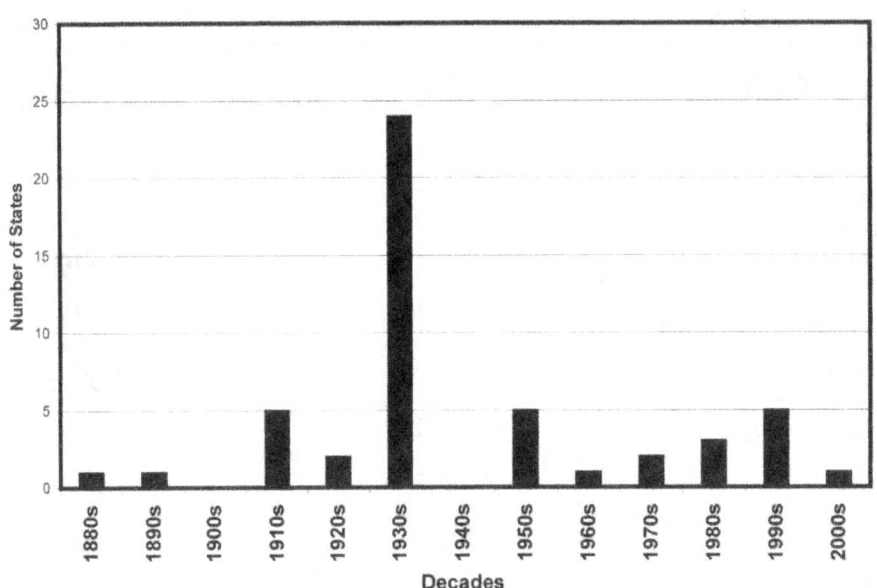

Figure 2. All time record high temperatures for 50 states (NOAA.gov)

The robust solar activity of the twentieth century caused a warming of the land and also a warming of the oceans. The 20th century warming correlated with the increase in solar activity as seen by the Solar Number and Carbon 14 proxies in Fig. 4.) The oceans cover 70% of the earth's surface and are the dominant influence on atmospheric temperatures. Atmospheric CO_2 levels rise as the sea surface is

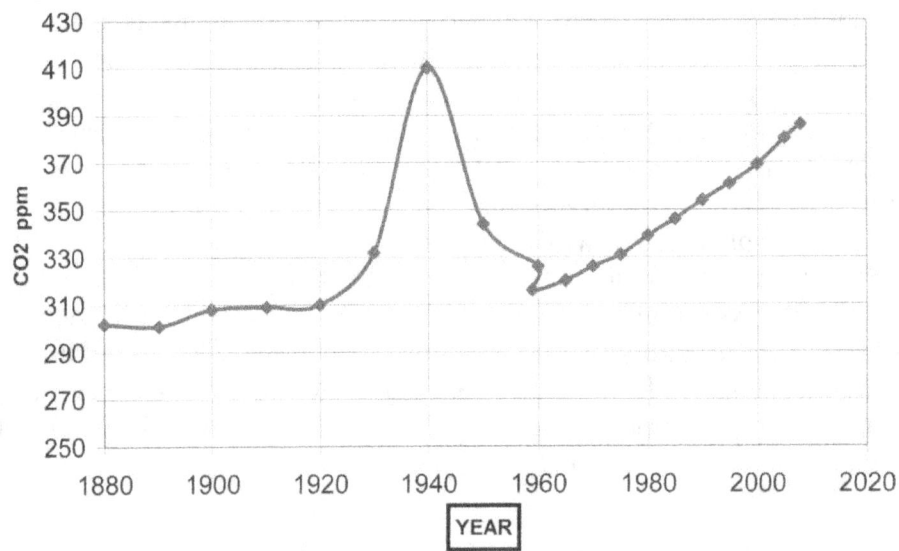

Figure 3. Atmospheric CO_2 concentration 1880–2005 (1880-1960 data; Beck (2007), 1959-2005 data; Mauna Loa)

warmed (Endersbee, 2008). Henry's Solubility Law, coupled with mass balances of carbon and its isotopes, show a 2–4% total increase in atmospheric CO_2 from pre-industrial times. Burning all our remaining fossil fuels, <u>cannot</u> <u>double</u> the CO_2, but only increase it by 20%. Beck (2007) cataloged 90000 chemical measurements of CO_2 in the 1800s, some as high as 470 ppm—greater than the current Mauna Loa value of 385 ppm. Beck's data exposed the UN IPCC's 280 ppm ice core values, supposedly measured during the 1800s, as false. IPCC's ice core measurements of CO_2 were incorrect due to their inability to correct for problems with gas solubility and the extreme pressures in glaciers. God rules the climate, not man.

Sunspot activity and δC¹⁴

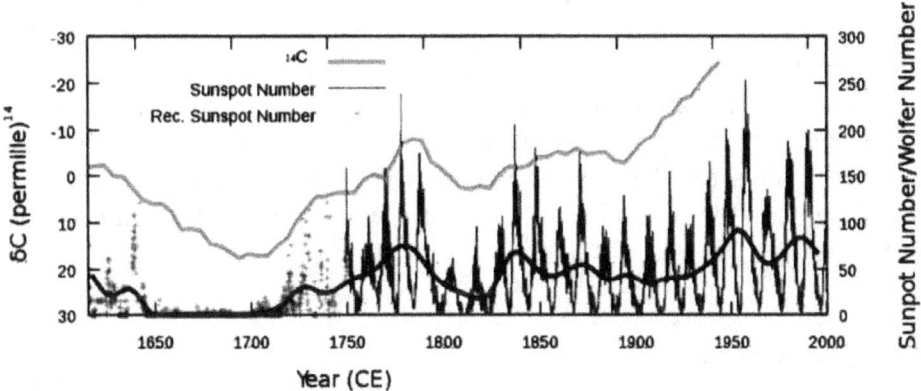

Figure 4. Proxies for Solar Irradiance are Sunspot Numbers and Carbon 14 Isotopes, years 1610 - 2000 (Wikipedia 2010)

When a cold bottle of soda is taken from a refrigerator CO_2 bubbles rise out of the glass as it warms. Similarly, CO_2 bubbles out of the oceans into the atmosphere if the ocean warms. Cooling oceans absorb CO_2 from the atmosphere. CO_2 is less soluble in water as it warms and more soluble as it cools. The warming during the twentieth century caused the oceans to emit more CO_2 into the atmosphere (Endersbee, 2008)

From 1880–2000, the global temperature increased about a third of a degree centigrade, primarily due to the increased solar activity seen in Figure 4. During this same period, a miniscule amount of atmospheric global heating of 0.5 W/m^2 was due to an increase of approximately 11 ppm of CO_2. This represents a 3% increase of atmospheric CO_2 due to the burning fossil fuel (Segalstad, 1996).

This corresponds to a tiny 0.05° C rise in temperature, using the climate sensitivity parameter of 0.1° C per W/m^2 (Kiehl and Trenberth, 1997). The temperature increase of 0.05° C is too small to be seen in the global averaged temperature measurements since it is much smaller than the normal yearly temperature fluctuations due to the earth's clouds and water vapor.

The climate sensitivity parameter of 0.1° C per W/m^2, agreed with eight natural experiments (Idso 1998). UN Climate models use sensitivity values 5–10 times higher than 0.1° C per W/m^2. Why, because it allows the UN to results to frighten people with exaggerated predictions of future global temperatures! Remember the UN's forecast in 1998 and August 2008, of impending planet disasters that never occurred?

Archibald (2008) assumed a theoretical rise in atmospheric CO_2 of 380 to 420 pp, in the next 20 years. Using the University of Chicago's MODTRAN facility, he obtained a 0.4 W/m^2 increase in global warming. Using Idso's 0.1° C per W/m^2 sensitivity value, he predicted a 0.04° C increase in temperature due to CO_2 green house effect (Idso, 1998). This is an insignificant rise in temperature. If a room temperature increased 0.04° C (0.07° F), a human would not notice such a tiny increase in temperature! Yet the environmental extremists scream the lie that man is going to burn up the planet and drown us all due to melting glaciers. Trillions of dollars can be poured into this non-existent problem by trying to eliminate the use of fossil fuels and CO_2 but it will accomplish nothing but wreck our economy.

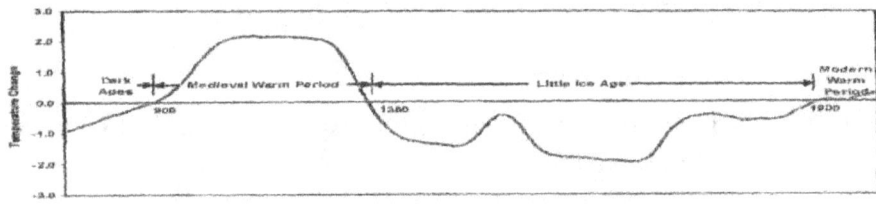

Figure 5. Medieval warm period & little ice age (IPCC 1995)

In 1995, the UN IPCC produced a chart, Fig 5, showing global temperature anomalies for the past 1300 years. This chart agreed with hundreds of scientific papers written on the Medieval Warm Period and the Little Ice Age which followed. From about 900 to 1350 AD, the earth was approximately 2° C warmer than now. During this time, about 5000 Vikings colonized Greenland, and many of the great cathedrals were built in Europe. The Chinese navy sailed in the arctic with little sign of ice. The Little Ice Age that followed lasted about four-hundred years (Soon & Baliunas, 2003). The bodies of the Viking colonists are now buried under Greenland's permafrost. Yet, in 2001, six years after the UN

258

published the Little Ice Age chart, the UN did an about face, and published a new radically different chart. It was named the Hockey Stick chart. It depicted a flat temperature for the one thousand years prior to the twentieth century, followed by a rapid rise of earth's temperature in the twentieth century. The UN blamed the rise on AGW. The man who conjured up this false chart was accused of "cherry picking" the data. The UN's hockey stick was broken in 2007 after scores of reputable scientists protested against this fraudulent "hockey stick" chart. Two scientists McIntyre and McKitrick, using one thousand years of proxy data uncovered the "Hockey Stick Lie." Hearings were held in Congress and the National Academy of Sciences. The UN hockey stick chart was labeled fraudulent, and designed to deceive the public and lawmakers. What do you expect from the UN ... the truth? No ... lies!

The green movement to limit or eliminate the burning of fossil fuels eagerly seized upon this UN AGW myth. As of now one of our political parties has joined hands with the environmental extremists and refuses to allow for oil & gas drilling in promising areas all in the name of controlling carbon. MIT Prof. Dr. Richard Lindzen, (maybe our top climate scientists) was an original member of the IPCC team (before he discovered the IPCC was a fraud, and quit). Lindzen stated, "Controlling carbon is a bureaucrat's dream. If you control carbon you control life!"

The most important greenhouse gas is water vapor. Its mass is fifty-four times greater than CO_2. "The first thirty feet of water vapor absorbs 80% of the earth's heat radiation. You can go outside and spit and have the same effect as doubling CO_2!" (Dr. Reid Bryson, late Director of Meteorology, University of Wisconsin.)

A hundred and fifty years ago, the atmospheric CO_2 contained 700 Gt of carbon (1 Gt = 1 billion tons), and the earth contained 7000 Gt of carbon in the form of fossil fuels. It is estimated that man has burned 1000 Gt of the original 7000 Gt. (Segalstad 1998). For water, at normal temperature, Henry's Law of Solubility dictates there will be 50 parts of CO_2 in solution, for one part of gaseous CO_2 above the water. Experimental measurements have shown that the residence time of CO_2 in the atmosphere is about 5 years. The corrupt UN politicians (without any proof) say the residence time is 50-200 years. Hence today, after one hundred and fifty years, the amount of CO_2 added by man to the atmosphere is (1/50) x 1000 = 20 Gt, and the increase in atmospheric CO_2 is (700+20)/700 = 1.03 or a 3% increase!! (Segalstad, 1998). The UN, using junk science, and mysterious fudge-factors, said the increase is 21%. Where is their proof ... they don't have any?

Segalstad (1998) developed an alternative method of determining how much of the atmospheric CO_2 is due to fossil fuels is by an isotopic mass balance of Carbon 12, C-12, and the heavier isotope Carbon 13, C-13. During photosynthesis plants absorb more of the C-12 than C-13. Ratios between C-12 and C-13 stable isotopes are commonly expressed as in permil by a so-called delta-13-C notation multiplied by 1000. CO_2 from combustion of fossil fuel have delta −13-C values of (−26 permil). Natural CO_2 has a delta-12-C value of (−7 perm). Keeling (1989) reported a 1988-measured atmospheric delta-13-C value of (−7.807 permil). Using a simple isotopic mass balance equation of [26X +7(1-X) = 7.807] produces an X value of 0.042. Hence the earth's atmospheric CO_2 is made up of approximately 4% CO_2 from the burning of fossil fuels. This is close to the 3% computed above by the alternate mass consumed method of Segalstad. Revelle & Suess (1957) using Carbon-14 data computed the amount of atmospheric CO_2 derived from fossil fuel combustion was 1.2 to 1.73 %. UN IPCC report state that at present, 21% of CO_2 is from fossil fuel burning! That is a lie!

Using Henry's Law, and <u>assuming all the remaining 6000 Gt of carbon in our fossil fuel reserves has been burned</u>, the increase in atmospheric CO_2 will be [{(700+ (7000/50)}/700 =1.2], a 20% increase over what the atmosphere contained back in the mid nineteenth century! (Segalstad, 1998) The corrupt UN with their junk science predicts a 170% increase. Even burning all fossil fuels (7000 Gt of carbon) will have no meaningful effect on global climate. CO_2 in the atmosphere cannot increase more than 20%. It cannot double!

The Earth receives about 1368 W/m^2 of radiative heat from the sun. The total amount of heat withheld is about 146 W/m^2, +/- 5 to 10 W/m^2 due to natural climatic variations. Clouds can reflect up to 50 W/m^2 and can absorb up to 30 W/m^2 of the solar radiation. Less than ½ W/m^2 ^{is} produced by anthropogenic CO_2, making it much smaller than the Earth's average "Greenhouse effect (water vapor, etc), which varies naturally within 96 to 176 W/m^2 (Segalstad, 2006).

The total internal energy of the whole ocean is 3.3×10^{27} Joules, about 3500 times greater than the total energy of the entire atmosphere, 9.4×10^{23} joules. The global climate is primarily governed by the enormous heat energy stored in the oceans and the latent heat of melting of the ice caps. From a thermodynamic heat balance, the small amounts of heat generated by anthropogenic CO2 (1/2 W/m ^{2}) could not possibly cause significant increases in sea level (Segalstad, 1995).

A study of fourteen hundred years of sea level data, found a 10-inch difference between the thermal expansion of the Medieval Warm period and the thermal contractions of the Little Ice age (van de Plassche). This makes the prediction of a twenty-foot rise in sea level by Al Gore and the UN, to be out of touch with reality.

One of the biggest global temperature drops ever recorded; occurred from January 2007–08. The drop in temperature was about equal to the net gain in average temperature for the twentieth century.

The UN and the Marxist Enablers

AGW is a hoax and has become the political agenda for the UN, the Democratic Party, and their environmental extremist supporters. There is no convincing evidence for it. The AGW hoax was initiated and nurtured by the corrupt politicians of the United Nations, the same people who colluded with Saddam Hussein to skim billions of dollars in their "Oil for Food" scheme! They have undiluted power to deceive the public. They are lawless, corrupt, anti-God and an utter fraud. They are using the delusion that, man's use of fossil fuels causes Global Warming, in order to frighten people into allowing them and their supporters to rule the world. Because <u>any group that controls carbon, controls the world</u>. AGW is a hoax, religion, junk science, and is worthless in its predictions of future global temperatures. It is as scientific as astrology!

Dr. Tim Ball the distinguished former climatology professor at the University of Winnipeg has stated it is "<u>possibly the greatest deception in human history</u>" Dr. Ball's life has been threatened a number of times because of his stand against global warming. The environmental extremists aim is to destroy the industry of western civilization. Many of these environmental alarmists are misanthropic fanatics. Man is bad … polar bears are good! Limit the number of people on the Earth! They will not be satisfied until we are all living in a hut, defecating in a bucket and cooking our food on dried dung.

After the 1989 fall of the Soviet Union, the Marxists, socialists, and anti-capitalists wackos were looking for a new way to reinvent themselves and find a new way to destroy America and Western economies. They found it in the evil UN's contrived "<u>man is destroying the planet by burning fossil fuel religion</u>" This new religion of the socialists and anti-consumerists oppose large corporations, global free trade, and economic growth. They hate oil, coal and gas companies. Thirty years ago they, along with Jane Fonda and her movie "China Syndrome," destroyed nuclear energy in this country. Our left-wing media, who love to publish gloom and doom stories about global warming, has joined hands with the extremists.

Marxist Maurice Strong and his associates at the United Nations hatched the rotten egg of Global Warming back in the 1980s. Former Vice-President Al Gore working the UN helped to spread the myth of Global Warming.

Al Gore is a politician, not a scientist. He made an "F" on his College Board physics exam and a "D" on the chemistry exam. He flunked out of Vanderbilt's divinity school, with five F's. He later transferred to Harvard where he took only two college natural science courses. He made a "D" in an evolution class, and a "C+" in an introductory science course. With this academic background, does anyone really think he could have written two cleverly crafted science-fiction books on global warming? Could Gore's friends at the UN have ghostwritten those two books for Al? Gore ducks all challenges to debate on AGW. It is reported he collects $250,000 for his colorful "dog and pony" slideshow on man's destruction of the earth by burning fossil fuels that cause global warming. He is a hypocrite since his home reportedly uses over twenty times the amount of electricity as an ordinary American. To add insult to injury, Al Gore was awarded a Nobel Prize in 2007. Finishing second was Irena Sendler, a nurse in the Warsaw ghetto in World War II by saving over two thousand five hundred Jewish children by sneaking them out of the ghetto until the Nazis caught her, tortured her, and broke both her legs and feet. She just recently passed away in her mid '90s. Couldn't the Nobel judges tell the difference between a candidate with true humanitarian courage and a politician preaching a scientific lie?

Strong and the UN set up the 1992 Rio de Janeiro conference entitled "The Earth Summit" It was attended by Vice President Al Gore. At the conference, Strong stated, "The Earth Summit will play an important role in reforming and strengthening the UN as the centerpiece of the emerging system of democratic global governance, i.e., a one-world government run by the UN. Strong and the UN set up a conference on global warming in Kyoto, Japan. All countries were urged to sign a treaty to reduce their CO_2 output in order to save the planet. China, India and the US refused to sign the treaty. Most of Europe joined, but have done little in the way of lowering their CO_2 output.

The National Review magazine, September 1, 1997 quoted Strong, "The only way of saving the world may be for industrial civilization to collapse, deliberately seek poverty, and set levels of mortality. Timothy Wirth, former US Senator from Colorado and president of the United Nation's Foundation stated, "We have to ride the theory of Global Warming even if it is wrong" Richard Benedict, former advisor to Kofi Annan stated, "A global warming treaty must be implemented even if there is no evidence of global warming" These guys know AGW is not true, and are deceiving the world by saying they are saving our planet. "The urge to save humanity is almost always a false front for the urge to rule" (H. L. Mencken).

In 1988, the corrupt UN politicians set up their political junk science department, the Intergovernmental Panel on Climate Control (IPCC). Its purpose was to frighten people of AGW, and have them beg for a government solution. There was no scientific evidence then or now of any significant AGW, so they had to create a masterful lie. Sir John Houghton, the first chairman of the UN's IPCC stated, "Unless we announce disaster, no one will listen"! Here is how they succeeded. As a smoke screen, qualified experts in science and climatology were hired to investigate if man has effected the warming of the earth. Here is the summary, the scientists wrote for the 1995 IPCC Draft Report:

1. None of the studies have shown any clear evidence of climate changes due to greenhouse gases.
2. No study has positively attributed any climate change to anthropogenic causes.
3. Any claims of positive detection of significant climate change are likely to remain controversial until uncertainties in the total natural variability of the climate are reduced.

This was not what the UN wanted! After the real scientists finished their report and went home, the UN politicians removed all three of the above quotes and inserted the following bold face lie in the final 1995 IPCC Summary Report for Lawmakers: "The balance of evidence suggests a discernible human influence on global climate" Because of this deceitful lie, many of the IPCC scientists quit, and threatened the UN with a lawsuit in order to have their names removed from the IPCC 1995 final report.

The UN political appointees have adopted a strange and unusual way of writing their IPCC reports. They first publish a *Summary Report for Lawmakers*. Then several months later they publish the Scientific Report so as to assure its consistency with the previous Summary Report. This is political indoctrination, not science! Real science projects complete the scientific work first and then write the final report. After the 1995 IPCC report, the lies, deceit and sleight of hand were repeated in 2001 and 2007 IPCC reports.

Another controversial figure in the fight for global warming is Dr. James Hansen. He has been Al Gore's global warming mentor for several decades, and is a media darling depicted as a non-partisan scientist. But his immense arrogance is dwarfed by his great dishonesty. He is the director of NASA's GISS Lab that keeps track of global temperatures. Several prominent scientists have accused Hansen of cooking the temperature books in order to make them appear to show that global warming has occurred. Over 9000 news stories have quoted his pro-"global warming" lies.

In 1988 Hansen appeared before Senator Al Gore's Committee and stated he was 99% certain the earth was warming due to man's burning of fossil fuels. He made a prediction to the Senators about how much the earth would warm up in the 1990s. His prediction was too high by 300%. In 2004, he publicly endorsed John Kerry for President, and then received a $250,000 gift from the charity of Kerry's wife, Theresa Heinz. On June 23, 2008, Hansen asked Congress to try

Big Oil leaders for the high crime of doubting global warming! In a moment of candor he did admit, he is willing to exaggerate science in order to get public attention. In February 2009, Hansen equated coal with death. Some responsible climatologists have called for his termination.

CO$_2$ Not Harmful to Man or the Earth

Obama's administration claims CO_2 is a pollutant. CO_2 is not a pollutant. It is the gas of life for plants, man, and animals. All plant life is sustained by photosynthesis, where CO_2 plus water plus the Sun's energy form carbohydrates plus Oxygen. Humans and animals breathe in oxygen and exhale CO_2. It sounds like an intelligent design!

It is not causing global warming or climate change, nor has it ever killed anyone. Our present atmospheric CO_2 level is 385 ppm. Historically the amount of atmospheric CO_2 has never reached a level where it is dangerous for humans. Humans are in danger when CO_2 concentration reaches 50000 ppm. Sailors in our submarines work in CO_2 levels of 8000 ppm with no ill effects. Crowded auditoriums, may reach 10000 ppm. The recommended threshold level in civilian workspaces for an 8-hour day is 5000 ppm. A typical office has 350 to 2500 ppm. Exhaled human breath is about 45000 ppm. The total carbon emitted by a human is about 2 kg/day. For six billion people on earth this amounts to 2.2 gigatons/year (1 gigaton = 1 billion tons).

According to the Mauna Loa observatory the present atmospheric CO_2 is about 385 ppm (parts per million.), but in times past it was as high as 2450 ppm. (Jaworoski, 1992a, 1992b). This contradicts the UN lie that CO_2 concentrations of 280 ppm existed from the beginning of the time to the late 1800s. Then man increased the atmospheric CO_2 in the twentieth century, due to his burning of fossil fuels. Beck (2007) criticized the authors of the UN lie, Callendar (1958) and Keeling (1989). Beck essentially said that "cherry picking" was involved, because CO_2 measurements were rejected if they did not fit the UN lie. Beck also stated that Callandar and Keeling examined only 10% of the available literature.

Zbigniew Jaworowski, MD, Ph.D., DSC (2007) is a CO_2 glaciologist who has studied glaciers all over the world. He has published many papers on climate, most of them concerning CO_2 measurements in ice cores. He strongly believes the CO_2 measurements used in the UN IPCC reports have been corrupted, and are false. He pointed out, "Drilling ice cores is a brutal system and a polluting procedure, drastically disturbing the ice samples" He also states that ice cores cannot be regarded as a closed system and used to measure CO_2 levels of air trapped in ice. He stated there are "more than twenty physical-chemical processes operating *in situ*, in the ice cores. In cold water, CO_2 is more than 70 times more soluble than nitrogen and more than 30 times more than oxygen" Liquid water is commonly in present in the polar snow and ice even at the eutectic temperature of -73° C" This phenomenon alone will greatly reduce the percentage of CO_2 in the air bubbles trapped in ice. The Knudsen effect, combined with inward diffusion, depletes CO_2 in ice cores exposed to drastic pressure changes (up to 300 bars, for ice buried in glaciers). The effects of increased solubility and extreme pressures, could explain the difference between Beck's chemical CO_2 measurements and the ice core measurements used by IPCC. Jaworowski noted that these effects were discovered only recently, many years after the ice-based edifice of anthropogenic warming had reached a skyscraper height. Jaworowski noted how Neftel (1985) et. al. fraudulently combined the CO_2 values of 328 ppm from ice deposited in 1890 and combined it with 328 ppm CO_2 values measured at Mauna Loa volcano, Hawaii, 83 years later. This fraudulent data curve was then published in the 2001 IPCC report, and is now part of the dogma of the AGW crowd. The discovery of this fraud makes it shockingly clear that pre-industrial level of CO_2 was the same as in the second half of the 20th century. This fudging of the CO_2 data 83 years apart has reinforced Jaworowski belief that AGW is a myth and human beings may be responsible for less than 0.01° C of warming during the last century."

Carbon Dioxide: Aerial Fertilizer

At present CO_2 is an endangered molecule. We need more CO_2 in order to increase crop yields! Award winning Princeton University physicist, Dr. William Happer has stated that CO_2 levels were much higher in the past and at present the Earth is currently in a CO_2 famine with atmospheric levels of only 385 ppm. (Testimony before Senate Environmental and Public Works Full Committee hearing, February 27, 2009). Ernst-Georg Beck (2007) reported CO_2 levels of 400+ ppm in the 1940s and early 1800s, which are higher than our present level of 385 ppm. This is a mystery the Global warming advocates cannot explain, because they have cooked the books on CO_2 levels prior to 1950, (Jaworowski, 2007).

Our atmosphere contains approximately 78% nitrogen, 19%oxygen, 1% to 3% water vapor, plus 1% trace gases, which includes 0.0385% CO_2. Carbon dioxide is vitally important because it is a component of all food, fiber and fuel. Now our U.S. government wants to tax it as a pollutant! CO_2 is the sole source of our food chain. Every item of nutrition we consume started out as atmospheric CO_2. Plants produce food from carbon dioxide and water in the presence of light by the wonderful mechanism of photosynthesis. Plants ingest CO_2 and release oxygen (O_2) to the atmosphere. Six molecules of CO_2 and six molecules of water (H_2O) are converted by plants into one molecule of carbohydrate plus six molecules of oxygen (O_2).

$$6H_2O + 6CO_2 \rightarrow C_6H_{12}O_6 + 6O_2$$

The resulting carbohydrates directly or indirectly, supply almost all animal and human needs for food and energy. Oxygen and some water are released as by-products of this process. Man has burned hydrocarbons in engines for the past one hundred and fifty years and wood since the time he was created, releasing CO_2 to the atmosphere. Decaying flora and fauna also release their stored carbon back into the atmosphere in the form of CO_2.

Humans and animals are fat, protein, and bone with the spark of life. Fat is a largely carbohydrate, protein is carbohydrates plus nitrogen, and bone is calcium phosphate, plus some carbon minerals. Notice the common thread … all life is carbon. CO_2 is not now, and will never be a pollutant. Someone must tell this to our Environmental protection Agency!

Carbon dioxide is an "aerial fertilizer" for plants, yet our government is talking about reducing atmospheric CO_2 by "Cap and Trade" (a.k.a. Cap and Tax) scheme and pumping CO_2 down old oil and gas wells. Of course that will cost us trillions of dollars and further bankrupt our country. They have already started pumping CO_2 down a hole drilled at the Ethanol plant in Greenville, Ohio! This is analogous to the pagan medical practice of bleeding patients.

Lower concentrations of CO_2 can kill plants. The U.S. Department of Agriculture (USDA) found there is so little atmospheric CO_2, that a field of corn in full sunlight consumes all of the CO_2 within a meter of the ground. If wind currents do not constantly stir up the air, the corn will stop growing! Plants start to suffer when the CO_2 level drops to 240 ppm and they die at 160 ppm (Darryl Smika, 2007).

Our government will endanger our food supply by reducing atmospheric CO_2. The best-kept secret, the green-extremists hide from the public in the global warming debate, is that plant life of planet Earth would greatly benefit from higher levels of atmospheric CO_2. Plants grow faster and larger because of more efficient photosynthesis and have a sharp reduction in water loss. Thousands of experiments have shown that higher levels of atmospheric CO_2 levels will result in increases in plant growth and food and fiber crop production. Other benefits include greater resistance to temperature extremes and better growth at low light intensities, improved root/top ratios, less injuries from air pollutants, and more nutrients in the soil as a result of more extensive nitrogen fixation.

The underside of plant leaves contain pores, known as stomata, which admit air into the leaf for photosynthesis, but they also are a major source of moisture loss. Higher CO2 levels partially close the pores; greatly reducing the plants water loss—a significant benefit in arid climates.

266

Thousands of experiments have measured increases in plant growth with increases in CO_2. For rice the optimal CO_2 level was found to be between 1500 and 2000 ppm. The US Department of Agriculture in Phoenix pulled together data from 800 scientific tests that doubled the CO_2 level and found an average of 32% improvement in plant productivity. Greenhouse-grown vegetables, including tomatoes, cucumbers and lettuce show earlier maturity, larger fruit size, greater number of fruit, and yield increases ranging from 10 to 70% averaging 20-50%. Cereal grains, including rice, wheat, barley, oats and rye show yield increases ranging from 25 to 64%. The food crops, corn, sorghum, millet, and sugarcane, show yield increases of 10 to 55%, resulting primarily from superior water use efficiency. Tuber and root crops, including potatoes, and sweet potatoes showed yield increases from 18 to 75 %. Legumes, including peas, beans, and soybeans, show yield increases of 28 to 75 %, with a spectacular increase in nitrogen fixation.

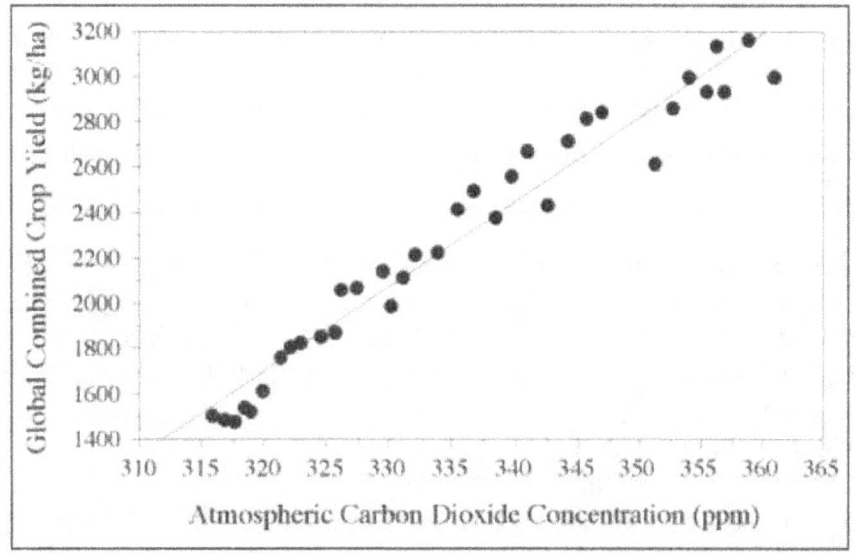

Figure 6. Rising atmospheric CO_2 increases crop yields (UN study graph from Climateresearch.com)

Figure 6 shows during the time period of 1958-96, that for every 1% increase in CO_2, there was a dramatic 8% rise in crop yield. The benefits of CO_2 enrichment in greenhouse yields was first discovered in Germany 100 years ago, and now it is widely used in Sweden, Denmark, Holland, Australia, Japan and the US Typically the nurserymen increase greenhouse CO_2up to 1000 ppm or higher by burning a small amount of liquid propane or natural gas, or they use Bottled Carbon Dioxide

Emitter Systems. CO_2 enrichment is more economical when greenhouse vents can be closed. Hence it is used most often in winter.

Trees and seedlings have shown remarkable growth responses to elevated levels of CO_2. At concentrations of 1000-2000 ppm, black walnut, sugar maple, oak, sweet gum, pine, and eucalyptus seedlings increased dry weight and leaf area by about 80%, height by 90 +%. The forestry department at Michigan State University produced plantable trees in months rather than years using CO_2 concentrations of 1000 ppm. The US Department of Agriculture found that in a two-year period with 650 ppm CO_2, orange trees produced 10 times more oranges than trees grown under ambient CO_2 level of 360 ppm. In 1992 the Finnish Forest Research Institute reported a 25-30 percent increase in the forest of Finland, Sweden, Switzerland, and West Germany between 1971 and 1990 due to increase in atmospheric CO_2 of 9 percent during the same period. A largely unknown phenomenon caused by rising CO_2 levels is that satellite studies have shown the Sahara desert has shrunk since 1980. Sand dunes are retreating, and vegetation ousting sand across a swath of land stretching 6000 kilometers (Pierce, 2000). Other important benefits of elevated levels of CO_2 include: 1) compensating for deficiencies in light during winter months in northern Europe, US, and Canada; 2) protection of plants against both extremely hot and cold temperatures; 3) increasing "biological nitrogen fixation" by legumes' 4) offering protection against air pollutants.

Fossil Fuels vs. Green Energy

Our government's aversion to nuclear energy, and coal is frightening. Coal provides over 50% of our electricity. Coal, oil, and natural gas built our modern economy. During a seventy-year period from 1850 to 1930, we went from horse transportation to planes, trains, and automobiles.

Until the mid 1800s the American life style was similar to "The Little House on the Prairie" with horses and buggies, out-houses, well water, candle or oil lamp lights, back breaking labor to raise crops and primitive health care. In 1859 oil was discovered in Titusville, Pennsylvania and coal was becoming more plentiful. Coal and oil are two of the greatest liberators of all times. They are energy miracles. Carbon-based energy has brought lower infant mortality, longer life expectancy. Electricity powers water to your home, runs our washing machines, air conditioners, telecommunications equipment, dental drills, X-ray equipment and a host of other medical equipment. The modern civilizations of America and Western Europe were built with the high energy density in coal and oil. To give an idea of the high energy density of coal, one ounce of it has enough energy to pull one ton of coal in a railroad car about two miles! Oil has about twice the energy density of coal. One ounce of oil can pull one ton by rail four miles. Oil and coal can economically pull themselves to northern homes in US and Canada and bring the heat of Florida and electricity to these homes.

Over 50% of our electricity is generated by coal. Unfortunately our scientifically and economically challenged government wants to replace the fuels that built our civilization with extremely expensive wind power, solar panels and biofuels. They cannot come close to doing the jobs of the miracle fuels, coal and oil.

How much more expensive is Green Power than fossil fuel power?

In July 2007, the author's home used 1900 KWH of electricity (generated by coal by his electric company). The author's electric bill was $169. The actual cost of the coal to generate 1900 KWH was approximately $28. If natural gas, oil or wood had been used, their approximate costs would have been $27, $48, and $21 respectfully (at 2007 prices). The cost to build a 10 KW windmill to generate the same amount of electricity per month would be approximately $40,000 - $60,000. The wind mill would require a 100 foot tower, with 22 ft. blades, and one acre of land. What about using solar panels? 733 sq. feet of 9 KW panels would cost about $120,000! The actual cost of wind mills and solar panels would be approximately double the quoted numbers above, because the wind doesn't always blow and the Sun doesn't always shine. At present Green Power is vastly more expensive than fossil fuel power. You can't run America on Green Power!

Large-scale wind farms are a waste of money. Everywhere they have been tried, they fail when the government subsidies are removed. They produce very small amounts of electricity at an enormous landscape and environmental cost to game birds and bats. I suppose they will provide some green jobs, by letting men pick up the dead birds. They only produce about 1/6th of their rated capacity due to the variability of wind. They produce zero energy when the wind is not blowing strongly enough. Hence, they need some type of fossil fuel or nuclear power stations to be kept on stream to pick up the load. They are normally located long distances from the user, which requires costly and unsightly transmission lines with large energy losses. Even when the wind is blowing, the fluctuations in the wind destabilize the electric grid and only a small percentage of its total power supply can be safely used. In winter, windmill blades ice up and throw chunks that can kill. They must be shut down and de-iced—more green jobs?

Vaclac Klaus of the Czech Republic reported that to replace their Temelin nuclear power plant by wind, it would take 7750 wind turbine power plants requiring 8.6 million tons of material, and would cover a 413 mile long line of turbines, 492 ft high, corresponding to the distance from Belgium to the Czech Republic. Hence they opted out of windmills. Due to from lack of sales,

England's last remaining windmill company that builds recently closed. Some more green jobholders are probably now unemployed.

Solar panels have many problems. Their efficiency is improving, but they are a long way from providing any meaningful portion of our nations electric needs. They are useful only in sunny climates and on a micro-generation scale (e.g., providing electricity for a hospital refrigerator in Africa). Larger solar collectors built in deserts suffer from extremes of temperatures, exposure to winds and sandstorms. Maintenance costs and transmissions line losses are high and reliability is low. One additional problem in areas of snow, you must shovel the snow off the panels—more green jobs!

Biofuels are for the scientific illiterate. Our government has forced biofuels upon us. Since the initiation of biofuels, the price of staple foods and agriculture land has doubled worldwide, and is causing starvation in places. Even if all of our corn (10.5 billion bushels in 2006) were converted to ethanol, it would only provide 6 percent of our oil needs. Using all of our soybeans would only provide 1.5 percent of our oil needs. More energy is needed to produce a gallon of Ethanol than is in a gallon of gasoline refined from oil! After Congress passed the biofuels bill, it was discovered that CO_2 emissions from the production of most biofuels and production costs were actually greater than that of gasoline! Adding used restaurant cooking oil to diesel fuel can present problems in cold weather. In January 2007, a diesel-powered bus in the middle of the Colorado Rocky Mountains could not start its engine because the fuel mixture congealed. The culprit was the 20% biofuels mixture. The people almost froze to death before being rescued. Governments need to have enough safeguards to be free of scientifically illiterate pressure groups. Using biofuels—forget it!

All of the rhetoric about a shortage of oil is baloney. America has 25% of the world's coal. We are the Saudi Arabia of coal. We can make oil and gas from coal using the Fischer-Tropsch technology used by the Germans in World War II. SASOL in South Africa has produced petroleum from coal for decades, since they have no indigenous petroleum supplies. The German synthetic oil and gasoline were so good that in World War II, as US General Patton's Third Army began outrun their supply trucks, they used the synthetic gasoline from German vehicles and raced ahead. A recent Royal Dutch Shell report indicated that when oil prices exceed $64 per barrel, it is economical to produce oil from coal. The term "we are running out of oil" is obsolete. We can make sulfur-free petroleum in any quantity and any grade we want! In addition, by using fast breeder reactors, the world has enough nuclear fuel to last for thousands of years! Our government needs to abandon their punitive tax schemes on fossil fuels that will kill our prosperity, and start developing coal to oil and nuclear energy programs. The climate extremists have a hatred of fossil fuels and wish to run this country on windmills and solar panels, but it can't be done!

Global Cooling

Does the atmospheric CO2 correlate with temperature? It should if AGW were correct. But Figure 7 shows falling temperatures and rising concentrations of CO_2.

This illustrates that the Obama administration is wrong when they say rising CO_2 levels in our atmosphere cause global warming. Our government has not told the American people that the Earth has been in a global cooling phase since 2000 (Fig.7), which may put millions of acres of croplands in the deep freeze. Our growing population may face a food shortage! Large increases in atmospheric CO_2 can help mitigate the coming food shortage. Stop taxing fossil fuels and stop pumping CO_2 underground. It is insanity!

Prior to the year 2000, Earth had undergone a hundred and fifty years of global warming, not due to man's burning of fossil fuel, but because sunspots and the Sun's magnetic storms were more active than they had been for hundreds of years. During this one hundred and fifty year period, atmospheric CO_2 increased as the warming ocean exhaled CO_2, since CO_2 is less soluble in warm water.

Many qualified climate scientists have noted the earth has been cooling for the past decade. Some are predicting this cooling effect is due to periodic cycles of the solar output and may continue for many decades. If this happens we are going to have less land to produce crops, since many millions of acres of land in the northern regions of North America, Europe, and Asia will be too cold to farm. To improve the productivity of the land that can be farmed we need as much CO_2 in the atmosphere as possible.

Solar cycle 23 is about to end (Fig. 8). It has lasted 13 years, which correlates with very few sunspots, weakened solar magnetic winds and colder weather. Normal solar cycles last eleven years. Shorter cycle lengths result in hotter weather, while longer cycle lengths result in colder weather. Archibald (2008) discovered that solar cycle lengths longer than eleven years reduce the average global temperature by $1.25°$ F per year. Since Solar cycle 23 is thirteen years old, predictions are the next decade to have about a $2.5°$ F global temperature drop. The last half of cycle 23 has switched Earth into a global cooling mode. The Earth experienced a similar dearth of sunspots during the period named "The Little Ice Age" (Fig. 5). This period lasted from 1300 AD until 1850 AD. There were three very cold periods during 'The Little Ice Age" that had extremely low numbers of sunspots and long cycle lengths: 1) the Sporer Minimum, 1425-1575 AD; 2) the Maunder Minimum, 1645-1715 AD; and 3) the Dalton Minimum, 1790-1820 AD. Almost half of the 550 year long "Little Ice

271

Age" were years of extremely low sun spot activity. These were periods where millions died due to famine and diseases.

Figure 7 shows the Earth has been in a cooling mode since 2002, yet atmospheric CO_2 increased during this period. How can this be? Our government insists increasing CO_2 causes the Earth's temperature to increase? This is another example of the scientific foolishness of the global warming theory. The originator of Figure 7 is meteorologist Joe D'Aleo, co-founder of TV's "The Weather Channel."

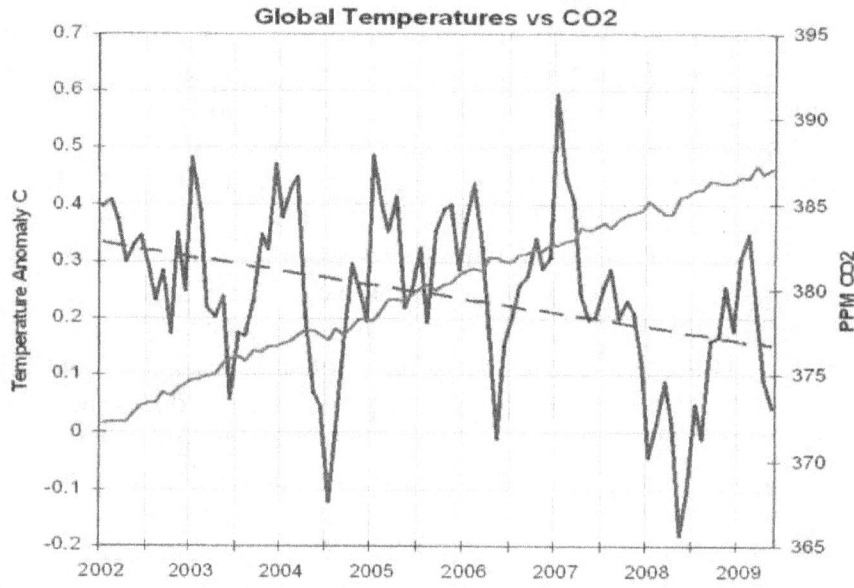

Figure 7. Global cooling has started (J. D'Aleo, www.icecap.us)

Based upon these past historic responses to low sunspot, many scientists now predict similar periods of severe cooling and famines for planet Earth that may last for decades. It would take very little cooling to destroy food crops of Canada, northern U.S, northern Europe, and Asia. Canada would welcome Global Warming! Archibald (2008) has computed the expected drop in global temperature due to the similarity between Solar cycles 22-23 and the solar cycles that started "the Dalton Minimum" of 1790–1820. He then computed the southern movement in the US hardiness zones.

The hardiness zone at the US-Canadian border is about $50°$ north latitude. It may move more than 260 miles south. <u>This could be bad news for Canada</u>. As of June 2009 the Canadians are having some problems getting their crops to grow in the cold weather. Predictions are for only a 50-mile movement south for the hardiness zone along the US southern border.

Our present world population is about six billion and some have predicted that it will grow to 9 billion in fifty years. A worldwide famine would certainly reduce that number. To minimize the deaths due to cooling, our government should be planning for ways to mitigate the effect of cooling on crop production and resulting food shortages. One obvious step is to stop the insanity of their war on CO_2.

During the last two years, we've had record-breaking cold temperatures and snowfalls, and glaciers have started advancing. June 2009 was the eighth coolest June in the last one hundred and thirty years. In early June 2009, snow fell in New Jersey and North Dakota. The lateness of spring, with six foot snow drifts have obliterated the breeding season for geese and other migratory birds over much of the Hudson Bay area. It was even worse than the cold spring of 1962. The past two winters have been so cold that hundreds to thousands of people froze to death in India, China and Afghanistan. Tourist filled a Russian icebreaker expecting to see evidence of global warming were shocked to experience global cooling when they became stuck for seven days in the Arctic ice. We need more CO_2, not less, especially if global cooling continues and freezes useable croplands in Canada, northern US and Europe. CO_2 is aerial fertilizer for plants and trees.

The poor will suffer the most. For the past one hundred and fifty years medical literature has documented that mortality rates are higher in the winter and lowest in the summer. In England 40,000 die every winter due to cold coupled with poor housing, insufficient insulation, ineffective heating and high fuel costs. 8000 more Britons die for each degree the cold dips below the winter average. This past winter was so cold that they expected that 1 in 12 seniors would die. This does not bode well for America if the global cooling continues and the massive Obama taxes on electricity, coal, oil etc. become law.

Governmental Insanity

In addition to rationing fuels, in hopes of reducing atmospheric CO_2, our government plans to pump CO_2 underground (sequestering). In fact they recently started pumping CO_2 underground at an Ethanol plant in Greenville, Ohio. It is extremely expensive to compress CO_2 into a liquid, drill deep holes and pump CO_2 into it. It is insane! It is completely worthless. The Department of Energy estimates it will cost $100 per ton to sequester CO_2 and they are considering sequestering 1.5 billion tons per year. Knowing the government estimates on the costs of new programs are usually too low by factors of 5 to 10, the total tax for this new program is likely to be over $ 1 Trillion! It takes a lot of energy to sequester CO_2. So additional coal must be burned to provide the electricity energy for the pumps to sequester CO_2. The bad news is for every ton of coal burned, three tons of CO_2 is produced and that drives the cost up even higher!

The EPA is considering a tax on the flatulence of cows, sheep and pigs! This could put many ranchers out of business. The EPA claims animal flatulence contains methane, which is causing global warming which will destroy our planet!

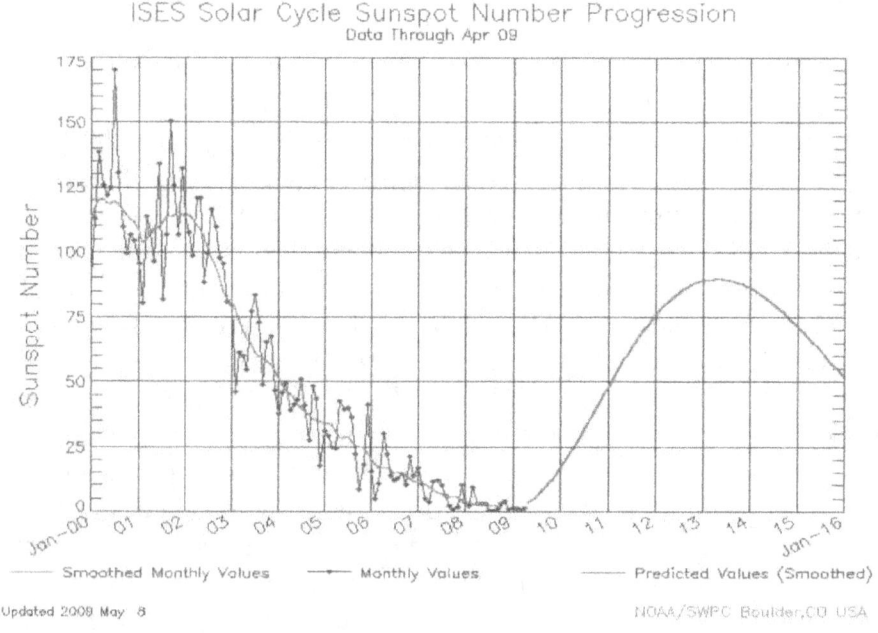

Figure 8. Solar cycle 23-24 progression (Archibald, 2008)

The proposed tax on dairy cows may be $175 per dairy cow, $88 on beef cattle, and $20 on each hog! If this happens we may have to import milk and meat! Unbelievable!

Our government plans to pay farmers for not growing crops, because growing crops adds more CO_2 to the atmosphere! They are also planning on paying foreign governments not to cut down forests so as to not add CO_2 to the atmosphere. Another government insanity is a full employment plan for lawyers: People will be allowed to sue the government if global warming harms them! Truth is stranger than fiction.

The Climate is Always Changing

Some people have suggested that our government is rushing full speed ahead into socialism by passing energy bills based upon the junk science of global

274

warming. Is our government so ignorant of science and history that they do not know that climate has always been changing. Consider the five climate changes that occurred from 1895 –2008:

1. Cooling 1895-1917
2. Warming 1917-1940
3. Cooling 1940-1979
4. Warming 1979-2000
5. Cooling 2000-??

The following was printed in the "Washington Post" on Nov. 2, 1922: Luckily for the American public in 1922, President Warren Hardin did not blame the global warming on all of those "Ford Model T" cars that were crowding our roads, or the coal that was heating homes and generating electricity. Harding did not blame fossil fuels and the "bad molecule CO_2" for melting icebergs and call for massive taxes on them. Fortunately he did the right thing…nothing! Eighteen years later, a global cooling period (1940-79) occurred. We can avoid an economic disaster now, if our government does nothing.

An article from the Russian newspaper *Pravda*, written in the spring of 2009 by Stanislav Mishin entitled, "American Capitalism Gone with a Whimper," accused President Obama of pulling the plug on the American way of life. Mishin wrote: "It must be said, that like the breaking of a great dam, the American descent into Marxism is happening with breath-taking speed, against the back drop of a passive, hopeless sheep, excuse me dear reader, I meant people … First, the population was dumbed down through a politicized and substandard education system based on pop culture, rather than the classics. Americans know more about their favorite TV dramas than the drama in DC that directly affects their lives. The final collapse has come with the election of Barack Obama. His speed in the past seven months has been truly impressive. His spending and money printing has been record setting, not just in America's short history, but also in the world. If this keeps up for more than another year, and there is no sign that it will not, America at best will resemble the Weimar Republic and at worst Zimbabwe. Prime Minister Putin, less than two months ago, warned Obama not to follow the path to Marxism, it only leads to disaster"

It appears that our government is becoming more and more radical and we are witnessing the collapse of the once greatest nation on Earth. Pray that God will once again smile on America. *"If my people, which are called by my name, shall humble themselves, and pray, and seek my face, and turn from their wicked ways; then will I hear from heaven, and will forgive their sin, and heal their land"* 2 Chronicles 7:14

References

Archibald, David, (March, 2008) "Solar Cycle 24: Implications for the United States", david.archibald@westnet.com.au.

Beck, Ernst-Georg, "180 Years of Atmospheric CO_2, Gas Analysis by Chemical Methods" Energy & Environment, Vol 18, No 2, 2007.

Callendar, G. P. "On the Amount of Carbon Dioxide in the Atmosphere," Tellus 10: 243-48, 1958.

Endersbee, L. "Global Climate Change has Natural Causes", Science, March 7, 2008.

Idso, Sherwood B. "CO2-Induced Global Warming: A skeptics View of Potential Climate Change", Climate Research, Vol. 10: 69-82, 1998.

Jaworowski, Z., Segalstad, T.V. & Hisdal, V.; Atmospheric CO_2 and Global Warming: a Critical Review, 2nd revised edition, *Norsk Polarinstitutt, Meddel*elser [Norwegian Polar Institute memoirs] 119, 76 pp. (1992 a).

Jaworowski, Z., Segalstad, T.V. & Ono, N. "O Glaciers Tell a True Atmospheric CO_2 Story"? Science of the Total Environment 114 227-284 (1992 b).

Jaworowski, Z., "CO_2: The Greatest Scientific Scandal of Our time" Science, March 16, 2007.

Keeling C. D., From, E., "Reassessment of Late 19th Century Atmospheric CO_2 Variations", Tellus, 38B 87 105, 1986.

Kiehl & Trenberth, "Earth's Annual Global Mean Energy Budget" Bull. Amer. Meteor. Soc. **78**, 197-208 1997.

Pierce, F. "Africans Go Back to the Land as Plants Reclaim the Desert", New Scientist 175, 21 September 2002, pp 4-5).

Revell, R & Suess, H. "Carbon Dioxide Exchange Between Atmosphere and Ocean and the Question of an Increase of Atmospheric CO_2 During Past Decades" Tellus 9, 18-27.

Segalstad, T.V., The Distribution of CO_2 between Atmosphere, Hydrosphere, and Lithosphere; Minimal Influence from Anthropogenic CO_2 on the Global "Greenhouse Effect", 1995, printed in "The Global Warming Debate", European Science & Environmental Forum, 1996.

Segalstad, T.V., "Carbon Cycle Modeling & Residence Time of Natural & Anthropogenic Atmospheric CO_2; on the Construction of the "Greenhouse Effect Global Warming Dogma", 1998, printed in "Global Warming: The Continuing Debate", ESEF, 1998.

Segalstad, T.V., "What is CO_2—Friend or Foe?" invited lecture KTH International Climate Seminar, Stockholm, Sept. 11, 2006.

Smika, Darryl, retired U.S. Dept. of Agriculture, Private Conversation with the Author, August, 2007.

Van de Plassche & van der Borg, Sea Level-climate Correlation During the Past 1400 yr., Free university Amsterdam & Utrecht Univ., http://www.fys.ruu.nl/-adejong/radiocar-bondating/Sea-level-climate_correlation.htm.

Author bio:

Edward Blick has three degrees from the University of Oklahoma, a BS in aeronautical engineering, MS aeronautical engineering and a Ph.D. in engineering science. He served four years in the US Air Force as a Weatherman. His industrial experience includes working at McDonnell Aircraft as an aerodynamicist on the Mercury capsule, our first space craft. He also was a space engineer for Lockheed Missile & Space Co. working on the "Polaris Reentry" missile.

He taught and performed research at the University of Oklahoma from 1958-2007 in four areas, School of Aerospace, Nuclear & Mechanical & Nuclear Engineering, School of Meteorology, School of Medicine, and School of Petroleum and Geological Engineering. He has written and published 135 technical papers, one book on the Bible and Science, and two engineering textbooks. He also served as the Associate Dean of Engineering. For the past 45 years he has enjoyed presenting slide shows on the "Bible vs. Evolution" and during the last four years, "The Global Warming Lie" You may email Dr Blick at: edblick@cox.net.

Section
Behavioral Science

Behavioral Science Introduction
Denyse O'Leary

Introduction

Neuroscience is not mainly about neurons; it is mainly about how we use them. The advent of new means of observing the mind at work, such as functional magnetic resonance imaging (fMRI) and electroencephalography (EEG) has greatly helped physicians and surgeons. It also became a focus for renewed debates about the mind and the brain. Is the mind just the dance of neurons in the brain? Is the brain the organ by which the immaterial mind communicates with the world?

How people answer such questions depends on their worldview. A materialist atheist such as genome expert Francis Crick announced with confidence on page 1 of *The Astonishing Hypothesis* (1994), that "You are nothing but a pack of neurons" On the other hand, Montreal neuroscientist Mario Beauregard and Toronto journalist Denyse O'Leary decided that the mind is an immaterial entity that communicates with the brain (*The Spiritual Brain*, 2007). Well, who is right?

The answer matters a great deal to you and your family. Are criminals responsible for their behavior? Neurolaw, a growing movement in law argues that they are not. How will that affect you, if someone you know should become a victim of crime? Darwinists teach that our brains are adapted for fitness, not for truth, with predictable results.

Darwinian thinking is reaching out to medicine as well. For example, does intensive care really help people? In 2010, British popular science magazine *New Scientist* published a piece in favor of "Darwinian medicine" which argued that it does not help much. That flies in the face of the established fact that countries that provide intensive care display dramatically longer life expectancies than those that do not. How might this affect your health care?

In her chapter, "Does your faith heal you? Yes, actually," Toronto journalist Denyse O'Leary outlines how those "nothing but a pack of neurons" theories of mind are not being confirmed by research evidence, despite massive effort. There is no credible materialist explanation of consciousness—the "I" interface between the mind and the brain—either, and none on the horizon.

O'Leary also talks about the importance of mental attitude to getting well. This is not a "New Age" belief, despite New Agers' attempts to co-opt it. Christians can refer to Luke 7:1–10 for an example of a Roman centurion who simply decided to ask Jesus to heal his servant, on the point of death, and he succeeded! One of her key messages is that a spiritual life is good for you, and research has consistently shown that people who stay in touch with God live longer—except in the unusual case where they think God is "out to get" them.

Now, one key area of Darwinian atheist materialism is the question of whether we are defined by our genes. This claim has a major impact on debates around homosexuality. Are people "born gay"? Is there a "gay gene"? Or is homosexuality a sexual and, possibly, lifestyle choice?

There is a distinction worth making here first: Some people are simply quietly homosexual, others participate in a widely advertised "gay lifestyle" The former usually just ask others to leave them alone; the latter may encourage gay lobbies, intent on using tax-funded school systems and other venues to promote a gay agenda.

The reason it all matters is easy to understand. If being born gay is the same thing as being born with hazel eyes, then people might reasonably conclude that marriage laws should be changed to permit same sex marriage, and children should be taught early in school that gay is okay, and perhaps even fun. Naturally, gay lobbies promote the "gay gene" view.

And with what consequence? David Williams sets out the consequences from a parent's and a teacher's perspective. And they are not pretty, especially because the gay outreach to the schools is tax funded and student attendance is mandatory. Students may be taken to gay weddings. Parents may be jailed for asking for their children to be opted out of the ideology. In society at large, people who sincerely want to lose their interest in same sex attraction may be unable to find a therapist who is allowed help them.

Denyse O'Leary is personally aware of at least one case in which a man left the gay community, in which he had lived for many years, and got married to a woman, and is quite happy with the arrangement. But these situations are not rare. The question is will people feel free to seek help to opt out of the gay movement or opt their children out of it, without conflict with the government? At this time, it is unclear, unless more people speak up.

Dennis Jernigan writes on the fact that no one is born gay. He speaks of his personal experiences eloquently, and points out that there is nothing unusual about the decision of hundreds of people with whom he has been in contact to just walk away from the gay lifestyle. Given that there is no serious evidence for a "gay gene," this chapter should give considerable reassurance to parents, teachers, and religious counselors who assist people who want to change their lives.

Emanuel Tundrea talks about the threat to us all from Internet pornography. He points out that the Internet is a mixed blessing. It has liberated hundreds of millions of people from dependence on the gatekeepers of legacy media, who often slanted their "news" toward materialism. On the other hand, it enabled pornographers to reach a much wider public, likely creating many more addictions. What happens, he asks, when a spouse gets a bill for pornography services? Good question and the answer could well be captured in seven letters: D-I-V-O-R-C-E.

The basic message of this section is, yes, we do have minds and we do have free will, and we are responsible for our choices. Read, enjoy and take heart!

Does Your Faith Really Heal You?Yes, Actually It Does!

Denyse O'Leary

Editor's note

Denyse is a Canadian journalist. We met online several years ago and have stayed in touch. She is also an evolution troublemaker extraordinaire and a good friend. She introduced me to an online group that deals with intelligent design and related issues. Through that group, I have made many friends around the world with similar interests including several who have written chapters in this book. As is the case for many of us who accept a supernatural creation, she has been attacked and harshly ridiculed for her faith. She has bravely stood her ground and defended her belief with extraordinary finesse. For that too I respect her and am happy to call her friend. She writes with uncanny clarity. Like a good teacher, she makes difficult topics easy to comprehend.

Introduction

A long, slow change has been occurring over the years in our understanding of faith's relation to health, which—despite much controversy and opposition—parallels a small but growing movement in neuroscience that affirms the reality of the human mind. The two ideas are closely linked. If the mind is an illusion created by the activity of neurons, it causes nothing. If it is an immaterial reality, it may well direct the brain through its focus of attention, and that can impact health.

Those who hold that the mind is an illusion insist that they have proven their case. As one put it, "Despite having been comprehensively discredited by philosophers and scientists, mind-body dualism is an infuriatingly sticky idea" (Gefter, 2009). But when an idea is infuriatingly sticky, it could just as easily be right as wrong. Often, there are good reasons for stickiness. What does the evidence suggest?

Here are five key factors to consider:

1. *Metaphysical naturalist theories of mind are not being confirmed in science.*

To explain this development, the author invited readers of a popular culture magazine, *Salvo,* to consider just for fun a timeline of epoch-making science news that truly supports the metaphysical naturalist perspective:

1995: A single gene switch is discovered that causes people to be gay or straight. Shortly afterward, the single switches that cause people to be faithful or unfaithful partners, and liberal or conservative voters are also found.

1998: Nobel Prize awarded for discovering the God gene, which causes people to be unthinkingly religious.

2000: A rogue module is found in the brain that controls the sense that one is an "I" rather than an "it" Everyone is really an "it" after all. The "I" delusion is a widespread, inherited—but treatable—glitch.

2007: Most people worldwide agree that belief in God is harmful because it leads to bad ideas like right and wrong, the harmful practice of accepting responsibility for one's own actions, and the even more deplorable practice of teaching children to do the same. Billions of people demonstrate, urging their governments to abolish traditional religions and take over the intimate details of their lives to reduce the harm done (O'Leary, 2009).

Obviously, none of this happened. What did happen? Among other things, genes did not provide a comprehensive explanation for human behavior; the hard problem of consciousness (the "I") proved very hard indeed; and most people worldwide still believe in God (or karma, which requires that the universe be organized in a meaningful way). Despite the enthusiasm of their supporters, the computer model of the brain and efforts to show that chimpanzees think like humans have fallen on hard times as well (O'Leary, 2007).

2. *The metaphysical naturalist theory of mind has no credible explanation of consciousness.*

Metaphysical naturalist theory requires that consciousness be an illusion created by neurons at work, but the explanations have been unsatisfactory, and sometimes implausible. Consider the problem of qualia, for example—general agreement on an overlapping range of meaning for concepts like "red", which have many associations, often opposite ones (Greenfield, 2005). Then how do we communicate, given that no one accesses anyone else's consciousness? But qualia are only one aspect of the "hard problem" (Phillips, 2004) of consciousness. Physicist Nick Herbert writes, "Science's biggest mystery is the nature of consciousness. It is not that we possess bad or imperfect theories of human awareness; we simply have no such theories at all. About all we know about consciousness is that it has something to do with the head, rather than the foot" (Radin, 1997).

That may be a bit of an exaggeration, but not much. It is not just that metaphysical naturalist theorists have no workable science model for consciousness but that they have no useful idea how to acquire one. Current efforts often amount to one of three things: simple appeals to promissory materialism (Edelman and Tononi, 2000), Karl Popper's phrase for the belief that metaphysical naturalist approaches will eventually yield answers, given an indefinite lease; claims that consciousness does not really exist, i.e. Francis Crick, "You are nothing but a pack of neurons" (Crick, 1995); or implausible models. Neuroscientist V. S. Ramachandran provides an example of the third, when he suggests that the concept of the self may have arisen from an evolutionary necessity to guess others' intentions: "Our brains were essentially model-making machines. We need to construct useful, virtual reality simulations of the world that we can act on. Within the simulation, we need also to construct models of other people's minds because we're intensely social creatures, us primates. We need to do this so we can predict their behavior. We are, after all, the Machiavellian primate" (Ramachandran, 2003). But surely this gets it exactly backward. We were, and are, sure that we are conscious, so we infer that others are, and we attempt to predict their behavior on that basis.

It doesn't help when cognitive psychologist Steven Pinker muses, "Our brains were shaped for fitness, not for truth. Sometimes the truth is adaptive, but sometimes it is not" (Pinker, 1997). Few scientists would consider their profession worthwhile if they could never be sure whether they had discovered facts or simply iterated the adaptations of their own brains. In any event, metaphysical naturalist doctrines of consciousness often sound as though their aim is more to preserve metaphysical naturalism than to account for consciousness.

3. *Third, the accepted model of the brain did not survive the Decade of the Brain (1990–2000).*

After seventy years of neuroscience proclaiming that the brain was static and brain cells could not regenerate, findings from the 1990s (Changeux, 1985) and onward have conclusively shown that the brain is highly dynamic and neurons can regenerate (*ScienceDaily,* July 2, 2001). As neurobiologist Michael Friedlander puts it, "One of the fundamental findings of the last decade is the plasticity of the synapses—their ability to alter their strength in response to experience and the context of a situation. As this happens, the synapses are actually changing shape—getting fat, getting short, becoming concave or convex, forming mushroom shapes. We knew this happened in the developing brain, but we didn't know that as adult brains think and learn it happens dynamically too" (Yount, 2003).

Neuroplasticity is of considerable interest in the treatment of strokes, for example, because it means that stroke victims are not necessarily consigned to involuntary paralysis (Doidge, 2007). But neuroplasticity is easier to account for if an immaterial mind directs the brain, according to a subjective "mind's I" consciousness of experiences. Wanting to explore the issues, the Nour Foundation, ECOSOC, and the Université de Montréal recently sponsored a conference called Beyond the Mind-Body Problem: New Paradigms in the Science of Consciousness (www.mindbodysymposium.com, 2008). The focus of the conference was on effective resolutions of the problem for medicine. The speakers included such well-known non-materialists (shorthand for metaphysical naturalism) as Sam Parnia, Bruce Greyson, Mario Beauregard, and Jeffrey M. Schwartz.

This conference prompted concern at Britain's *New Scientist* magazine, which has crusaded for metaphysical naturalist interpretations of mind for years. A few weeks later, *New Scientist* published a story that implied that the conference was somehow associated with the Seattle based intelligent design think tank, Discovery Institute (Gefter, 2008). Most observers doubt that the UN and the Discovery Institute are on such good terms.

The Université de Montréal's Mario Beauregard patiently wrote *NS* a letter pointing out that, "Most participants in the 11 September symposium "Beyond the Mind-Body Problem: New Paradigms in the Science of Consciousness" at the United Nations were medical doctors or neuroscientists who work with them. We do not question materialist models of the mind-brain complex merely for ideological or political reasons. We want to move beyond them because we have not found them adequate explanations of mind-brain interactions, nor do they point to useful treatment plans" (Beauregard and Schwartz, 2008).

It might be worthwhile here to review the recent history of the mind-brain-body relationship, for medical science purposes. For much of the twentieth century, the problem was simply avoided. No consensus emerged from the many theories on offer (Edelman and Tononi, 2000). Mid-century behaviorist psychology ruled out the discussion of mental events as such. Behavior was to be studied in terms of stimulus and response, dismissing the question of mind (Skinner, 1971). Later, with the development of cognitive psychology in the 1950s, the computer became the favored model for mind. Strangely, in many quarters, there was only limited discussion of the serious conceptual problems of this "machine" model. Artificial intelligence enthusiast Ray Kurzweil spoke for many when he predicted in *The Age of Spiritual Machines*, "The machines will convince us that they are conscious, that they have their own agenda worthy of our respect ... They will embody human qualities and will claim to be human.

And we'll believe them" (Kurzweil, 1999). Another model gained popularity in the 1970s—the "Machiavellian primate" or "98% chimpanzee" approach—which sees the human mind as a souped-up version of primate minds, and assumes that the illusion of consciousness or self arose as an adaptation during the course of evolution" (Ramachandran, 2003 and Smith, 2005).

Models that dismiss the human mind as an irrelevance or illusion can be understood in relation to the inspiring history of nineteenth and twentieth century success in medicine. For example, vaccinations prevent diseases and antiseptics kill microbes irrespective of patient beliefs. Ignoring the patient's mind came to seem "scientific" By the 1930s, the Index Medicus contained no reference to the effect of mental states on physiology (Benson and Stark, 1996). In the 1960s, physician Herbert Benson had a hard time persuading colleagues that mental stress could contribute to high blood pressure. When he encouraged patients to meditate, mentors warned him that he was risking his career (Benson and Stark, 1996). In reality, however, all eliminative models have serious conflicts with evidence. The view that they are the only "scientific" way of viewing the mind was due for a challenge, and an early, significant challenge came from medicine.

Most of us have experienced the impact of the mind changing the brain in a commonplace situation known to physicians as the "placebo effect" A woman in pain and discomfort from the flu makes an appointment and, while sitting in a doctor's waiting room, she begins, embarrassingly, to feel better. Is she neurotic or delusional? Maybe not. It could be the placebo effect, the significant healing effect created by a sick person's belief in a powerful remedy. In this case, the first remedy is her belief that the doctor knows what to do and will give her a prescription. She has been told by trusted friends that he is a good doctor and comes highly recommended. This immaterial information enables her mind to start the healing process even before the prescription is given, let alone taken— but only after the conviction is established that it *will be* given, because she is sitting there in his office.

For millennia, doctors have given placebos, knowing that they often help when all else fails. Doctors are also taught a reassuring manner that itself functions as a placebo. Indeed, J. N. Blau has written, "The doctor who fails to have a placebo effect on his patients should become a pathologist" (Thompson, 2005). A new drug's effectiveness is routinely tested against placebos in controlled studies, not because placebos are useless but precisely because they are useful. Placebos usually help a percentage of patients enrolled in the control group of a study, perhaps 35 to 45% (Greenberg, 2003).

In 2005, *New Scientist*, hardly known for its support of non-materialist theories of mind, listed "13 Things That Don't Make Sense," and the placebo effect was number one on the list. It is easy to see why it would not make sense in a closed metaphysical naturalist system. In one study, for example, the tremors of Parkinson's disease (Colloca, 2004) were eased by a placebo (saline solution). It was established that neural activity associated with tremors declined as the symptoms decreased, so the patients were not simply confabulating that they felt better. The belief that they had received a powerful medication had triggered the release of dopamine in their ailing brains.

Other studies of Parkinson's disease have shown similar results. Raül de la Fuente-Fernández and colleagues reported in 2001 that "our results suggest that in some patients, most of the benefit that is assumed to be obtained from an active drug might derive from a placebo effect" The researchers had observed from PET scans that the placebo effect in Parkinson's patients was mediated through activation of the damaged nigrostriatal dopamine system (Fuente-Fernández et al., 2008). The significance of the placebo effect is that the patient's belief, or focus of attention, triggers the effect. In short, not only does the mind change the brain (and body) in these cases, but, in the Parkinson's studies noted above, the changes were observed by neuroscientists. A medically useful model of the relations between the mind and the brain can assume that the mind is real.

4. Treatment methods that assume the reality of the mind work.

Non-materialist neuroscience is gaining ground primarily through useful treatment plans. Berkeley neuropsychiatrist Jeffrey Schwartz's mindfulness-based treatments for obsessive-compulsive disorder (Schwartz and Begley, 2003), for example, or Vincent Paquette's demonstration that phobias can be cured through focused mental effort (Paquette et al., 2003) are examples, but others are being explored. For an aging population dealing with chronic, long-term disorders, this is welcome news. There may be no "magic bullet" for such patients, but they may benefit from awareness that their focus of attention makes a difference. There are exceptions, of course; it does not appear to work with cancer, except to improve pain control and appetite (Temple, 2003), or with Alzheimer, if declining cognition robs the patient of the ability to expect a proven painkiller to work (Neergaard, 2005).

A brief look at Jeffrey Schwartz's work offers an insight into mind-based treatment. Obsessive-compulsive disorder (OCD) is a neuropsychiatric disease marked by distressing, intrusive, and unwanted thoughts (obsessions) that trigger an urge to perform ritual behaviors (compulsions) like constant, abrasive hand washing. As a practitioner of Buddhist mindfulness, Schwartz saw OCD as a good candidate for a non-pharmaceutical approach to treatment. That is because OCD sufferers are not delusional. They actually *know* that their beliefs are mistaken and their compulsive activities are useless. But they do not know how to stop them. Yet, giving in makes the sufferers worse over time. The more they give in, the more persistent the beliefs and behaviors become. It is as if their brains have been hijacked. During most of the twentieth century, OCD was little understood and considered very difficult to treat.

Schwartz used neuroscience techniques to identify the cause of the disorder. Specifically, the cause is most likely a defect in the neural circuitry connecting the orbitofrontal cortex, cingulated gyrus, and basal ganglia, from which panic and compulsion are generated. When this "worry circuit" is working properly, we worry about genuine risks and feel the urge to reduce them. But, Schwartz found, when that modulation is faulty, as it is when OCD acts up, the error detector can be over activated. It becomes locked into a pattern of repetitive firing. The firing triggers an overpowering feeling that something is wrong, accompanied by compulsive attempts to somehow make it right. But the attempts only create the demand for more attempts because new attempts cause the circuit of neurons to grow.

He then developed a four-step program (Relabel, Reattribute, Reassign, and Revalue) to help patients identify and reassign OCD thoughts, until they diminished in severity. Schwartz was not simply getting patients to change their opinions, but to change their brains. Subsequent brain imaging showed that the change in focus of attention substituted a useful neural circuit for a useless one. For example, it substituted "go work in the garden" for "wash hands seven more times" By the time the neuronal traffic from the many different activities associated with gardening began to exceed the traffic from washing the hands, the patient could control the disorder without drugs (Schwartz and Sharon Begley, 2003). The mind was changing the brain.

Here, as with the placebo effect, the active ingredient is the focus of the mind. With the placebo effect, the focus is on a belief; with OCD treatment, it is on the deliberate choice of an alternative activity.

5. *Faith or spirituality is good for one's health.*

British theologian Alister McGrath asks, countering Richard Dawkins, "So just what is the experimental evidence that God is bad for you? Recent empirical research points to a generally positive interaction of religion and health" (McGrath, 2005).

Why have so many doubted this fact, now widely attested in studies? One confounding factor was the way in which early research into spiritual influences on patients was conducted. For example, patient histories that ask merely for a religious affiliation cannot distinguish "intrinsic" from "extrinsic" faith. Pioneer sociologist of religion Gordon Allport defined intrinsic faith as internalized experience; extrinsic faith expresses group membership. This distinction is critical where health is concerned, because identified health benefits come mainly from intrinsic faith (Sabom, 1998). In addition, sophisticated instruments for measuring attitudes were rarely used in the early research on faith and health. Also, study samples were often unrepresentative of the general population (Levin and Koenig, 2005).

Dr. David B. Larson (1947–2002), an epidemiologist, psychiatrist, and also a devout Christian found considerable bias in diagnostic criteria as well. The *Diagnostic and Statistical Manual of Mental Disorders (DSM-III)*, for example, used many case examples that characterized religious patients as "psychotic, delusional, incoherent, illogical, and hallucinating," suggesting a general psychopathology that misrepresented clinical experience. When that edition of the manual was in use, only 3 of 125 medical schools in the United States provided any instruction on the relationship between health and spirituality—in a nation where roughly one third of the population claimed to have had a religious experience. Despite his sudden, untimely death in 2002, Larson played a key role in helping to revise the *DSM-III*, to eliminate the (perhaps unintentional) anti-religious bias. And, thanks in part to his work with the Templeton Foundation, nearly two-thirds of medical schools now offer course work relevant to spirituality (Levin and Koenig, 2005).

In the 1980s, together with Jeff Levin and Harold Koenig, Larson pioneered an evidence-based approach to the relationship between faith and health. He developed a "systematic review" method that avoids selection bias by looking at every quantitative article published during a given number of years in a single journal. This method provides a comprehensive survey of findings that is both objective and replicable. In *The Faith Factor: An Annotated Bibliography of Clinical Research on Spiritual Subjects*, Larson, Dale Matthews, and Constance Barry conducted a detailed review of 158 medical studies on the effects of religion on health, 77 percent of which demonstrated a positive clinical effect (Sabom, 1998). Of course, one must qualify. One study shows that people who believe that God wants to punish them with illness do worse than people who believe that God wants them to get better (Pargament et al., 2001). Few clergy would be surprised to hear that.

Clearly, it makes a difference whether belief creates hope or despair. But does it make a difference what God one prays to? Dale Matthews, a physician associate of Larson, notes: "While science has demonstrated that being devout provides more health benefits than not being devout, we haven't shown that being a devout Christian will make you healthier than being a devout Buddhist" That does not make theology irrelevant (Sabom, 1998); it implies rather that the health effect of faith derives less from specific doctrinal beliefs than from resulting positive mental states. The research results do not, of course, substantiate extravagant notions that "faith surely heals," let alone that medicine is superfluous. They demonstrate only that choices in mental attention are important in maintaining and restoring wellness.

In recent years, the discussion has become much more focused. The question, "Does spirituality make any difference?" is giving way to "Under what circumstances does spirituality make a difference?" Some interesting recent research includes evidence that patients often want their doctors to know about their spiritual beliefs and take them into account (McCord, et. al., 2004), as well as evidence that doctors themselves are more likely to have spiritual beliefs than academic or research scientists (Curlin, et. al., 2005).

But isn't this dualism?

The non-materialist perspective is often derided as "dualism" If so, what's wrong with dualism, apart from the fact that it is fatal to metaphysical naturalism? We know only a small part of nature, so we cannot know all it contains, permits, or can tolerate. In practice, metaphysical naturalism has the effect of dismissing evidence for the reality of the mind or revelation from God, among other things, without the need to take the contrary evidence seriously. If we decide in advance that something simply cannot be true, then no evidence for it can be accepted. A materialist explanation that is clearly questionable must be accepted instead. One thinks of the many efforts of evolutionary psychologists to "explain" faith and spirituality (Boyer, 2001; Dennett, 2006; Wilson et. al., 2003; Wilson, 1998). Whether considered adaptive or not, the fundamental message is that faith and spirituality are not the result of revelation. But the only way one could know that is by positing in advance that there is no source of revelation. Or that methodological naturalism requires science to work with unprovable speculations about the Old Stone Age instead of considering the evidence in the present day.

Evidence is precisely the issue. The argument is changing, because the weight of the evidence does not consistently support metaphysical naturalist views of the mind. The frequently heard dramatic claims from its proponents, expressed in books and articles, are not evidence. And, as metaphysical naturalism is a total system, even one consistent contrary example would be fatal.

The role of the new atheism

A reader may well ask, doesn't the steady stream of "new atheist" books endorsing metaphysical naturalist perspectives show that the world is converting to their views? It is true that in recent years they have poured forth from the presses, to rave reviews in key papers. One reason is cultural. A literary agent told this author recently that decisions to sign books typically originate in Manhattan or similar venues, where the authors' views are taken to be what all reasonable people believe. The rave reviews from similarly sympathetic sources generate brisk sales. But, in a country where the Pew Forum on Religion and Public Life estimates that only 1.6% of the population are atheists, rave reviews can do only so much (McCane, 2009). Bryon R. McCane, Professor of Religion at Wofford College, asks, "Has something gone wrong with the new atheism? For awhile, it was really on a roll. Several best-selling books aggressively attacked religion, calling it a "delusion" (Richard Dawkins), and a "spell" (Daniel Dennett) that "poisons everything" (Christopher Hitchens). Bill Maher's movie "Religulous" warned that humankind must get rid of religion or die. New atheism looked like the wave of the future. But not anymore. "Religulous" got mixed reviews and disappeared quickly. Rebuttals to Dawkins, Dennett and Hitchens have appeared, culminating with Karen Armstrong's new book, *The Case for God*. Sales of atheist books have fallen off the charts, literally. Months have gone by since one appeared on the best-seller list (McCane, 2009).

McCane argues that the new atheists made several key errors. They vastly overestimated their own numbers, incorrectly assuming that the unchurched were mostly atheists. He feels that they also overestimated the influence of books on "science and logic" But their biggest mistake, in his view, was to be openly intolerant of religion in a society overwhelmingly committed to religious tolerance.

He is surely right about the intolerance. The public quickly realized that the new atheists offered no new message from atheism. Their key innovation was an attempt to make virulent detraction of theists respectable among scholars. Such detraction is hardly new, but formerly it was the village atheist's approach, not the scholar's. Despite new atheist claims, most observers doubted that a better society would result from all this venting of hostility; rather, it suggested an underlying insecurity.

One element is missing from McCane's analysis: growing evidence that the materialist point of view in general is a poor fit with reality. Large numbers of well-educated people know or suspect this fact. The "new atheist" books are *not* a good advertisement for either science or logic. Increasingly, they have the aura of a fad.

Fads fade. Popular media is not, of course, science. But by the time a concept is familiar enough to feature in popular media—be it the dangers of a high fat diet or the Big Bang theory—we can assume it is taking hold. And metaphysical naturalist doctrines of the mind are beginning to be scrutinized with more suspicion than formerly. For an example of the change, consider that in 1995, Michael D. Lemonick offered a typical "mind as illusion" pronouncement in *Time* Magazine: "Utterly contrary to common sense … and to the evidence gathered from our own introspection, consciousness may be nothing more than an evanescent by-product of more mundane, wholly physical processes (Lemonick, 1995).

But in 2009, Sharon Begley offered a blistering attack on many key favorite theses in evolutionary psychology in *Newsweek*. Her approach was picked up by David Brooks in *The New York Times* (Brooks, 2009). And this development was noted in a *Scientific American* podcast with supportive rather than alarmist comment (Nicholson, 2009). Not all evolutionary psychologists may consider the mind an illusion. But their discipline is based on the assumption that behavior is the outcome of evolution—either successful selfish genes or their free riders. Efforts are made to explain a remarkable variety of behaviors in this way (O'Leary, 2009). For example, in one recent study involving lost wallets, the authors concluded that an evolutionary program explains why people return lost wallets. They discount assumed traditional drivers like rational judgment, free will, and moral choice, concepts which assume that the mind is not an illusion (Devlin, 2009). However, the fact that the evolutionary psychology thesis can even be questioned in the public square without provoking an angry backlash suggests a change in the wind.

Are we in the midst of a changing paradigm? It's hard to be sure. There have been many serious critiques of metaphysical naturalism over the years; one thinks of John Eccles, Wilder Penfield, and Alister Hardy if troublesome evidence accumulates, newer, younger scientists may be less willing to accept the doubtful explanation than older ones whose careers were fostered by it. However, prediction is very easy, except where the future is concerned.

Denyse O'Leary is co-author of *The Spiritual Brain: A neuroscientist's case for the existence of the soul* (Mario Beauregard and Denyse O'Leary, HarperOne, 2007).

References:

Amanda Gefter, Review of What's Next? edited by Max Brockman, *New Scientist* 25 August 2009 Magazine issue 2722.

Denyse O'Leary, "Deprogram—Trust Brain: Thinking About Thinking Might Do a World of Good, Even at the UN", *Salvo 8* Spring 2009.

For a discussion of these areas, see Mario Beauregard and Denyse O'Leary, *The SpiritualBrain*, (San Francisco: Harper One, 2007).

Amy Butler Greenfield wrote a 300 page book on the various, contradictory meanings of "red" and their significance, *A Perfect Red: Empire, Espionage, and the Quest for the Color of Desire* (New York: HarperCollins, 2005). One could as well write a book about "green" or "blue."

The term "hard problem" is generally recognized. For example, Helen Phillips writes in *New Scientist* ("The Ten Biggest Mysteries of Life", September 4–10, 2004) "Why should the activity of a mass of neurons feel like anything? Why does pricking your finger feel like pain? Why does a red rose appear red? This has been dubbed the "hard problem" of consciousness."

Quoted in Dean Radin, *The Conscious Universe: The Scientific Truth of Psychic Phenomena* (San Francisco: HarperSanFrancisco, 1997), p. 265.

In *A Universe of Consciousness: How Matter Becomes Imagination* (New York: Basic Books, 2000), Gerald M. Edelman and Giulio Tononi provide a good example.

Francis Crick, *The Astonishing Hypothesis: The Scientific Search for the Soul* (New York: Simon & Schuster, Touchstone, 1995), p. 258.
V. S. Ramachandran, Reith Lectures, Lecture 5 2003; available online at http://www.bbc.co.uk/radio4/reith2003/.

Steven Pinker, *How the Mind Works* (New York: Norton, 1997), p. 305.
See, for example, Jean-Pierre Changeux, *Neuronal Man: The Biology of Mind,* trans.
Laurence Garey (New York: Oxford University Press, 1985), p. 282: "Of all the organs of the body, the nervous system is unusual in that its total number of cells is fixed at birth. Any neurons that are destroyed are never replaced.... The possibility of restoring function is quite high in the young but gradually declines with age"
See, for example, "Manipulating A Single Gene Dramatically Improves Regeneration In Adult Neurons: Finding May Lead To New Approaches For Treating Brain And Spinal Cord," *ScienceDaily* (July 2, 2001).

Quoted in Kathleen Yount, "The Adaptive Brain," UAB Publications, Summer 2003. This article,which won the Robert G. Fenley award for basic science writing (2004), provides a good (and relatively nontendentious) discussion of some implications of neuroplasticity for medical treatment.

For an account of the implications of neuroplasticity in the treatment of stroke victims, see Norman Doidge, *The Brain That Changes Itself* (New York: Penguin Group, 2007). Much paralysis was found to result, not from irreparable brain damage, but from "learned helplessness"; the failure to use a limb diminished its neural networks. In many cases, the process was reversible with prompt treatment.

September 11, 2008. Most of the proceedings are online at http://www.mindbodysymposium.com/Beyond-the-Mind-Body-Problem/New-Paradigms-in-the-Science-of-Consciousness.html

"Materialism" is often used as shorthand for metaphysical naturalism. Non-materialist neuroscientists, for example, might be considered "non-metaphysical naturalist neuroscientists"

Amanda Gefter, "Creationists declare war over the brain", *New Scientist*, 22 October 2008.

"Nonmaterialist mind", 26 November 2008 by Mario Beauregard and Jeffrey M. Schwartz, Montreal, Canada, and Los Angeles, California, US Magazine issue 2684, p. 23.

19. In their book on consciousness, (*A Universe of Consciousness: How Matter Becomes Imagination*, New York: Basic Books, 2000, p. 110), Gerald Edelman and Giulio Tononi provide a list—which they emphasize is not exhaustive—of theories that have tried to account for the relationship between mind and brain, including Spinoza's dual-aspect theory, Malebranche's occasionalism, Leibniz's parallelism and doctrine of preestablished harmony, identity theory, central state theory, neutral monism, logical behaviorism, token physicalism, type physicalism, token epiphenomenalism, type epiphenomenalism, anomalous monism, emergent materialism, eliminative materialism, and functionalism (various types).

B. F. Skinner, *Beyond Freedom and Dignity* (New York: Knopf, 1971), p. 198.

Ray Kurzweil, *The Age of Spiritual Machines* (New York: Penguin, 1999), p. 63.

See, for example, V. S. Ramachandran, Reith Lectures, Lecture 5, 2003; http://www.bbc.co.uk/radio4/reith2003/ and David Livingstone Smith, "Natural-Born Liars," Scientific American Mind 16, no. 2 (2005):16–23.

Herbert Benson and Marg Stark, *Timeless Medicine: The Power and Biology of Belief* (New York: Scribner, 1996), pp. 116–17.

Benson and Stark, 1997, *Timeless Medicine*, pp. 99–100. The question of the relationship between high blood pressure and mental stress is still debated, but the prejudice against considering the possibility has largely evaporated.

Quoted in W. Grant Thompson, *The Placebo Effect and Health: Combining Science and Compassionate Care* (Amherst, MA: Prometheus, 2005), p. 49; originally from S. Wolf, "The Pharmacology of Placebos," *Pharmacological Reviews* 11 (1959): 689–74.

Gary Greenberg, in "Is It Prozac or Placebo?" *Mother Jones*, November/December 2003 gives this figure in "Is It Prozac or Placebo?" Similar figures are found elsewhere.

F. Benedetti, L. Colloca, E. Torre, et al. "Placebo-Responsive Parkinson Patients Show Decreased Activity in Single Neurons of Subthalamic Nucleus," Nature Neuroscience 7 (2004): 587–88.

R. de la Fuente-Fernández et al., "Expectation and Dopamine Release: Mechanism of the Placebo Effect in Parkinson's Disease," Science 293 (August 10, 2001): 1164–66. They write: "Our observations indicate that the placebo effect in PD is mediated by an increase in the synaptic levels of dopamine in the striatum. Expectation-related dopamine release might be a common phenomenon in any medical condition susceptible to the placebo effect. PD patients receiving an active drug in the context of a placebo-controlled study benefit from the active drug being tested as well as from the placebo effect"

Jeffrey M. Schwartz and Sharon Begley, The Mind and the Brain: Neuroplasticity and the Power of Mental Force (New York: HarperCollins, Regan Books, 2003).
V. Paquette et al., "'Change the Mind and You Change the Brain': Effects of Cognitive-Behavioral Therapy on the Neural Correlates of Spider Phobia," Neuroimage 18.2 (February 2003).
R. Temple, "Implications of Effects in Placebo Groups," *Journal of the National Cancer Institute* 95, no. 1, 2–3 (January 1, 2003).
Lauran Neergaard, "The Placebo Effect May Be Good Medicine" *Pittsburgh Post Gazette*, November 30, 2005.

Jeffrey M. Schwartz and Sharon Begley, The Mind and the Brain: Neuroplasticity and the Power of Mental Force (New York: HarperCollins, Regan Books, 2003), pp. 54–90.

Alister McGrath, *Dawkins's God: Genes, Memes, and the Meaning of Life* (Oxford:Blackwell, 2005), p. 136.

35. Michael Sabom, *Light and Death: One Doctor's Fascinating Account of Near-Death Experiences* (Grand Rapids, MI: Zondervan, 1998), p. 82.

Jeff Levin and Harold G. Koenig, eds., *Faith, Medicine, and Science: A Festschrift in Honor of Dr. David B. Larson* (New York: Haworth, 2005), pp. 16, 140.

37. Ibid., pp. 15–16, 140-43.

Michael Sabom, *Light and Death: One Doctor's Fascinating Account of Near-Death Experiences* (Grand Rapids, MI: Zondervan, 1998), pp. 81–82.

K. I. Pargament et al., "Religious Struggle as a Predictor of Mortality Among Medically Ill Elderly Patients," *Archives of Internal Medicine* 161 (August 13/27, 2001): 1881–85.

Quoted in Michael Sabom, *Light and Death: One Doctor's Fascinating Account of Near-Death Experiences* (Grand Rapids, MI: Zondervan, 1998), p. 126.

G. McCord, Valerie J. Gilchrist, Steven D. Grossman, Bridget D. King, Kenelm F. McCormick, Allison M. Oprandi et al., "Discussing Spirituality with Patients: A Rational and Ethical Approach,"*Annals of Family Medicine* 2 (2004): 256–361.

Farr A. Curlin, John D. Lantos, Chad J. Roach, Sarah A. Sellergren, Marshall H. Chin, "Religious Characteristics of U.S. Physicians," *Journal of General Intern Medicine* 20 (2005):629–34.

See, for example, Pascal Boyer, *Religion Explained: The Evolutionary Origins of Religious Thought* (New York: Basic Books, 2001); Daniel C. Dennett, *Breaking the Spell: Religion as a Natural Phenomenon* (New York: Viking, 2006); Wilson, David Sloan. *Darwin's Cathedral: Evolution, Religion, and the Nature of Society* (Chicago: Univ. of Chicago Press, 2002); E. O.Wilson, *Consilience: The Unity of Knowledge* (New York: Random House, 1998).

Bryon R. McCane, "A kinder, gentler American atheism", The Pew Forum on Religion and Public Life (July 19, 2009). This article can be viewed in its entirety at:

http://www.wofford.edu/newsroom/woffordinthenewscontent.aspx?id=58190.

McCane, Ibid.

Michael D. Lemonick, "Glimpses of the Mind," Time, July 17, 1995.

Sharon Begley, "Why do we rape, kill, and sleep around?" *Newsweek*, June 20, 2009.

David Brooks, "Human Nature Today", *New York Times*, June 25, 2009.

Christie Nicholson, "Questioning evolutionary psychology" *60-Second Psych*, July 17, 2009 Excerpt: "Well it's pretty cool when we can see scientific viewpoints turning, slowly of course. Also known as a paradigm shift. Right now, it appears evolutionary psychology is under mainstream fire"

Denyse O'Leary, "Dissecting the caveman theory of psychology", (*MercatorNet*, August 10, 2009).

Hannah Devlin, "Want to keep your wallet? Carry a baby picture", *TimesonLine*, July 11, 2009. The experiment involved offering a variety of pictures in dropped wallets, and the highest proportion of returned wallets included a baby picture. "According to Dr Wiseman the result reflects a compassionate instinct towards vulnerable infants that people have evolved to ensure the survival of future generations. "The baby kicked off a caring feeling in people, which is not surprising from an evolutionary perspective," he said. " The conclusion the journalist drew was that the return of the wallet "depends rather more on evolution than morality."
Evolutionary psychology has always been symbiotic with popular journalism, though that may be changing.

Wilder Penfield, *The Mystery of the Mind: A critical study of consciousness and the human brain* (Princeton University Press, 1975); Alister Hardy, *The Spiritual Nature of Man*. Oxford: Clarendon, 1979; John Eccles and Daniel N. Robinson. *The Wonder of Being Human: Our Brain and Our Mind*. New York: Free Press, 1984.

Thomas Kuhn, The Structure of Scientific Revolutions, 2d ed.(Chicago: University of Chicago Press, 1970), p. 90.

Author bio:

Denyse O'Leary, was born in Saskatchewan and is a Canadian journalist, author, editor, and blogger. She has worked for major publishing houses. Her most recent books are Faith@Science (Winnipeg: J. Gordon Shillingford, 2001), **By Design or by Chance** (Minneapolis, Augsburg Fortress, 2004), and co-authored with Mario Beauregard as lead author, **The Spiritual Brain** (San

Francisco: Harper One, 2007). Her father was educated by Fr. Athol Murray, in the depths of the great Depression.

Although Denyse never finished high school, she was given early admission to a local University, due to high scores on their test, after two free courses. She intended to become a teacher and earned a four year honors degree in English Language and Literature. After writing her MA thesis the journalism bug bit her and she has remained active in that area. She is perhaps one of the best known evolution troublemakers, including lots of online activity. Check out the following:

http://www.google.ca/#hl=en&source=hp&q=Saskatchewan+River&aq=f&aqi=g 10&aql=&oq=&gs_rfai=&fp=933c5a55b2a4dbb7.
http://www.thecanadianencyclopedia.com/index.cfm?PgNm=TCE&Params=A1A RTA0007158.

Also, do have fun with "The Pirate of the Saskatchewan" - a guy trying crime in Canada and not making it work:
http://www.google.ca/#q=Pirate+of+the+Saskatchewan&hl=en&prmd=v&source =univ&tbs=vid:1&tbo=u&ei=N-UJTIOOB8L68AabmLiMBw&sa=X&oi=video_result_group&ct=title&resnum= 1&ved=0CB0QqwQwAA&fp=933c5a55b2a4dbb7.

Apparently, the Mounty was always after him. Watch for my upcoming co-authored book: *Christian Darwinism* by William A. Dembski and Denyse O'Leary (Broadman and Holman, 2011), on "Why theistic evolution fails as science and theology." It's not as abstruse as it seems. Either you believe in survival of the fittest (Darwin) or you believe in a God who loves you as you just as you are (Jesus). Incredibly, some people want to have it both ways.

Check out her blogs
The Spiritual Brain: A Neuroscientist's Case for the Existence of the Soul by Mario Beauregard and Denyse O'Leary (San Francisco: Harper 2007)
Search inside: http://tinyurl.com/yo6b6r Book's blog: The Mindful Hack http://mindfulhack.blogspot.com/

By Design or by Chance: The Growing Controversy on the Origins of Life in the Universe by Denyse O'Leary (Augsburg 2004) Search inside: http://tinyurl.com/2un7tg Book's blog: The Post-Darwinist: http://post-darwinist.blogspot.com

Check out her newest blog:

Colliding universes, http://collidinguniverses.blogspot.com (If there are zillions of universes, all your nightmares are real and are coming to get you at some point.) You may contact her directly at: oleary@sympatico.ca.

Internet Pornography A Threat to All
Emanuel Tundrea

Editor's note

The Internet is a mixed blessing. Truly the world is at our fingertips with the touch of a mouse. We can communicate with someone half way around the world as easily as we can someone across town. It provides an endless resource for study, research and of course for meeting people. The Internet also has a dark side. Witchcraft and occult websites abound. Pornography is also available by the click of a mouse.

To most men, a woman's body is the most beautiful thing God created, but we are commanded in Scripture not to lust after another woman. Jesus gave a stern warning about this very issue. ***"But I tell you that anyone who looks at a woman lustfully has already committed adultery with her in his heart"*** (Matthew 5:28, NIV). I learned early on that as a man I cannot even look at girlie magazine pictures. For me it takes a great deal of "washing by the Word" to erase such images. This is an even greater problem for our young people today. Sex is blatantly presented on TV, movies and even commercials. This rise is discussed in the following essay. Sadly, some in our culture say pornography is merely a form of art. Certainly the female form has been drawn, painted and sculptured from the beginning of time, but for most of us, it is not art and can lead to lust which is sin in the eyes of God. Let me say it differently, pornography as art is yet another sacred cow. Christians must have no part in it.

Let me share one other thought. I feel honored to have someone from the former Soviet Block to be a part of this effort. As a young boy growing up on a farm in western Oklahoma with my mother and grandparents, I remember my grandfather telling me the leaders of Soviet Union and the United States disagree on many things, but we have nothing against the people of Russia. As it was then, so it remains today.

In view of the terrible risk Internet pornography poises, I strongly recommend the Internet filter by American Family Radio. Here is a link ... check it out: Christian Internet Filter providing Online Protection for your family-AFO Internet Safety and Web Content Filters.

Introduction

Eastern Europe has experienced fifty years of communism when liberty was just a dream. What do you think would be the first attraction, the first thing a man from the East would like to buy right after the wall between West and East

303

went down? It is amazing to discover that the answer is a pornographic magazine (Hensel, 2008). We are now after twenty years of freedom and the ways to become addicted to pornography are so many, so easy available. How to answer to these new challenges of our freedom? How to deal with the power and pleasure of pornography? What are the consequences? How to set on a path to free from pornography which seems to be a new enslavement of our freedom?

If I would ask now at a youth meeting in Romania adolescents and even primary school kids on how many of them have an e-mail address, their faces would obviously show how cumbersome my question is, for they ALL have one. If in the same way I would ask how many of them have yahoo ID or a Face book account, I would still get the same reaction. Social networks are among the most attractive way to spend time and one of the main motivations of young people to spend time "online."

I have taught at both high schools and universities and one of the first things that youth do when meeting someone is to exchange virtual IDs. And then, each one will brag on the number of girls that are "connected" to their virtual ID. And there was no difference between youth from the urban area and those from the rural one. Some might expect the later to have a limited access to the Internet and new technologies, but such is not the case.

The Internet is indeed a fascinating domain that can be described using two perspectives. One of them is that of Apostle Paul, a man who I think never imagined at that time any idea about what the Internet is today. But he was a man of amazing wisdom who through his writings has left behind so many instructions to find answers for the problems that our society is currently facing. His warning certainly applies to our "connected" society today. *Have nothing to do with the fruitless deeds of darkness, but rather expose them. For it is shameful even to mention what the disobedient do in secret. But everything exposed by the light becomes visible, for it is light that makes everything visible. This is why it is said: "Wake up, O sleeper, rise from the dead, and Christ will shine on you" Be very careful, then, how you live, not as unwise but as wise, making the most of every opportunity, because the days are evil* (Ephesians 5:11-16, NIV). Do not think of light as the sun or even as today's optic fibers. The Apostle Paul refers to light as our testimony. Let us see the Internet through these filters: "goodness, righteousness and truth." Apostle Paul also warns us that as Christians we are to "expose them"

That is my motivation for writing this chapter, although I'm fully aware of the fact that I will not be able to cover all the questions and issues raised by the Internet, but I would like to make sure that we are people who "do not participate in unfruitful things" neither for the individual or for the society, and who moreover prosecute these things which we just highlighted.

We have all heard of Bill Gates, the wealthiest man on the planet. It was he who said: "the way you collect, manage and use information will determine if you are a winner or loser" The way that we, in the 21st century, as young individuals gather information, use it, disseminate it, and interpret highway of information, will determine if we are winners and losers.

Why have I chosen this affirmation? It is because I believe that it holds a tremendous value for us as Christians. You will not be able to dissociate yourself from the Internet; most of the jobs in the near future will depend on it, as we will also not be able to live apart from the reality of this global network. Nevertheless, they way you handle this issue will determine if you are losers from the righteousness point of view and cleanness of your minds or if you are winners! We are winners!

What is the internet? I didn't plan of presenting the history of Internet (I don't think it would be relevant for this topic), but I have thought however of looking at a few aspects which reflect its spread. What would you say is the degree of spread of the Internet in a country which does not top among the richest in the world?

First of all, the internet is the most free zone, one of the spaces where people can travel and speak with no fear, with no natural inhibition! If you want to visit America, you need to travel to our capital, you need to apply for a visa, you need to pay for it and you need to have money to buy an airplane ticket in the positive circumstance in which you get acceptance for the visa. As for the Internet, you need only one click to visit Dubai, a single click to visit the beaches of Egypt, or a single click to see how Hawaii looks like. It's an absolute unrestricted access for the entire world's youth. If you have watched the recent TV news, you have probably seen a program initiative in Bangladesh for electronic information spread. Teenagers who have nothing to eat can surf the Internet. It's a free area in which you can interact with anyone who is connected, from anywhere. Please note the danger of having others getting in our private lives.

On one hand, using the same internet, you can find the Bible, sermons in all possible formats (text, audio and video), wonderful worship programs, but on the other hand there is a lot of pornography. Almost every evangelical church in Romania that exceeds 100 members is broadcasting using live streaming over the internet and you can watch the program! I somehow fear the moment in which any teenager having an iPhone who will dislike the way the choir sings at his church will manage a few clicks and be transported to watch a church's program in the US. I fear this is the direction we are currently going!

The internet is a platform open to any possibility and to anyone's opinion through posts. However, this platform is used for either Scripture, Christian resources, or anything else as it is unlimited, uncontrolled and unrestricted. See how Al-Qaida is posting frequent threat messages, how pornography extends so much because it is a completely unrestricted information zone. What is also interesting is the fact that the largest number of users is not located in America but rather Europe (larger population) and Asia who is catching up.

The amount of electronic information on this platform is tremendous. The multiplying measurement units for memory are GB, TB, and PB (Petha-Bytes). At very this moment, the Internet (excepting image, audio and video file) defined just by the unformatted text volume (which can be accessed with no login/access restriction) is estimated to be approximately 100,000 Petha-bytes. We also have an abundance of connection solutions (wireless or mobile networks, even in the rural area) that are becoming less and less expensive. More and more parks, universities, high schools, coffee shops, truck stops and restaurants offer wireless hot-spots for free in order to attract and retain customers. It is obvious the trend is for having easy access to Internet. There are people who can follow from anywhere their own home-based webcams, from construction sites, from their laptop or even using their mobile phones. If technology has become so cheap, why not incorporate the mobile phone's chip in a fridge or in the heating system of a home to visualize or to control them from distance.

If I tell my grandfather that I can walk on the street and at the same time talk to him over the phone, his reaction is: *"Eh, these are science-fiction stories!"* This is why I dare to say that the way we manage these technology tools to connect to the internet will determine if we are winners or losers!

In Eastern Europe, the Internet offers among the best work positions and successful carrier paths (it's clean, and enjoyable, and is considered to be second in the wages top after pharmacists). Today, there are fewer and fewer jobs which you can get that will not require you to know how to send an e-mail and to work in MS-Office – these have become basic requirements as much as writing was in the early times. Usage of Internet has become an aspect which is not subject for employers as it is assumed that any new employee knows about it.

The Internet offers a tremendous amount and variety of pleasure. You want to find good jokes? Didn't anyone send you a joke recently? If you write in a search engine the word "jokes", you find over 80 million websites which have jokes sorted in categories and you could spend years reading those in your favorite category. In the last 10 years countless online games have been developed. A lot of them can be played remotely using the network or as part of social networks (Facebook, hi5) and these also attract a large number of users.

Then, is there anyone who has not viewed YouTube at least once? This is a rhetorical question! Thus, the list of Internet tools which are offering entertainment is extensive. We can say then that the Internet is probably the largest entertainment platform at this moment and it seems there's no substitute in the future. If in the past you had to have money to indulge yourself with sins like pornography or vices like gambling, but today the Internet offers them almost for free! In the past, a lot of teenagers, even if attracted by these temptations, could not afford them and this was a good protection. Today, unfortunately, these are so easily available. It's with deep regret that I say today: I have met teenagers so caught up in Internet backgammon, online sports bet, online poker and I have asked them: "Do you really have the money for this?" Can you think of a way in which they attracted them? It all started from only pennies. It's the risk of having things available so easily.

I would emphasize two more aspects before closing this point. The search engines, besides the basic abilities, they also offer the possibility to analyze the search frequency of a certain key word using geographic location, language and many other criteria. Thus, I requested a similar analysis for the following key words: "*Jesus, God, porn*" on Google Trends and the difference between *Jesus and God* on one hand and *porn* on the other hand is astonishing.

Thus this analysis raises the following questions: Is the Internet a black zone or a white one? Is it something good or bad? Should we not connect to it because it contains so many dangers or should we connect just because it contains good and useful things? And even if we are looking for good things, isn't the danger too high? These questions are being asked by youth, families, and the whole society.

I also tried to find the reasoning of the Bible on this subject, because many have said: "The Bible does not hold any answer on this topic!" I dare to say that it's not the first time when God surprises me how clearly He makes a statement regarding this dilemma – the Bible offers absolute liberty to the man! You heard me right: absolute liberty! But before exploring the solution lets analyze in the next section the problem of pornography. Now we zoom in from the general Internet platform to the pleasure and power of pornography which it supports.

The Pleasure and Power of Pornography

Sexuality is such a wonderful gift for us and yet also causes so much pain and struggle in many people's lives. A young man came to see me for counseling. He had been married for about a year and he thought that being married will solve his problem of pornography addiction. Much of his dismay he found out that within a year he had been pulled back into it again. When he was in early years of school colleagues brought to him pornographic magazines and the fascination of that age was enormous for what he was looking at. Then he gradually went into Internet pornography and then began to even visit a striptease club in the city where he lived. He wanted to be free, but under pressure he could not resist. He said that when the thought of pornographic images came to his mind he was like on a bob-sled run.

There are many areas where we can easily become addicted. Once you are on the hook—whether is an addiction to alcohol, drugs, smoking or sex it is very hard to stop. The word addiction comes from the Latin term "*addicere*" means to surrender our rights to someone, to give your self up to the power of something. Therefore, you become the captive of a substance or a sexual image or experience. We think of addiction not only for a substance, like heroin, cocaine, alcohol, but for any habit which soon moves to a compulsion and then to an addiction.

An addiction is characterized by a repeated craving and you begin to lose control on behalf of that thing. You cannot stop doing something even though you know it is something which damages you and the people whom you love.

Most Internet addiction is sexual addiction. You know the power of this addiction. We hear about the multiplication of the number of Internet sites which are available to people. It becomes an incredible problem for many professors and teachers which deal daily with our children and youth in schools and universities. Many of them live a double life: very successful and promoting morality in society, but in their private life could not stop using pornography.

The Pleasure of Pornography

Sex, obviously, is a wonderful gift from God (Genesis 2:24). As I will address the negatives of sexuality it is important to affirm the goodness of the sex as it was a gift from God. Therefore, sex has an enormous potential for good because it is a gift from God and for evil because we live in a fallen world. We live in a erotically stimulated generation of all times because of the availability of pornography. So many people are so visual, whether is movie on Internet or television. The amount of pornography that is available is easy to find with the click of a button. It is all there prepared to enslave you.

We live in a culture that says: W*hy not? If it feels good, just do it*. It is a self-gratification culture when I do what I want when I want to. The art of delayed gratification is something we are losing in many aspects of life.

Therefore, sex is everywhere in the culture and yet Christians still have a hard time talking about this problem. We need in the church and in our families education to learn to be able to talk about it both in affirming the positive good that God has given us in our sexual lives and the creativity in that, but also helping each others to avoid the danger that is in it because it has such power. What we find is that people easily can become addicted and I would argue that in all of us there is the potential for this addiction.

The people that are more vulnerable to become addicted often find ordinary relationships difficult. Intimacy in relationships is hard to them (due to the individualistic society we live in); therefore they attempt to go into relationships in the wrong way. We will explore this more in the next section.

Ordinary normal relationships take hard work, long term commitment and you do not get instant gratification. You have to work at a good relationship to have the best sex in marriage for example. It gets time to get to know someone, to build relationship. As compared to the use of pornography, it is a sort of quick fix (Winter, 2008). You can get the pleasure and even the orgasm without having involved in the difficulty of building relationship. You can have instant gratification, release without responsibility. Therefore, the physical release pleasure, which is an important part of the whole sexual pleasure, is there without the responsibility.

That is why the Internet pornography addiction is sky-rocketed. It is because of the *availability* and *accessibility*—you can have it quick, easy, at no cost—you do not have even to go out and buy a magazine, you do not have to go in a video-store and risk to be seen renting a sexy-video; you can do it at home in privacy.

Then it is the *anonymity*—you can remain totally anonymous. It gives the impression that nobody sees you because you are all alone in your room.

Finally, there is the *safety* of using pornography. It seems that you are not doing any harm to anyone else and not even to yourself. There is no risk of sexual transmitted diseases or anything like that. Therefore, it is a very large temptation.

There are an increasing number of girls using pornography, but the figures are so alarming among the boys between 11 and 19 years old. The building blocks of pornography addiction are fantasy, then pornography and then masturbation (Winter, 2008). They are central ingredients.

Fantasy is a lot easier to deal with than reality. One may have more control over fantasy than a real person. Reality is difficult, unpredictable and more untidy. For many people, the factor of power and control is decisive, especially for men who have felt powerless in the presence of women earlier in their lives (Archibald, 1995). They may have had a very powerful or critical mother. Therefore, it is the idea of controlling reality with fantasy; men controlling women.

It is also an element of reward. Many sex addicts use sex as a reward. When they succeed in something, they reward themselves with pornography. The opposite is also true: if they had a bad day and they think "*I deserve something better than this!*" will use pornography as a reward.

The Power of Pornography

The power of this addiction is also seen in the addictive nature of the sexual orgasm. It is one of the most pleasurable experiences known to mankind. The more someone associates fantasy to pornography, pornography to masturbation and to orgasm, the more you fix that cycle in your brain. When someone becomes addicted to a particular lifestyle we find that the brain begins to change. The chemical and even the structure of the brains changes and that make it harder to go back and liberate. The more someone uses pornography, the harder is to backtrack because it is a physical change there.

It is not only the pleasure and the power of orgasm, but it also becomes a means to channeling emotional tension (Arterburn 2000). Many young men and women living in unhappy and dysfunctional homes discover masturbation and find its pleasure. Therefore, every time they are unhappy they use it and it becomes a channel to release emotion.

If a young man is addicted to hard drugs (like heroine) at age 13 and he remains addicted for ten years. At 23, emotionally he will be still around the age of 13 because he learnt to deal with uncomfortable emotions like sadness or depression by having a fix from the drug. When he comes off he needs help to grow up. He has to deal not only with the drug temptation, but also to deal with uncomfortable emotions.

The same is true with the pornography addiction. If someone has used pornography for ten years they are not really learning to deal with their emotions in a healthy way. Once they come off there is a huge battle helping them to do that.

For people who were abused and this built up a feeling that they are dead inside, pornography is a way of feeling alive. For a moment it gives a sense of life or at least feeling something rather than nothing.

In this context, pornography can be seen as a mood altering or self medication (*I medicated myself with pornography*). Some people are more vulnerable than others. The risk is bigger for those who were abused emotionally or physically or they were not loved at all when they were children. Those who were raised in a loving family do not have such a desperate need for love and affection; do not have such a pit in their soul that needs to be filled by something.

Patrick Carnes made a study among 1,000 fairly-severe pornography addicts and found out that 97% have been emotionally abused, 74% have been physically abused and 81% have been sexually abused (Carnes 1991). The more abuse has been experienced, the more likely is to be hooked into pornography addiction. In Romania, I believe that the main reason is not abuse in the sense of hurting physically or active abuse, but in the sense of abandoning: isolation, lack of love and lack of communication between teens and their parents, working abroad during their children's adolescence. The legacy of divorce is seen in this area also. Children that were abandoned by their parents are more vulnerable.

I observe that teens which are in a family which communicates with them and provides a loving environment, are still facing the risk to be hooked into pornography, but it is not such a deep addition and they can be free easier. They do not have such a desperate need for love, but they still have the curiosity and peer-pressure to experience pornography.

There is also a cultural change which advocates more and more that pornography is morally acceptable. There are even married women which assert that they use it in order to find out what men want. Then they become hooked somehow themselves.

However, men fantasize about sex more than women do. Men tend to be much more visual and pornography is mainly visual. Also, for men sex and relationship are more split. They separate them more than women do. Women need the romance and the relationship. On the other hand, men confuse love and lust more often than women do (Anonymous 1992). This is the same thing as separating sex and relationship.

In this respect, women tend to be more addicted on reading romance novels, romantic stories. This is a paradigm which is changing because of the highly erotic society we live in.

Pornography Addiction Evolution

As time passes the addiction to pornography evolves and the addict tends to become more tolerant. Tolerance here means that the people build up a bigger need of heroin to have the same buzz. People using pornography experience this type of tolerance: they need to increase the type of sexual stimulation to get the same good feeling. They need more images, then video and there is a progression of exciting stimulus (Arterburn, 1993). It seems that in the last years the greatest command is for hard-core pornography and even sex with children (Kirk, 1995). It is really a progression in what people need.

Besides tolerance, another feature is withdrawal symptoms. Young people feel unhappy and begin to crave and need next "fix." These people spend more and more time on the Internet searching for images that will stimulate them. They spend more and more time planning elaborate rituals for masturbation with pornography. It is not just looking, but this takes the form of a schedule they crave for and it takes more and more of their time.

The research on the effects of pornography shows a desensitization of conscience (Donnerstein n.d.). The more you give yourself to something, the more you have to harden your conscience to deal with your shame and your guilt. You begin not to feel it so much anymore. Soon you begin to deceive yourself, tell yourself lies. For example: *I can give this up anytime. This is not a problem to me. I am not addicted.* This is typical to any addiction.

Then you begin to deceive other people when they start questioning you. You start to lie when you have to explain where you spent your money, where you were for the last hour, do you ever look to pornography on your computer. Deception, cover-up becomes a lifestyle.

This gets into total addiction. The more it progresses the more it becomes like a poison. It gradually kills them. They do not see it because it is not very visible at first, but afterwards they start to see the world from the perspective of porn-spectacles (Laaser, 2004). It is horrific to think that a whole generation of young people, between the ages of 11 to 19, is learning about love and about intimate aspects from the Internet pornography. Because it is so available a huge percentage of young people are taking their life lessons from Internet pornography. They come home from high-school and most often their parents are out working and the youth are alone for 2 hours before their parents came back. The first thing they do is to go straight into chat-rooms with their friends. In this way they start to see the world through porn-infected-eyes and the women become devalued. Women become just objects for my use. It is a much distorted view of reality in their conscience.

It is also a strange sense of false intimacy (Schaumburg, 1997). Just a screen or an image gives someone a strange sense of intimacy which is clearly false. For many young people this becomes more real than real relationships.

The research also shows that there is an increasing narcissism, self-love that people experience when they use pornography. It is sort of making love with yourself. When we think of the difference between lust and love: lust means to take what you want and love is giving pleasure to another person (as a result you get pleasure for yourself, too). Lust is about taking, love is about giving. Because lust becomes the master of the heart instead of love, a kind of heartlessness begins to appear: a hardening of the heart to force maybe a real partner to do things that someone saw in pornography even if he/she does not want that. He has an expectation of it and it is much less sensitive to relationships.

The final step is to develop a lack of commitment to long-term relationships, to marriage, to family, to having children. It is an increasing desire that my needs are met rather than me caring for other people or seek the good of the next generation.

Pornography's Painful Consequences
The previous section showed how pornography poisons the whole life and as a result causes a lot of pain. First, it involves secrecy in your life: hiding, denial, deceiving. Then it involves a split of life: a professor may teach morality

312

and require morality in studying from his students, but in his private life he is immoral. Just like a doctor who is an alcoholic or is smoking. He manages to keep going, but in his spare time when he does not manage his patients he is thinking where to smoke his next cigar. It happens the same way with the sex addict: in his spare moment he thinks where he can get his next "fix." This is a typical split life.

Maybe for some time the conscience of this people is dull, but there are moments when they become aware of their state and self-hate is rising. They hate themselves for what they are doing and why they cannot stop: *"I am a disgusting person before God and other people"* Self-loathing and shame drive people on a vicious-circle: they go back to masturbation or pornography to dull the conscience again for a while. All this becomes a repetitive cycle (Roberts, 1999).

The pain may be felt in financial area. When your spouse gets a phone bill with a hundreds of dollars because of a phone-sex service, it ruins the relationship and it is incredibly painful (Kirk, 1995). There are people who were fired because at lunch time or even during work time they used pornography. In this case, the relationship with God is affected because of the sin, the self-loathing and shame is great, the marriage is stroked and there is also a financial consequence.

Pornography and Scripture

1 Corinthians 6:12: *"Everything is permissible"*—but not everything is beneficial. *"Everything is permissible"—but not everything is constructive"*

Have you ever heard someone that tries to promote standard of living, or tries to promote the direction for a society or tries to provide answers to teenagers, and say: "You know what? Are you interested on whether the Internet is bad or not? All things are allowed!" Except the Bible, I have never heard a treaty in the world which states: "Everything is allowed!" Anything you want is rightfully yours—all things are allowed. Any organization, no matter how liberal it would be, it ought to have some rules of conduct. You have certain rights, but you also have some obligations. This is why the Bible is amazing in its statement: **"All things are allowed!"**

If we compare this topic to the extraordinary experiment that God did with man in the Garden of Eden, we can then say that God did fulfill His promise towards man. He said: "I give you ruling power over this earth, I'll tell you what to eat or not, but I want you to know that you are free. Trees are there, I'm asking for your willing obedience. However, you are free to walk (*"browse"*) wherever you wish on this earth, and do whatever you want. Just keep in mind that there is a forbidden tree because it can bring you death"

I think this statement is as much as valid over the Internet, even if, so as to help us chose right between good and wrong in this area, God has given much wisdom to the writers He has inspired. One example is Apostle Paul who has left

313

a few filters of wisdom so as to understand this issue: 1 Corinthians 6:12a: *"Everything is permissible"*—but not everything is beneficial. *"Everything is permissible"*

This is the first filter which I propose for you to consider in the time that you decide to browse the Internet. When you decide to enter a website, the very moment in which you click a certain button, on an advertisement, or you read an e-mail, accept an IM invite or start an IM conversation, before you do all that, ask yourself: "Is this useful to me? Does it help? Is it enhancing you knowledge? Is it healthy for your spiritual life? Is it keeping your righteousness? Is it useful for your time?" Use this filter very seriously: "Is it useful to me?" 1 Corinthians 6:12b: *all things are allowed, but nothing should set rule over me*.

This is another important thing to consider: Have you become addicted to the Internet? Online games? Instant Messaging? Do you foresee that the addiction that the Internet can cause is starting to develop in you? Do you become irascible when the Internet connection is low?

In the times of our parents a lot of them were talking during sleep. I just got a joke on my e-mail which says that nowadays it seems the issue has evolved and Internet addicts move their fingers during sleep! The aspect of Internet addiction is very important. You develop and identity crisis whenever someone uses an ID which is similar to yours? Whenever you see a person offline – do you get upset? Do your virtual relationships seem more real and deeper than actual relationships? Have you got to a point in which you are so open to share things over IM and communicate intimate things? Do you communicate more with friends you don't see rather than your parents or real friends? Some have reached such a degree of addiction that if they are woken in the middle of the night, a strong drive to devour Internet and pornography takes control of them, giving up their sleep.

If something which is considered important by your family or friends takes place, does your Internet addiction cause you to become nervous and upset because you must give up some computer time? If you have unexpected visits from close friends, do you become unhappy? If you are aware that you hold responsibility but however your thoughts can't be control and fall under Internet pornography , can you manage to discipline yourself and hold yourself responsible?

How many times have you told yourself that you connect to Internet just to "*relax*" and when relaxed to take on more important things? Didn't you deceive yourself? How many times have you said: "*it will be just 5 minutes*" and stay online half of night? Internet creates addiction, and in the context pornography is a lethal drug both for your mind and body. It can also destroy your soul.

The last filter that I would like to mention from the wisdom of Apostle Paul is found in 1 Corinthians 10:23b: *"Everything is permissible"—but not everything is constructive.* Be honest towards your own evaluation of the time spent on the Internet and ask yourself if that actually helps you grow. How constructive are the things you read on the internet? How constructive is it whenever you are provoked over IM? Did you observe that your capacity to focus at an abstract level and time allocated to a particular study has dropped lately? Have you noticed how bored you are at courses, reading a book or following a seminar? Do you need frequent breaks? Do you prefer during those breaks to rather stay in front of your PC and engage in pornography than going for a jog? Are you all the time attentive of your browsing experience and prefer to browse *"in Private?"*

Install these three filters and I think these will take you on the way of healing: Is it useful? Has it taken control over you? Is it constructive for you? Don't forget: you are free! If you are an adult, it is very unlikely that someone will verify your browsing history. Most will not be controlled by anyone. This is why I think that the Internet has a huge capacity to hold evil and it has destroyed numerous lives, but it also hold a tremendous potential for good!

Path to Healing

The previous sections spoke about the power, the pleasure, the poison, the pain and the answer of the Scripture. I believe that the Scripture is the best start to heal. What is the prescription that the Scripture gives us and then some common grace wisdom to help people be free again.

As with any addiction, the first thing is for people to come to their sense and realize that this is something bad and that it has a grip on them (Winter, A Time to Run n.d.). The first step is to confess and to admit that they cannot take this battle on their own. It is important to acknowledge that you are powerless in front of this addiction and you need something bigger than yourself to deal with it. At this point I have great news that you have the promise of God in 1 John 1:9 **"If we confess our sins, he is faithful and just and will forgive us our sins and purify us from all unrighteousness"** This is the beginning of the path.

Sometimes people come to this understanding themselves, but most of the times the addiction is so powerful that people need the love and care from someone close to them. When the spouse discovers such addiction in her partner she rightly becomes very angry. This shocks sometimes and this is needed in order to realize that this causes a lot of hurt to the spouse and it is profoundly wrong. Confrontation is important and sometimes people will admit from the beginning or deny.

When someone acknowledges his problem and calls for help what sort of ways we can help? Many times this is a big challenge, because people come to us searching for solution. This is tragic to hear, but we need to love them, accept them and be glad that he/she told us. It is like the Father of the prodigal son who welcomes him home. We need to model grace and forgiveness in love, but confronting the truth and the reality of the addiction, the sin. The question of what to do is still there.

Therefore, next important step is that the addicted to pornography man to confess his sin to his spouse or parents (his closest family) in order that he will develop accountability in his relationships which will care for him for the next steps. This will probably crush down the view of people around, but will create an environment where those around will care and he will be monitored until the addiction is vindicated. It is a lot of sorrow and grief for the spouse who will face the question *"why am I not good enough for him? When those dirty images from his mind will delete? Probably never"*

Then Richard Winter suggests a period of 90 days of total abstinence from all forms of sex (Winter, The Struggle of Sexual Purity 2008). This is a tough decision, but to bring a man to understand that he does not need pornography to live or to deal with his depression or anxiety or anger it is important to *"un-alcoholize."* There are better and healthier ways to deal with problems than pornography. It is a time which gets someone out of his addiction and helps him see how to live a normal life, sensible again to normal desires and limits. Of course, people will find hard to keep to abstinence, but here again we must come alongside with care. It is an important lesson to learn how to have sex in a correct way.

For someone who is not married, in most cases this will mean abstaining from masturbation and maybe harder will be abstaining from fantasy and waiting pure for fulfillment in marriage.

This shows the importance of education about sex in order to communicate people that pornography has such a wrong view on women, on love, on intimacy. Basic teaching on this subject is an important part of their re-education. There are wonderful books that you can access (check the bibliography at the end) and help along.

It is also important to be realistic about lapses. Lapses will occur. If the incidence of sin is lower than at least there is a progress and this must be an encouragement to continue the battle.

Fantasy is particularly difficult to deal with and most of the people say that this is harder to deal with and get rid of than anything else (Winter, The Struggle of Sexual Purity 2008). There are various behavioral techniques that can be used. For instance, people can learn the art of distraction. If fantasy is starting in my mind, than I have to find other thoughts to bring in to play. Just as a small child who is crying, if you can manage to distract him he will stop very quickly from being upset. Of course, this is more difficult with adults, but if they realize the need, they can welcome other thoughts. It can be a physical exercise (like going for a run) or a mental exercise. All of them are forms of self-discipline.

Another thing that I found helpful as I interacted with people is to let the craving go. When a thought comes (and I find this in my mind, too), a thought that is tempting to do something that I know is unhelpful to me, rather than engage with the thought, start to fight with it and to say to myself *"I know I should not do that, it is very strong, but I do not want to do it"*, to just let it go. It is not grabbing the temptation and fight with it, but ignoring it. If you can resist from craving about ten seconds, it will very quick start to diminish its power. It is not long you have to wait for the thought to die.

Others say that they need a mentor or someone they can call and just talk on the phone for five minutes and then the eye of the storm is gone. Many people say that this was the most helpful tool. The problem is that you need someone you can call day or night. To be available is a huge gift to someone who needs deliverance from pornography addiction.

Some people use pain. For instance, they wear a plastic on their wrist and just pinch when fantasy is tempting.

Another mental exercise is to ask people to remember what the worst moment during their pornography addiction was. Maybe it was when his wife found out about his addiction. It is a terrifying remembrance. Others may confess other painful experiences, most of them will be related to exposure of shame. In this way, the people can be encouraged to bring back in their mind those shameful feelings every time temptation returns. It is about helping people to think about consequences again, where the temptation leads. Help people think where they will be in five years time if they continue to give in pornography. What will your life be like if you continue to use pornography? Will you have your wife? Will you have your job? What impact will be on your children, what sort of father are you? Think of consequences on long-term instead of short-term fulfillment.

Then we can think how will be like if you stop using pornography. It is also the positive reasoning. If I stop, how my relationship with my wife, my children, my friends, with God will be like?

At the next step is important to set boundaries. A good approach is to discuss all the details of their struggle: where are the dangerous points of your life? Many of my youth have to get rid of the TV's because just its presence was too much for them. To have a TV connected to cable programs is far too dangerous if you struggle with pornography. Then you have to get rid of the home Internet access so that you can only go online when you are in a public place (in a library, in the office at work, ...) Having a computer always in a public place at home is crucial for youth. From my pastoral experience, I believe it is a suicide to give to your teenager Internet access or TV or DVD player in his bedroom. If it is in a public place, it is a better training to learn self-control. But for a person who has gone much further than that and became addicted is important to get access only in further places in town to avoid access so easy. In case of certain pornographic clubs nearby, than it is maybe better to have access from home, but from a public place, not the private bedroom.

Setting boundaries and finding ways of blocking the temptation are together part of the solution. Blocking the Internet access is of a particular importance. There are two other helpful approaches. First, it is helpful to install a firewall or an adult filter and ask your spouse or a good friend to store the password. A safe blocking agent is maybe the ultimate solution. A second approach is to use a solution which sends the links you visit to another partner. In this way, when you browse the Internet you realize that your wife or best friend will receive a report on all the sites you visited. It is all about having an accountability partner. It does not block your access to sites (that can be a real problem), but it makes you responsible and help you realize you can win through accountability. If your accountability partner is not your spouse (which is wise not to be because it is most of the times a too much burden for her), a possible consequence is that your friend partner may be tempted to visit those dirty sites himself.

This external control may be extended to your bank account if you feel you are in danger of spending money on pornography.

Usually, people who travel a lot, who stay in hotels, are very vulnerable. In these cases, a solution is to try to stay with friends. I much prefer to stay in a tiny apartment with friends, than to be alone in a hotel. It is better to just stay away from that temptation.

Therefore, the need for honesty with me and my spouse, my friends, the need for accountability and support and also the need for counseling are major areas to healing and going in the right direction. Small groups of friends are a great support to deal with the culture around us, but sometimes it is needed specialized counseling in order to deal with past hurting, guilt or inferiority complex. This applies to deeply damages people which for long time were pornography addicts and which became so due to other abuses or terrible circumstances in the past. Because of that, these people need not only to deal with their battle against pornography addiction, but also other deeper issues in their lives where they have been sinned against. All of us are both sinners and also sinned against.

Nevertheless, in a small group these people start to experiment acceptance, grace, accountability and different relationships. They are learning to share intimately, not sexually intimately, but to share their feelings and thoughts with other people in ways there maybe never done before in their lives. They are learning how to relate in a more mature way. That is very healing as well.

Radical and Personal Applications

I would get closer to the end by recommending some practical aspects: Be very careful towards relationships that you're being suggested! Please note: The Internet is not a platform on which you make or search friends! The internet is a platform which you use to engage in already existing relationships. New relationships are formed and entertained only if there is a link in the real life (even if I never met face-to-face the editor of this book, there is somebody I know personally who recommended him to me). It is a virtual space which should function just as a useful instrument in developing already initiated relationships and tested in real life. Connect just with people which are real-life friends.

We already have too many sad stories which show many that have paid a high price for not listening to this advice! Don't try to justify yourself by saying that you verified discussion partner by asking for his ID and picture! I think that that we are all aware that it's no point in being so naive. Don't try to make friends in the virtual world because you will be deeply hurt. There were so many girls that survived rapes with drug dealers' mafia, and many had started from a game on the Internet. Don't think that you are so wise that in the snare of this virtual world in which there is no police, rules or morality, you will be more agile than all others and will know how to stay out of trouble. Sooner or later you might fall into their net and the consequences are not to be ignored as we have just mentioned above. Don't look for new friendships on the Internet.

Be careful as to what sites you are visiting! God has given us an extraordinary memory and especially in the case of the young generation it translates into a visual memory which is tremendously developed and awakens very early. This is why the screen and now the Internet have become so popular, due to their power of image and the fact that we so easily can remember images.

Recent studies in the micro-processor area try to model the human brain: memory and rationing (Orr n.d., Wikipedia n.d.). Thus, a lot of studies are being carried as to the way that associations are being made in our mind between images, films and ideas. For example, our brain associates a lot faster an image to a person rather than reading that person's name and identification. The conclusion states the fact that we are not in the spot in which we can have the courage to say that we can play with pornography on the Internet or even just browse the Internet without some clear principles and filters. Scientists behind the pornographic-profile websites study the human psychology in parallel with the development of technology and are targeting creation of a customized catch for any psychic, any personality or any online behavior. So be careful what websites you visit.

If by any mistake you click on a pornographic link, try to close on spot as those images will be immediately imprinted in your mind, whether you want it or not. Your awareness should include suspicious websites. We are called to stay away from anything that seems bad: 1 Thessalonians 5:22 *"Avoid everything that seems evil."*

I have a friend who from his student years was passionate of movies on PC, but in his passion he was drawn into downloading pornographic movies over from the Internet. Meanwhile, he realized the consequence and today he wants to be clean, but he is permanently haunted by those vile images. In his student year I discussed many times his problem and said to him: *"My friend, try to set a restriction on you. Don't try to see how much longer you can continue without getting sensual, but how far can you stay from it and be protected"* In one of his honest moments he told me: *"I decided more than a thousand times that if I watch a movie, I will fast forward and jump over the lustful scenes and change the channel or switch to another website. But the reality is that I never had the power to change the channel or close the movie file. If I'm alone in my room, and I need to be honest now, I don't have the power to set any filter. The addiction is much too strong"*

Avoid entering those websites; because once you're in, you will not have the power to close it right away. It's a catch which is too powerful for us. If you detect something rotten, close it in the first second. If you don't, you will not have the power or freedom to do it. It's unrealistic and a lie if you say you will close it later!

Careful to the content of the messages! Check who is it that you're getting the messages from. If the sender is not known to you and the subject sounds provoking, ignore the message. Don't try to enjoy the messages that are coming from unknown senders as you might get contaminated. Would you do it in your real life? Would you go out on the street and enter people's courts and house or institution just because you would find them interesting? If you would walk on the street, would you go into a building just because someone that you don't know is inviting you in and promises you amazing experiences?

And finally: don't lose your time! The Internet is an extraordinary platform of information but which can eat up a huge amount of time! May it be in school or church, if a sermon is longer than "x" minutes, you already look out for your watch. On the Internet you can't realize how "n+1" minutes have gone by. The Internet catches you and captivates you. After the first click, you start from one idea and then you find so many things which will make you forget about time or any other responsibility.

My parents had no idea about the Internet or computers for that matter. However, I admired so much their way of setting some very practical filters born out of pragmatic life principles. For example, they decided the following: "young man, you got free time, you're young and we understand that it is normal for you to have hobbies, but consider first cleaning the house, groceries and homework. And only then you will have your time!"

Establish your own priorities well, and as you come from school you say: "I will relax a bit, and only then I will work," you will not realize how the hours will slip through your fingers. Then you will be hungry and you will want to eat. As you eat, you will think of watching a sexual movie as it's more interesting to eat this way. After that you remember that you needed to look up for something on the internet and another hour will go by. Then, have your siesta and will feel again the need to relax. In this way, time management is ruined. Look to prioritize your responsibilities in the right way and only when these have been respected, and in case you still got some time, use it for the internet, without forgetting to use the filters.

Another important rule that my father set out for me and for which I have appreciated him very much was that the computer and access to the Internet was not in my bedroom. Don't take the risk of leaving the computer in your bedroom. Put it somewhere where everyone has access, in the most public place. And I come back to the advice of not considering yourself strong. It's not worth it to assume such a risk by damaging your mind, even if you ask for forgiveness afterwards. If you're honest, you will obtain God's forgiveness, and even your partner's forgiveness, but the consequences over your mind will remind for the rest of your life.

The Lord Jesus made things very clear: Matthew 5:29 ***"If your right eye causes you to sin, gouge it out and throw it away. It is better for you to lose one part of your body than for your whole body to be thrown into hell"***

How many times can you lust in your life? Just two times by taking the Bible's standard after which you have lost both eyes. The Internet is much more aggressive then you think and temptation does not play fair.

I will end with the example of Daniel who was a man that at the age of adolescence, took a radical decision for the times that we are living in: *"I don't want to taste!"* Daniel had realized that if he would not taste the food from the king's table, even if he would not get drunk, or dance with Nebuchadnezzar's dancers, he would rather make sure there would be no instance to give course to any temptation. Daniel thought: *"I don't want to taste from anything that is part of Nebuchadnezzar's party, because I realize that if I will taste it, I will be caught in the snare!"*

Recently in Romania, there was a TV commercial for chocolate which said: "dare for just one!" Do you know what the idea behind is? "Take just one piece and then stop!" and for sure we know that after a piece, another will follow, and another, and another … Things are the same with the Internet. Just one click and then minutes of fall will follow. Set these filters for you: ask yourself if it is useful, if it is addictive, and if it is constructive and may God make us all victorious!

Bibliography

Anderson, Kerby. *Pornography.* Edited by www.probe.org. quoting Final Report of the Attorney General's Commission on Pornography, ed. Michael McManus Probe Ministries. Nashville: Rutledge Hill Press, 1996.

Anonymous. *The War Within.* Leadership Magazine, http://ldolphin.org/lust.html, 1992.

Archibald, Hart. *The Sexual Man.* Thomas Nelson, ISBN-10: 0849936845, ISBN-13: 978-0849936845, 1995.

Arterburn, Stephen. *Addicted to "Love": Understanding Dependencies of the Heart : Romance, Relationships, and Sex.* Vine Books, ISBN-10: 0830733760, ISBN-13: 978-0830733767, 1993.

—. *Every Man's Battle: Winning the War on Sexual Temptation One Victory at a Time.* WaterBrook Press, ISBN-10: 1578563682, ISBN-13: 978-1578563685, 2000.

Carnes, Patrick. *Don't Call It Love.* Bantam, ISBN-10: 0553072366, ISBN-13: 978-0553072365, 1991.

Dictionary, Latin. http://www.italatin.com/English_to_Latin_Dictionary_A.html (accessed June 02, 2010).

Donnerstein, Edward. *The Pernicious Effects of Pornography.* http://www.ankerberg.com/Arti-cles/practical-christianity/PC0607W2.htm.

Hensel, Jana. *After the Wall: Confessions from an East German Childhood and the Life That Came Next.* PublicAffairs, ISBN-10: 1586485598, ISBN-13: 978-1586485597, 2008.

Kirk, Jerry. "A Way of Escape" *Leadership Magazine*, 1995.

Laaser, Mark. *Healing the Wounds of Sexual Addiction.* Zondervan, ISBN-10: 0310256577, ISBN-13: 978-0310256571, 2004.

Orr, Genevieve. *Computation in the Brain.* http://www.willamette.edu/~gorr/classes/-cs449/brain.html (accessed 06 01, 2010).

Roberts, Ted & Hayford, Jack. *Pure Desire: Helping People Break Free from Sexual Struggles.* Regal Books, 1999.

Schaumburg, Harry. *False Intimacy: Understanding the Struggle of Sexual Addiction.* NavPress, ISBN-10: 1576830284, ISBN-13: 978-1576830284, 1997.

Wikipedia. *Neural Network.* http://en.wikipedia.org/wiki/Neural_network (accessed 06 01, 2010).

Winter, Richard. *A Time to Run.* Edited by Covenant Theological Seminary. Chapel Message. http://www.resourcesforlifeonline.com/audio/3999/ (accessed 06 01, 2010). "The Struggle of Sexual Purity" *European Leadership Forum.* Eger, Hungary: http://www.euroleadershipresources.org/resource.php?ID=288, 2008.

Author bio

Emanuel Tundrea was born in Romania under communism, but in a Christian family. His father suffered for his faith and well understands the price for being a pastor and distributing Bibles. His childhood in this context and his parents' faith had a great impact on him to devote to Christ.

In June 2001 he graduated Computer Science, Software Engineering Specialization at, "Politehnica" University in Timisoara, Romania and in January 2009 he completed his Ph.D. in Modelling Software Product Lines as a cotutelle at Institut Universitaire de Technologie de Nice et de la Cote d'Azur (CNRS/UNSA) and as a researcher at Laboratoire Informatique Signaux et Systemes de Sophia-Antipolis, France.

Since 2001 he is a lecturer at "Emanuel" University of Oradea, Romania (www.emanuel.ro) and now he is part of the Center of Research on Information and Communication Science to Promote Evangelical Values. Now he is pursuing a Master on Pastoral Theology at the same university. He is a member in the Scientific Network at European Leadership Forum (www.euroleadership.org) and a fellow of Cambridge Scholars Network.

He is also serving as a pastor of Emanuel Baptist Church of Timisoara (the town with one of the largest student community in Romania). This is one of the greatest joys to see people coming to faith. Therefore, he is engaged in proclaiming the Word in the church, but also in public debates in universities in Romania and in the media on Intelligent Design, moral choices, academic freedom and Bible criticism. His experience among students was the motivation to write this chapter in order to deconstruct one of the seductions of our anti-God education of our century and present the power of God who can transform and set captive free.

He is married with Nadia and they have two boys: Nathanael who is five and Titus who is two years old. His prayer is "to understand the times and to know what to do in this part of the world" (1 Chronicles 12:32) under the guidance of God. His email address is: emanuel@emanuel.ro.

Homosexuality and the Public Schools
A Parent's and a Teacher's Perspective
David Williams

Editor's note

As both a parent and public school teacher, David provides insight into the way the homosexual lifestyle is presented to our young people today. He is also keenly aware of what scripture teaches on this important topic. Increasingly, the hands of our educators are tied and they are no longer free to discuss ethics and morality. In many schools teachers are forbidden from hugging a hurting child to console them. Where will it end?

> *Although they claimed to be wise, they became fools…Because of this, God gave them over to shameful lusts. Even their women exchanged natural relations for unnatural ones. In the same way the men also abandoned natural relations with women and were inflamed with lust for one another. Men committed indecent acts with other men, and received in themselves the due penalty for their perversion. Furthermore, since they did not think it worthwhile to retain the knowledge of God, he gave them over to a depraved mind, to do what ought not to be done* (Rom 1:22-28, NIV).

Introduction

As I begin this chapter, I must say up front that I am not a scientist or a theologian, just a man who has raised four sons and four daughters and have now been teaching in the public schools for ten years, following a twenty-two year military career. In today's world the words homosexuality and gay are ever present on TV, at school, work, or in the news. Science and psychology appears split down the middle on whether the behavior is inherent or the result of a complicated series of life events and choices. I have done some research on human sexuality to get the facts on the issue at hand. Most public schools follow guidelines in a pamphlet entitled, "Just the Facts about Sexual Orientation and Youth," published by the NEA (National Education Association) and mental health professionals. This publication presents homosexuality in what is known as an essentialist viewpoint, which sees sexual orientation as a fixed, inherent trait.

However in my research I found another view, often called constructionism or developmentalism. This viewpoint of sexuality sees one's orientation as a mix of nature and nurture subject to change through personal reflection in the context of environmental factors. There are also thousands of people who have left the homosexual lifestyle who prove this viewpoint.

Movements are in place in politics, the workplace, and our school system to promote and support homosexual rights. Even the religious world has no consensus on the matter. I do not plan to cite and quote experts that support either viewpoint, just my observations over the past thirty-five years.

One may ask, why any interest at all in this matter? It all began with a question from my youngest son about some very aggressive girls at his high school asking for signatures on a petition about gay marriage. It was the year a ballot measure to amend the Oklahoma Constitution to allow marriage between a man and a woman only was to take place. I thought that may have been the reason. A few days later my son brought up the petition telling me it was for student council consideration of a new student organization, a gay straight alliance (see the student "how-to" to start a GSA on the GLSEN or ACLU website). The name of the club was to be "Open Arms" Its purpose was very unclear.

As a member of the Christian Educators Association International I was aware of GLSEN and what a gay straight alliance were all about. The stated purpose was to promote tolerance and understanding between gay and straight students which would prevent bullying at school. My son passionately asked for me to get involved even at the risk of retribution on him at school. Through a series of events which included letters to the editor of our local newspaper, a television interview, a community and schoolwide information blitz, and much prayer from the community, the students themselves voted the club down. At no time did anyone bash or demonize the students. We only asked that both views, essentialist and developmentalism, be presented to the students before the vote. Here is where the sacred cow nature of gay activism in the schools is very evident. In our case when both views were presented to the students and the community, the need for a gay straight alliance became a mute point. A majority agreed that to support a high school age student in his self-identification as gay is not a positive thing to do. Ex-gays would agree (see Exodus International, PFOX, and First Stone websites).

Since this incident I have found myself in the role of helping others look at gay activism in their schools in a healthy, respectful manner but not at the expense of the truth. The First Amendment Center also published guidelines for dealing with this matter in a balanced, open way in a publication called Sexual Orientation and the (http://www.firstamendmentcenter.org/about.aspx?id=16611) Public School.

Focus on the Family has also developed a website to help students, parents, and schools respond in a positive way (http://www.truetolerance.org/).

326

The push to normalize homosexuality in schools is very strong in our current culture through gay pride days and think pink anti-bullying days in elementary schools. The reality is that through the freedom students have gay activists usually push an all or nothing approach, a sacred cow not to be challenged.

I have family and co-workers that identify themselves as gay. Secondly, as both a military officer and teacher, professionally I have had to respond to the issue. And finally, most recently as a parent, the issue at school and our community's response to it. This book is about sacred cows and one thing is evident, the viewpoint that homosexuality is inherent has the cultural upper hand and to point out that truth may lie in another direction has serious consequences for the one stating the facts.

Am I biased on this issue? Maybe so. I was raised in a Christian home that held the principles of the Holy Bible as truth and I consider myself born again through faith in Jesus Christ. If there is a bottom-line on this issue one must ask the question "Is there any such thing as right and wrong?" Does each person choose what is right and wrong? Why do courts across the land make contradictory decisions? Is there a transcendental truth outside of us that overrides one's personal opinions and feelings? If truth is found in each individual then the need to read further is not necessary.

Near the end of my military career I had the privilege to teach a moral development class to inmates in a military prison. The issue of right and wrong and what is truth was central to the course. Most of those in prison are very aware that personal standards of right and wrong and an individual's feelings about truth can be wrong based on a higher standard of right and wrong held by a court of law. Could it be that such transcendental truth exists for us all here on earth? This is something to seriously ponder. If such truth was built into the system for our benefit it must be worth pursuing. To me this is what one today may call a Biblical world view and I must admit this is what I adhere to.

As we look at ourselves and those we love, we may note that we all have a different propensities to do harm or good to ourselves and others. For some it may be alcohol or drugs, others sex drives of all types, still others overeating or gambling. Are these inherent and not subject to our will to be chosen or embraced? Why wouldn't same sex attraction fall into the same category? For the many I have met over the years that have left a gay lifestyle, all say it is a very complex situation that developed over time and was reinforced by others along the way.

As a parent and a teacher I have observed that each child is unique and special with strengths and weaknesses. Behavior has a great deal to do with the complexity of the family in which they live and the influences of caregivers, media and peers. Teens are in the process of figuring things out for many years. It is at this time of confusion and lack of personal fear that many slide into behaviors that give immediate gratification that turn into a lifestyle. This could be anything from promiscuity with members of the opposite or same sex to pleasure from drugs or alcohol. All it takes is others (adults or peers) to provide the support and opportunity to continue the behavior.

To me this is the time of life that youth need to be given all the facts about the consequences of choices, not just one; hence our sacred cow. As a middle school teacher I hear the phrases, "that's gay'" or "that's so gay," or "are you gay?" a lot. They all have a negative connotation and are a poor choice of words! On the other hand those who believe homosexuality is an immutable trait are pushing hard to stop the bullying such phrases lead to for youth who have self-identified as gay at a very young age. This is the dilemma found in homes, schools, and worksites around the nation. What is a workable solution? The trend today is to consider sexual orientation (does it, in fact exist as a civil right?) a protected class to the point of granting the status of legal marriage. But what about those who believe that it is wrong behavior and can be resisted and overcome? What happens to *those* kids and adults at their schools and the workplace? Do they lose their rights? Evidence is mounting that is the case. The same tactics used by those advocating gay rights are now being used to protect and litigate on behalf of ex-gays and students who have religious objections to homosexuality. Politicians and teachers (me included) have experienced extreme persecution or job loss for speaking publicly against using the local school to normalize homosexuality.

In states where gay marriage has become legal, children from kindergarten up are being exposed to same sex relationships as a positive good with no opt out provision. Gay teachers have taken their classes to their weddings. A dad in Massachusetts was even jailed when attempting to get opt out privileges for his kindergarten son. These are stories that do not regularly appear in the news. I questioned a GLSEN (gay lesbian straight education network) conference in my state on the job and was not rehired at the end of the year. Ex-gay ministries and reparative therapists across America are beginning to fear that they may soon be outlawed. This is one sacred cow that has the potential to hurt many in the name of tolerance. Is this the only way to go?

The situation looks bleak but there is hope. A backlash of sorts is beginning to occur. While no one supports gay bashing or bullying a reasonable approach is possible. In some communities where dialog on this issue has been held without name-calling and hostility a compromise has worked. In some cases the sacred cow has been put out to pasture. The first step is to agree that there is a debate on the origins of homosexuality and also whether the school is the place to hold the debate. Those who have left the gay lifestyle must be given the same rights and privileges to speak as those who remain in it. Those students who believe the behavior is wrong must be given the same respect as those who don't. Teachers and staff can operate from a position of neutrality, not one of support for either perspective, much like they do on matters of religion as agents of the state. Yes, there is hope to send this sacred cow back to the herd with the rest of its peers.

Reference websites:
GLSEN How to start a GSA http://www.glsen.org/cgi-bin/iowa/all/news/record/-2226.html.
ACLU How to start a GSA
http://www.aclu.org/lgbt/youth/28336res20070209.html
Exodus International http://www.exodus-international.org/.
Parents and Friends of Gays and Ex-gays PFOX http://pfox.org/default.html.
First Stone http://www.firststone.org/.
Here is an excellent website from Focus on the Family that gives information about how to address gay activism in schools.
http://truetolerance.org/.

Author bio
David Williams and his wife, Deborah, currently live in Oklahoma where he teaches high school math in a public school. They have raised eight children, and now enjoy nine grandchildren. He grew up as the son of a U.S. Navy dad who traveled extensively during his early years through high school. He served both in the U.S. Navy and U.S. Army and retired in 1995. He is a Gulf War veteran. At the encouragement of his wife and family he completed a Bachelor's degree in elementary education in 1998 and took additional courses in math and science to teach at higher levels. His teaching assignments have included elementary, middle school, and now high school. He was awarded the U.S. Department of Education No Child Left Behind American Star of Teaching award for the state of Oklahoma in 2006. He also serves the educational community as the South Central Regional Director of Christian Educators Association, whose primary focus is to encourage, equip and empower educators according to Biblical principles. His experiences in the private sector, the military, and in the classroom as both a parent and teacher provided the motivation to write the included chapter. Email David at: davidmwilliams77@gmail.com.

No One is Born Gay

Dennis Jernigan

Editors note

I have been a huge fan of Dennis Jernigan's music and followed his ministry for over 20 years. I also know his wife and nine beautiful children. God has called him to an unusual ministry…helping homosexual men and women find freedom. Our Creator is STILL setting the captives free.

For several years he held a monthly night of praise in Oklahoma City and I seldom missed a service. Perhaps part of the connection I had with Dennis is that I was sexually abused by a pedophile stepfather as a boy from fourth grade until seventh grade when he went to prison for molesting over 100 young boys. The most difficult part of that whole ordeal was others thought I was "tainted" and I was ostracized for spending time with other boys as a teenager. As it was with Dennis, I was ashamed of this dark chapter in my life. I told no one. Time passed and I was a university professor teaching mostly pre-nursing students. I had NEVER shared my childhood sexual abuse with no one, but I felt God wanted me to share it with one of my classes. It was very difficult, but I did it in obedience to what I felt strongly that my Creator wanted me to do. Of course I did not go into detail, but mentioned the sexual abuse I had experienced. It was like opening a flood gate. Over 85% of the students in that class had been sexually abused. Some openly told of their experience in class, others came into my office, closed the door and shared with me. One girl said she had been sexually abused by her cousin. I tried to consul her by telling her that such abuse often comes from a relative or trusted family friend. She began weeping and said, "Dr. Smith, you don't understand. My abuse was from a female cousin" Although rare, sexual abuse of women does sometime occur by other women.

Although a terrible experience for anyone who has experienced sexual abuse, as scripture teaches, *And we know that all things work together for good to those who love God, to those who are the called according to His purpose* (Romans 8:28, NKJ). Notice it say ALL THINGS, not just those things that seem good at the time. Perhaps the best example was young Joseph being sold into slavery … on order for God to save a nation. So it was with my situation, I turned inward, fell in love with God's Creation and became a scientist.

Over the years I took three of my university students who were lesbians to learn of his testimony and hear him sing. Two of the three were set free. I have the utmost respect for him, his family and his ministry and am deeply honored that he has written the final chapter in this book. Thank you Dennis and may the

God of Creation continue to protect and bless you, your family and your ministry. You are a true friend. It is time to put this sacred cow out to pasture.

Introduction

I used to struggle daily with unwanted same sex attraction—unwanted homosexuality. From my earliest recollections I felt drawn to other males. Many circumstances came my way that only seemed to reinforce those feelings. When I was five years old an adult male confronted me in a sexual manner. As a child I was very emotionally sensitive, artistic, and musical. The other boys at school in my formative years seemed to relish in reminding me how much of a fag I was. In my college days, a friend and mentor—married with children, Christian, and community leader—made a sexual advance and I was convinced this was my lot in life. The only problem with that is that I became more miserable than ever. When I got to the end of my rope, God met me there with a new identity and the power to change my way of thinking. To my great dismay (but not to my surprise), the world has begun to think in an upside-down manner, calling what is righteous ridiculous and what is perverse normal and acceptable. The Bible speaks of this very thing. ***Woe to those who call evil good and good evil, who put darkness for light and light for darkness, who put bitter for sweet and sweet for bitter*** (Isaiah 5:20, NIV).

Once I began to understand God's true plan for my identity, I began to think in a way I had never thought before. No longer was I one trapped in bondage (homosexuality). Now I was a NEW CREATION with the power to put off my old way of thinking and the power to receive and PUT ON a new way of thinking. This statement will probably produce a lot of controversy, but this is how I think of myself: I do not consider myself a recovering/former/ex gay. I consider myself a new creation. The slate of my mind is being erased and the old thoughts are being replaced with new thinking. What I have discovered in the process is that when I change my thoughts, my attitudes change.   When   I change my attitudes, my behaviors change. When I change my behaviors, my perspectives change. When my perspectives change, I see life from a vantage point that homosexuality NEVER afforded me. What is the bottom line for my recovery? God loved me right where I was…but loved me enough to not leave me there! Oh, and by the way, I have been married over twenty-six years and have been blessed with nine children … and have never once delved back into my old life. (My story in its entirety follows at the end of this chapter).

Until 1973, homosexuality was considered a mental disorder by the American Psychiatric Association (APA). In the early 1970s—after many years of protests by the pro-gay movement—homosexual activists campaigned against the DSM (Diagnostic and Statistical Manual of Mental Disorders) classification of homosexuality as a mental disorder, protesting at APA offices and at annual meetings from 1970 to 1973. In 1973 the Board of Trustees of the APA voted to remove homosexuality as a disorder category from the DSM...and thus began our slide down the slippery slope toward a Romans 6 mentality. With that one decision by such a powerful group of therapists, man became the central focus in the matter of homosexuality and effectively began to remove hope for change from the realm of possibility for many men and women.

Let's cut right to the chase. My belief (and experience) and my observance (having personally talked with HUNDREDS of men and women desiring to walk away from unwanted same sex attraction) is that the facts of truth do not bear witness to the current and popular conventional wisdom of this age concerning homosexuality. Let's look at truth for a bit.

Lowest Common Denominator

Sex is for procreation. Would everyone agree with that statement? Do I really need to say more? Sex, in its uttermost and ultimate form is for the express purpose of reproducing ... for making babies. Of course, man in his wisdom, believes everything is about him and life is all about getting the most pleasure ... leading mankind to pervert and co-opt the things of God in such away as to eradicate the reality of having to face his sinful nature. Perversion is simply taking something God intended for a holy purpose and using that something in an unholy or unnatural way. Sex is pleasurable ... but its ultimate purpose is to reproduce the human race. Everything we do to preserve life was made pleasurable by God to insure that mankind would continue to thrive. We need water to survive ... so He made it pleasurable to drink. We need food to survive ... so He made food taste good and smell good, making it a pleasurable thing to do. We need to produce the next generation for our species to survive ... so He made sex pleasurable.

We pervert our desires when we substitute something else to meet that need or when we go outside the boundaries God established for our own good. Alcoholics do not want to face their weakness. (As a side note, 10-15% of Americans identify themselves as alcoholics yet we do not consider this normal, do we? Why are we showing such preferential treatment to those who want to impose their desire that homosexuality be seen as a normal sexual behavior when only 1-2.8% of Americans identify as exclusively homosexual?) Overeaters or gluttons do not want to face their weakness. Homosexuals do not want to admit their weakness. Overcoming any addiction or disorder or sinful habit requires work...blood, sweat, and tears. Many give up simply because they cannot withstand the pressures fighting such a battle brings up. I can say this honestly: knowing Jesus intimately and trusting Him for my identity has been worth every struggle and battle I have had to endure. Victory and subsequent freedom are simply worth it. And as we have heard from the military world: freedom is not free.

Consider what the inerrant Word of God says.

Professing to be wise, they became fools, and exchanged the glory of the incorruptible God for an image in the form of corruptible man and of birds and four-footed animals and crawling creatures. Therefore God gave them over in the lusts of their hearts to impurity, so that their bodies would be dishonored among them. For they exchanged the truth of God for a lie, and worshiped and served the creature rather than the Creator, who is blessed forever. Amen. For this reason God gave them over to degrading passions; for their women exchanged the natural function for that which is unnatural , and in the same way also the men abandoned the natural function of the woman and burned in their desire toward one another, men with men committing indecent acts and receiving in their own persons the due penalty of their error. And just as they did not see fit to acknowledge God any longer, God gave them over to a depraved mind, to do those things which are not proper, being filled with all unrighteousness, wickedness, greed, evil; full of envy, murder, strife, deceit, malice; {they are} gossips, slanderers, haters of God, insolent, arrogant, boastful, inventors of evil, disobedient to parents, without understanding, untrustworthy, unloving, unmerciful; and although they know the ordinance of God, that those who practice such things are worthy of death, they not only do the same, but also give hearty approval to those who practice them (Romans 1:22-32 NASB).

The FACTS about Homosexuality

Homosexuals are born gay is NOT true! Let's look at the science.

• Quite simply, THERE IS NO GAY GENE! Even openly homosexual researchers have come to that conclusion. In 1996, a research team of five led by Dean Hamer at the National Cancer Institute released a study that attempted to link homosexuality with a specific region of the X chromosome. Dean Hamer made the statement "...environmental factors play a role. There is not a single master gene that makes people gay" He went on to say, "I don't think we will ever be able to predict who will be gay"[1]

• A well-known brain study of 1991 by Simon Levay tried to find the differences in the hypothalamuses (a very small portion of the brain) of both homosexual and heterosexual men. Levay, who was one of the researchers and himself a gay activist, offered criticism of his own work: "It's important to stress what I didn't find. I did not prove that homosexuality is genetic, or find a genetic cause for being gay. I didn't show that gay men are born that way, the most

334

common mistake people make in interpreting my work. Nor did I locate a gay center of the brain"[2]

• Clinical professor of psychiatry at the Albert Einstein School of Medicine and past president of the National Association for Research and Therapy of Homosexuality, Dr. Charles Socarides, argues that since psychologists and ministers have treated homosexuality with success, the genetic cause theory must be suspect.[3]

• We have all heard bits and pieces from the famous 'twin study' conducted in 1991 by psychologist Michael Bailey and psychiatrist Richard Pillar. But the truth of their own study flies in the face of conventional wisdom. This study attempted to show that homosexuality occurs more frequently among identical twins than fraternal twins. The interesting result is that the study actually provides support for ENVIRONMENTAL factors versus genetics. If homosexuality was a part of the genetic code then both identical twins would have been homosexual 100 percent of the time…but that was not what they discovered![4]

• Prominent researchers Bruce Parsons, William Byne (psychiatrist with a doctorate in biology), Richard Freidman (psychiatrist), and J. Downey each came to the conclusion that there is no evidence to support a biologic theory, but rather that homosexuality could be best explained by an alternative model where "… temperamental and personality traits interact with the familial and social milieu as the individual's sexuality emerges"[5]

My own experience? My emotional sensitivity and artistic/melancholy outlook caused me to see the world and my experiences in ways that caused me to believe I was not like the other boys I was growing up with. Combine that with a very real and deep need for affirmation and approval from my father and mix in a lack of knowledge concerning what it meant to be sexually masculine and by the time I was 11 years old I was fully gripped by a belief that I was somehow different (that is an overly simplistic overview of my journey into homosexuality, but accurate).

Ten Percent of the Population is Homosexual…I Don't Think so!

Due to the error-filled Kinsey report of the 1950s, the popular belief is that 10% of our population identifies themselves as life-long homosexuals. True scientific research reveals that the reality is that only 1.4 to 2.8% of our population identifies themselves as life-long homosexuals. A series of recent studies from 1989 through 1993 all show similar figures for the real proportion of exclusively homosexual individuals in America: about 1.4% to 2.8%.

• A 1994 study entitled the *National Health and Social Life Survey* conducted by researchers at the University of Chicago and the State University of

New York at Stony Brook found that only 2.8% of the men and 1.4% of the women surveyed said they thought of themselves as homosexual or bisexual. [6]

Homosexual Relationships are just as Normal as Heterosexual Relationships…At least That's What Conventional/Popular Wisdom Says!

Reality and facts tell a quite different story from the popular trend in television and cinema that paints homosexual relationships and behavior as the new normal. Reality is that the average male homosexual has hundreds—HUNDREDS—of sex partners in their lifetimes. The median number of partners for homosexuals is FOUR TIMES HIGHER than for heterosexuals.[7] In his study of male homosexuality in *Western Sexuality: Practice and Precept in Past and Present Times*, M. Pollak says that "few homosexual relationships last longer than two years, with many men reporting hundreds of lifetime partners"[8] In a study involving 2,583 older homosexuals published in the *Journal of Sex Research*, Paul Van de Ven and colleagues found that only 2.7% claimed to have had sex with only one partner.[9] In yet another study, it was found that 24% of gay men had over 100 partners, 43% of those studied had over 500 partners, and more shocking still, 28% of gay men had over 1,000 partners! That does not sound normal to me.[10]

As with any besetting sin or addiction or abnormal behavior, we try to convince ourselves that what we are doing is normal and acceptable. I tried this in my own life and became more miserable than ever … trying to replace what I knew to be true (and convicting) deep inside with alternative ways of thinking. We cannot replace the right order of God's universe with altered/perverted ways of thinking. Even common sense tells us that a man's body was created for a woman's body and vice versa. A sad commentary in our world is that the squeaky wheel (gay community) gets the oil (attention and altered world view for the masses) … and the rest of the population—97% of us—resort to the easy road of simply accepting the new way of thinking rather than to fight it.

If the truth be told, most homosexuals I talk with do not want to be homosexual. That fact alone should ring loud and clear. Dr. Ariel Shidlo of Columbia University published the results of a study on "internalized homophobia" among *homosexual* persons. Let that sink in. *Homosexual* persons. He found that a significant number of homosexuals surveyed held *negative* views/attitudes toward their *own* homosexuality and toward other homosexuals! Fifty-three percent of homosexuals agreed with the statement "homosexuality is not as satisfying (good) as heterosexuality," while another thirty-seven percent agreed that "homosexuality is a sexual perversion"[11] Wow! What does that tell you?

People can try all they want to try and promote homosexuality as an equally satisfying and normal lifestyle option, but even members of the homosexual community would have to admit that such a view is false. Reality is that gay couples adhere to a very different moral standard than straight couples. A prominent conservative gay author, Andrew Sullivan, says their moral standard is one in which "a greater understanding of the need for extramarital outlets" exists.[12] It should be noted, also, that two researchers who avow to be a homosexual couple have concluded that gay relationships between men rarely survive if they are not open to outside sexual contacts.[13]

Homosexuals Are as Normal as Heterosexuals … Really?

I will not belabor the point, but those who identify as homosexual are statistically at a higher risk of drug use, suicide, depression, and other emotional problems.[14] In addition, in the October 1999 issue of *Archives of General Psychiatry*, research on the relationship between homosexuality and mental-health problems found that men with same-sex partners were 6.5 times as likely as their co-twin counterparts to have attempted suicide.[15] "Homosexuality in America: Exposing the Myths", a report by Richard Howe exposes the very intentional lies promoted by the gay agenda. Howe writes, "The slick and somewhat successful public relations campaign hides the true face of the homosexual lifestyle. This can be demonstrated from two (indisputable) sources: former homosexuals who reveal what the typical homosexual lifestyle is like and videotapes of other homosexuals in certain public gatherings"[16] Reality is that all men have fallen short of the glory of God. One sin – be it homosexual or heterosexual in nature – separates us from God. In order to get back to the true identity God has placed deep within the being of every male and female we must return to Him, repenting of our sin and asking Him to reveal to us our *true* identity.

What I Know Now

If I had listened to all my gay friends, I would never have left the homosexual lifestyle/mentality behind. I have been told "you'll be back" to which I say, "Why would I return to something that would require me to be less than who God has called me to be?"

Many have told me that my story is made up … that a real homosexual would never be able to truly leave the lifestyle. The truth is that if I were going to make up a story I would make up a much better one than mine! Why make up a story that has damaged my image in the minds of so many? Why make up a story that has been a true detriment to my musical career (I actually see myself as a minister and a song receiver rather than an artist or a musician) making it difficult, at times, to even provide for my family because even "the church" hesitates at "exposing" my life to their people because "they just do not have that problem" in their church. My story is my story … and it is true … and the reason I share is because I do not want others to go without knowing that freedom is possible. Psalm 107:2 reminds us that the redeemed are to "say so" If we who have been redeemed and set free through a relationship with Christ do not "say so," how are those in the same bondage going to know there is hope for a way out?

Once I came face to face with a God Who loved me right where I was but loved me enough to not leave me there, I followed Him with my whole heart. I decided that I would no longer believe a community (gay) that used me and used me and called it love…but would rather follow after the One Who loved me righteously and wholly for who I am … and not for how much sexual pleasure they could use me for. Reality was suddenly revealed to me in this manner: I could either choose to follow mankind whose sever and utter focus was on mankind as the answer and highest authority OR I could follow after the One Who created mankind. My reasoning? When I have a problem with a computer or piece of technology of some kind I do not go to the local ice cream shop to find the solution. My first response is to search the manual and then to call the manufacturer. That is the sure way to fix a problem. Why do we see our spiritual life any differently? I have God's manual—His Word—and I have direct access to my Manufacturer via my relationship with God through Jesus Christ!

My healing was instant and yet has been a process. In one instance of faith I was given a brand new identity. In the 29+ years since that moment I have been walking toward Jesus relationally while He helps me rip off the old clothes of death from my past life of sin, revealing my true identity as a man in the process. My past failures do not define me. No present circumstance defines me. No temptation defines me. We do not get to choose what we are tempted by … but in Christ we are given the power to choose what is righteous. No person, place, event, or law defines who I am. Only One has that authority. My God defines who I am … and I choose to believe that and to walk in that. And, yes, I have had many from the gay community tell me that I am brainwashed…and they are absolutely correct! I have had my mind cleansed by the transforming power of relationship with Jesus Christ and now think with the mind and walk with the attitude of Christ … through temptation or lack of temptation!

Overcoming homosexuality—or ANY sin—is a battle of the mind. Over the past twenty-nine years I have fallen so many times into stinkin' thinkin' … but rather than give up and cower in defeat, I have chosen to recognize that I am, indeed, in the battle of my life. I grew tired of being the victim who was sexually violated as a boy or the victim who was sexually used and thrown away by others along the way. I decided that since Christ was victorious over sin, death, and hell … then SO WAS I! I decided I was Victor rather than Victim … and watched the walls of my own prison fall in the process.

The bottom line is this: **NO ONE IS BORN GAY**. My belief, my experience, scientific facts, and most importantly, my God tell me otherwise! I was gifted of God in artistic and emotional ways. The world told me these gifts were effeminate. I had a need to hear my dad tell me he loved me. My perception was that I was not lovable nor was I acceptable to my dad as a fellow male. Due to the sexualization of my mental disorder and skewed outlook on life, I came to the conclusion I was born gay. Reality and God's truth say something different, though. Every man is born into sin. It is our sinful nature that separates us from God. Our only recourse is through repentance … acknowledging that He is God and WE ARE NOT … turning from our sin and turning to God…becoming new creations in and through Christ. When we, by faith, are born again we become someone brand new. The process of becoming that new creation is quite simple. I walk and talk with the Lord and choose to believe what He says about me … putting off the old ways of thinking and putting on the new ways of thinking. Changing our thoughts leads to a change of attitude. A change of attitude leads to a change of behavior. A change of behavior leads to a hopeful, abundant life where TRUE intimacy is discovered in a living, growing relationship with the Creator.

My Story
We Have Believed a Lie!

Before I begin my story, you must know that I desire to bring honor to my earthly father and mother as well as to my heavenly Father. The reason I share the things I am about to share with you is because I believe many people will be able to identify with what I have "gone through" My greatest desire is that you would come to know the Father even more intimately than I have. Because we are all born sinners we all have some very basic needs. Yes, we have physical needs. But I'm referring to the many emotional and spiritual needs we are born with. Little children gain their identity through their father. I can remember being a little boy and desiring my daddy's approval and acceptance for every area of my life. Being a father of both boys and girls myself I can see not only how my sons need me to help them realize "who they are" but my daughters as well. One of my daughters may "do" her own hair and come to my wife, Melinda, and ask how it looks. But it takes dad's stamp of approval before she will really believe that it looks acceptable. And isn't that the way it should be with our heavenly Father? I desire to gain my worth and acceptance from my heavenly Father and who He says I am. As a father, I desire to nurture my children in such a way that they do not become dependent upon me but will be able to transfer their deep needs to their heavenly Father. I realize I will never be perfect as a father, husband, worship leader, or person. But my Father is perfect--in every way! My healing has come and will continue to come as I seek an intimate and life-giving relationship with Him.

I was born in Sapulpa, Oklahoma. Soon after my birth, my parents moved to the farm my grandparents (Samuel Washington and Myrtle Mae Snyder) had built ... the farm where my father was raised. We lived three miles from the small town of Boynton, Oklahoma (Pop. approx. 400) where my brothers and I attended school. The Lord gifted me from an early age to play the piano. By the time I was nine years old I was regularly playing for the worship times at First Baptist Church. This was also the church my grandfather Herman Everett Johnson had pastored. This was the church where my parents, Samuel Robert Jernigan and Peggy Yvonne Johnson, had met. My father had also "led singing" here from the earliest I can remember. When I was about six or seven years old, my grandmother Jernigan moved back to the farm in a trailer next to the old farmhouse where we lived. And each day after school I could be found at my grandmother's house practicing piano conveniently forgetting about my chores.

It was through my grandmother Jernigan that the Lord taught me to play the piano. Since we lived so far from any town with a music teacher, I had to learn to play by ear. My grandma was very patient with me and taught me how to "chord" for "church playing!" It was also my grandma who told me there was more to a relationship with Jesus than getting saved. She once told me that she would know my grandpa Jernigan when she got to heaven because the Lord had told her his "new name in glory!" I was in awe! God spoke to my grandma ... but I could never hear him speak to me. Needless to say, I grew very close to this godly woman. It would be many, many years before I would begin to realize the full impact that she was to have and is having on my life.

An early (and defining) memory that helped set me on the road toward homosexual identity happened one Sunday when I was about ten years old. My brother and cousins were playing on the church steps before church began, oblivious to the conversation of the men who were standing near us. My ears tuned right in, though, when their topic of conversation used words like "fagging" and "queering" because I knew they were talking about ME even if they did not know it! These were the men who taught me about God. These were the men whose image God had been painted in my mind! So guess what I thought GOD thought about me?! I thought He hated me ... and this thought tortured my soul for far too many years!

Even though I equated my musical/artistic abilities and emotional sensitivities to being effeminate, God used those abilities to save me from torment and to actually preserve my life and sanity. When the boys at my school would mock me and tease me, I could get lost at the piano when I got to grandma's house after school...and no one could hurt me there. I also had the ability to dream the same dream each night. In my dreams I lived on the *USS Enterprise* and my dad was Captain James Tiberius Kirk. Each dream had the recurring theme of me being captured by aliens and my dad, Captain Kirk, swooping in at the last second and destroying my captors. At the end of each dream, my mother (Doris Day!) would sing "*Que sera, sera! Whatever will be, will be. The future's not ours, you see! Que sera, sera!*"[17] Even in my despair, the Lord was using the gifts and sensitivities he had built into me (which I assumed were effeminate in nature) to preserve and protect me!

My relationship with my parents, from talking with many others over the years, was quite typical for my generation. We were not an affectionate family. While I did feel affection from my mother, I never remember receiving physical affection from my father or among my brothers and myself. My daddy was very hard working. We were not poor. ... but we were not rich monetarily. In addition to working the farm, my dad was employed by a utility company and eventually worked as a mechanic for many years. Since I have gotten older, God has reminded me of many ways my father expressed affection and love for me as I was growing up. My problem was not my father. My problem was that I believed a lie. Once Satan got his foot in the door of my heart, any rejection - no matter how big or how small was perceived as a lack of love from my dad (or whomever I felt rejected by at the time).

One incident in particular was used to set me on the course toward identifying myself as a homosexual. One day (when I was five years old) while playing with my brothers and cousins at a family-owned business; I needed to use the restroom. Running into the public facility nearby, I had not noticed the older man standing behind me until I had finished my business. As I turned around, he dropped his pants exposing himself to me. I was at once terrified yet captivated by what I saw. I had never see an adult male in that way so a natural curiosity overcame me ... but the fear took over when he asked me if I wanted to touch him "there." I ran for my life ... but stopped short of telling anyone else what had just happened. In that moment I began to hear a little voice in my head saying, "Why would that man ask you to do that? Something must be wrong with you, Dennis" And yes, a little boy can think that way ... because I certainly did. This was to set me on a pathway of identifying myself as being 'different' than other boys and somehow "less than a boy" In those days I could not even bring myself to put my hands in my pocket as is common for young men ... because I did not think I deserved to be labeled a boy ... and I certainly did not see myself as masculine. In those early days I began to focus solely on myself and tried desperately to manipulate how others perceived me.

Looking back, I realize that I was a very selfish child. From the earliest I can remember, I found it hard to believe anyone loved me. I felt worthless. Since I didn't believe anyone loved me, I couldn't really receive love. What I did discover, though, was that if I did something well, people would like me. So, I tried to be the best in whatever I did: schoolwork, basketball, music, etc. But I became so frustrated because no matter how well I performed, it never seemed to be good enough. I was very miserable and felt all alone (even though I wasn't alone!). Sports and grades weren't giving me any hope ... neither was music. Because I made choices based upon how or what I perceived people thought of me, I became a very selfish person ... usually at the expense of others and most often as the expense of my little brothers. What people thought was so good, my outward performance, soon began to hide the deepest hurts and failures of my heart. And I must add that my daddy and mama never missed one single event I was involved in while growing up, this should have spoken volumes to me. Still I chose to believe a lie.

Now I need to tell you about what I consider to be the most painful part of my life, a part I tried to hide. Since I felt so rejected, I allowed it to permeate every part of my life. What I didn't realize was that Satan was lying to me, all the while trying to keep me from God's plan for my life. This included the sexual part of my life. In this area I felt so ashamed and afraid of rejection that I became even more selfish and perverted in my way of thinking. As a boy I needed a role model to show me the way to manhood. But because I felt rejected by the main man in my life I, in turn, rejected him and began to yearn for intimacy with a man in perverse ways. Because of this wrong thinking I came to believe I was homosexual. It must have begun early in my life because I remember having those feelings for the same gender at a very early age. I hid this from others through high school and through my four years at Oklahoma Baptist University even though it wasn't hidden from those I had relations with. I might add that even though I was involved in homosexuality through my college days that I still regard that time with fondness. It is in looking back that I can see the awesome and mighty hand of God ministering His love to me in the midst of my sin and confusion. Because of my lack of musical training while growing up, my musical studies at OBU were like learning a whole new language. To be able to actually read and write the music I could see or hear was like a whole new world opening up to me. This would be very valuable later in my life as I began to express my heart and my feelings in song.

A major turning point in my life occurred my senior year at OBU. Desperate to hide myself from others, I lived alone the first semester of my senior year. During that time I thought of myself as a worthless worm-of-a-guy … not worth another's time or effort to get to know. So, imagine my surprise and delight when an older man—a husband and father and Christian leader in the community—who would ask me about my studies each week and take me out for a Coke and pray with me about my troubles and concerns, befriended me. I had never felt I could trust someone as much as I trusted him, simply because he invested so much time and energy into getting to know me. I felt valuable again. I will never forget the moment of honest confession when I shared my deepest, darkest secret with him. It was as if the entire weight of the world accumulated over the past twenty-one years of my life had suddenly been eradicated! I felt free and light and affirmed all at once! That feeling lasted for a few moments … until my friend, my mentor, made a sexual advance.

I went home to my little apartment after that encounter and turned on my gas stove and did not light the flame. As I lay down in the floor I remember thinking this would be best for everyone (me focusing on me, deciding what was best for me … selfish to the last second!). After a few minutes I begin having thoughts like, "Are you prepared for eternity? Do you know what's really waiting for you out there? Are you truly ready for this?" I became so terrified that I quickly turned off the gas and made this statement to myself: "This is just the way I was born. I will stop fighting it and simply BE who I am" I lived that way from that night in 1980 until well after my graduation in 1981.

Upon my graduation from OBU in 1981, God began to move in supernatural ways that even I couldn't see! One of these instances was a simple music concert. A group called The Second Chapter of Acts was going to be in concert in Norman, Oklahoma, and I knew that I was supposed to go. By that time in my life I was looking for anybody who was real, someone who had a real walk with the Lord. Among Christian musicians, I was looking for more than entertainers. So, I went to their concert. I knew by the words they said and the music they sang that these people were genuine, and the message was born out of times of desperation in their own lives. I needed hope. As I listened to Annie Herring speak and sing I was overwhelmed by the love she spoke of. This was the love I had dreamed of but still couldn't believe was available to me! So I listened very intently with great expectation—until she came to the song *Mansion Builder*. This song caught my deepest attention because of the simple phrase, *"Why should I worry? Why should I fret? I've got a Mansion Builder Who ain't through with me yet?"*[18]

All of a sudden she just stopped in the middle of the song and said, "There are those of you here who are dealing with things that you have never told anyone and you are carrying those burdens and that's wrong—that's sin and you need to let those hurts go and give them to the Lord. We are going to sing the song again and I want you to lift your hands to the Lord—and all of those burdens that you are carrying, I want you to place them in your hands and lift your hurts to Him" This was all new to me—worship and praise. I had always thought before that this was just an emotional response that didn't really mean anything. But you know what it did for me? As I lifted my hands, God became more real to me than I had ever imagined! The lifting of my hands was more than a physical action. My hands were an extension of my heart! I realized that Jesus had lifted His hands for me—upon the cross. I realized that He truly was beside me and that He was willing to walk with me and carry me and just be honest with me. And I could be honest with Him! At that moment, I cried out to God and lifted those burdens to the Lord and said, "Lord Jesus, I can't change me or the mess I've gotten myself into—but you can!" And you know what? He did change me!

At that time I acknowledged the fact that I was totally helpless and I turned everything in my life over to Jesus—my thoughts, my emotions, my physical body … and my past. Basically, I took responsibility for my own sins and yielded every right to Jesus—my right to be loved, my right even to life. Because of my choice to sin, I deserved death and hell—and that's where Jesus came in. At that point, something wonderful began to take place in my life … I began to hear the Lord speak to my heart—"Dennis, I love you. I have always loved you! Dennis, you are my child—I love you no matter what. Dennis, I will always love you!" It was then that I lost the need to be accepted or loved by others because I realized Jesus would love me and accept me no matter what, even when I was rejected by others! It was also at this same time that those sexually perverse thoughts and desires were changed … and He began to replace them with holy and pure thoughts about what sexual love was all about. You see, the sexual drive is a creative drive and Satan knows that if he can pervert that drive, he can kill and pervert God's creativity in us.

This all seems to fit in place for me now. For when I was about nine years old, I felt the Lord speak to me that I would someday have a large family of my own … with nine children! I thought, "Lord, You must be crazy. How can I have children if I have homosexual (unnatural) desires?" Do you see what Satan was trying to do? Not only is God blessing me with a wonderful marriage and many children, He continues to pour out His music in my heart. It is out of the gratefulness of my heart towards the Lord that I will have all the children He will bless me with and I will never stop singing praise to His name. The secret—the key for me is knowing that Jesus loves me and that I need Him desperately more every day … and realizing that He wants to change me--to change my heart—

every day. My desire is to come into His presence (lay myself on the altar) that He might change me into His own image. You see, when I was nine years old, Jesus began calling me to Himself. On September 8, 1968, I asked my mother how to be saved. She explained the plan of God's salvation—that we were all sinners and that we deserved to perish in hell. I was saved that Sunday afternoon and baptized that same evening. I believe that I was saved when I was nine years old, but because I looked and perceived my heavenly Father through my own perverted image of my earthly father, I couldn't fully receive all He had in store for me—like acceptance and forgiveness. It is so amazing to me that He loved me enough to preserve my life the way he has in this day and age of promiscuity, perversion, and sexually transmitted diseases like AIDS. One thing that kept me going during the early years of my life when I felt like giving up and living in sin, was the fact that Jesus kept calling me. If He was God then there was truly hope for me! The most precious thing of all is that He loves me with all His heart … and that's how I want to love Him. Because of this relationship with Jesus, my healing has been and will be a continual process … until the day I die and can see Him face to face!

Another major point of change for me came during this same time in 1981—yet another divine setup! A close friend found out about my past. I knew I would be disgraced and rejected now! When he confronted me, I ran from the house and continued to run until I could run no more. At that point, I simply cried out to God to speak to me. At the same, my eyes were directed to look into the darkness of the evening sky where I was drawn to a puffy white cloud floating above. This cloud looked like an old man with a beard and outstretched arms. Near this cloud was a smaller cloud in the shape of a lamb. As I watched, the bearded man engulfed the little lamb in His arms. I knew immediately that God was speaking to me … that this was what He wanted to do for me in this time of need. I then had the grace to return and "face the music" But that's not what happened! This friend was a true friend. He told me he loved me and was willing to stand with me as I walked through this time of deliverance in my life. And you know what else happened? God began to bring others into my life who were willing to love me unconditionally and to walk with me through the trials of my life—no matter what--for my complete healing.

In 1983, God called me to marry my wife, Melinda. I assumed that since I considered myself to be healed that there was no need to share my past with her. But I soon realized that I was really still trying to hide—which meant I still carried a burden and that I was still more concurred with what man thought of me than what God thought of me. Soon after we were married, the babies started coming! And with the babies, the added pressure of responsibility to deal with the real issues of total healing in my life. Hiding the truth would keep me from the healing God wanted for me in my life.

Because I hid these things from others, my relationships could never truly be what God wanted them to be—because in true love there is no fear. I was always afraid to tell anyone because I thought no one would love me. Why am I telling you now? Well, on July 13, 1988, I realized God wanted to take the greatest failures and weaknesses of my life and make them my greatest strengths—and that Satan wanted me to keep them hidden so he could use them against me. But like the prostitute, Mary Magdalene, I realized that to hide those things kept me from fellowship and freely loving the One I loved the most--Jesus. Not only this, but if I confessed my past freely, Satan would have no ammunition against me. So here's what I did. In July of 1988, I shared what I just told you (in a much more brief way!) with my church … and something beautiful took place. People began to come out of the woodwork that had been hurting just like me and even more so, men and women who were involved in homosexuality (sodomy), women who were abused by their fathers, those who had been raped and never told anyone, and even those who had abortions, etc. As they confessed their sins and hurts, Jesus was able to begin healing all their past. On that day, I publicly laid down my life and my reputation to serve Jesus in an awesome way. However, I want my life to be broken and poured out life the perfume Mary Magdalene used to wash Jesus' feet even though they said she was foolish. I want to lay down my life and reputation for others just as my Lord Jesus did for me. Imagine that—the perfect King of the Universe humbled Himself and gave up all His power and glory because He loves me! I can do no less!

Since the day I first shared my past publicly, God has called me to tell others what He has done for me—to lead and call others into intimacy with Jesus through the avenue of music and worship. It was after such a time of sharing in my hometown of Boynton in 1989 that I began to realize the true depth and extent of God's great love for me and the calling upon my life and the role of my grandmother Jernigan's vision and prayer upon my ministry. After leading worship at the Boynton Community Center, one of my grandma's old prayer partners said to me, "Isn't it wonderful how your grandmothers prayers have been answered?" Amid feelings of shock and tears of joy, I asked, "What prayers?" And she answered, "Didn't you know? Your grandmother told me how she would stand behind you as you practiced the piano at her house each day and would ask God to use you mightily in His kingdom to lead in music and worship! And He has answered her prayers!"

Your circumstances, your sins, your wounds, etc., may all be different than mine, but the answer is still the same—Jesus. You may have been sinned against and wounded very deeply. For those times you are not guilty! If you have

346

been used or abused in any way, you can be healed. Do not receive the false guilt that Satan would try to put on you because of circumstances that were beyond your control. I urge you to deal with your own heart and the things you were (and are) responsible for—like attitudes, actions, thoughts, and feelings! There is hope for the hurting. If you are like me, you may need radical surgery. Surgery may take more time than it takes to put a Band-Aid on a wound. But surgery generally gets to the cause and doesn't just cover up or pacify the symptoms of the wound. If you are willing, you can get to the root(s) of your sin(s). I urge you to get to the root of and deal with whatever you may be facing. I've been there and found the way out, and I must share my story—the story of Jesus with those who are hurting. Aren't we all hurting in one-way or another?

The bottom line is this: I can't make it one day without the Lord. I ask Him to fill me with His spirit day-by-day and moment-by-moment and to lead me. You see, we are all helpless and in need of a Father to care for us. And He is the Father Who will never leave us or forsake us. He is the Father Who enjoys our presence more than we could ever enjoy His! I am no longer afraid of what others think of me (at least I'm asking the Lord to help me in that area!) Please pray for me and my family as we seek God's direction for our lives. I love you.

In His Love and Grace, Dennis.

Update

Dennis Jernigan lives with his wife of 26+ years, Melinda, in Oklahoma on the farm where they raised their nine children. He currently serves on the Board of Directors for the world's largest ministry to those who struggle with unwanted same-sex attraction, Exodus International. Each year Exodus related ministries receive over 350,000 calls for help from strugglers as well as family members and friends of those who struggle.

Dennis has recorded over 35 Christian/Worship/Ministry recordings and has many songs that are sung daily somewhere in the world wherever the body of Christ meets in worship. For recordings, books, and other resources by Dennis Jernigan, go to www.dennisjernigan.com.

Books by Dennis Jernigan

1. *Giant Killers.* Dennis' personal journeys through the defeat of the giants in his life.
2. ***What Every Boy Should Know … What Every Man Wishes His Dad Had Told Him*** Dennis' timely message to fathers and sons who desire to preserve a godly heritage of what masculinity really is.

3. ***Daily Devotions For Kingdom Seekers*** daily encouragement to those who desire to see and live life from the King's point of view. Only for those who desire true victory in their lives!

4. ***This Is My Destiny*** Jernigan's personal journey toward discovery of who he is in Christ.

5. ***Victim to Victor*** a continuation of *Giant Killer* specifically to help get those who struggle with sin or bandage of any kind to the place of victory.

6. ***Help Me to Remember*** a book to help lighten the journey of grief.

7. ***A Worshiper's Guide to the Holy Land*** DJ's personal memoir of his travels in the Holy Land (written with friend Chuck King; includes a CD of worship music!)

Recordings by Dennis Jernigan

Kingdom Come
I Cry Holy
Hands Lifted High
Giant Killers
This Is My Destiny
I Surrender
Daddy's Song
Break My Heart, O God
I Belong to Jesus, Vol. I & II
I Will Trust You
Songs of Ministry
And many others!!!

Booking

Dennis Jernigan is a much-sought-after speaker, worship leader, and minister. To contact him for possible ministry in your church or group, contact his booking coordinator at 1-800-877-0406 or by email at booking@dennisjernigan.com.

Online Ministry - Worship from the Living Room

Watch and listen live as Dennis leads worship from his living room! To join in worship visit:

http://www.livestream.com/dennisjernigantv

You can join Dennis LIVE via webcam on the first Monday of each month at 9:00 PM CST!

Endnotes

Much of the information was gleaned from the resources available through www.lovewonout.com as well as www.exodusinternational.org and is used with permission.

[1] P. Copeland and D. Hamer, *The Science of Desire,* New York: Simon and Schuster, 1996.

[2] "The Innate-Immutable Argument Finds No Basis in Science", "In Their Won Words: Gay Activists Speak About Science, Morality, and Philosophy" By Dean Bryd, Ph.D., Shirley E. Cox, Ph.D., Jeffery W. Robinson, Ph.D. http//www.narth.com/docs/innate.html, 30 September 2002

[3] Charles W. Socarides, interview, in "New Gene Study: Homosexuality Inborn?" p. 9

[4] "The Innate-Immutable Argument Finds No Basis in Science", "In Their Won Words: Gay Activists Speak About Science, Morality, and Philosophy" By Dean Bryd, Ph.D., Shirley E. Cox, Ph.D., Jeffery W. Robinson, Ph.D.
http//www.narth.com/docs/innate.html, 30 September 2002

[5] "The Innate-Immutable Argument Finds No Basis in Science", "In Their Won Words: Gay Activists Speak About Science, Morality, and Philosophy" By Dean Bryd, Ph.D., Shirley E. Cox, Ph.D., Jeffery W. Robinson, Ph.D.
http//www.narth.com/docs/innate.html, 30 September 2002. William Byne and Bruce Parsons, "Human Sexual Orientation: The Biologic Theories Reappraised" Archives of General Psychiatry 50, no. 3.

[6] *Sex in America: A Definitive Survey*, Robert T. Michael, John H. Gagnon, Edward O. Laumann, and Gina Kolata, Little, Brown and Company, Boston, 1994, p. 176

[7] Whitehead, N.E.; Whitehead, B.K. (1999): *My Genes Made Me Do It!* 1994

[8] M. Pollak, "Male Homosexuality: Practice and Precept in Past and Present Times", ed. P. Aries and A. Bejin, translated by Anthony Forster, NY: B. Blackwell, 1985, pp. 40-61.

[9] Paul Van de Ven et al., "A Comparative Demographic and Sexual Profile of Older Homosexually Active Men", *Journal of Sex Research* 34 (1997): p. 354. Dr. Paul Van de Ven conversation with Dr. Robert Gagnon on September 2000.

[10] "Survey Finds 40 Percent of Gay Men Have Had More Than 40 Sex Partners", *Lambda Report*, January/February 1998, p. 20. A. P. Bell and M. S. Weinberg, *Homosexualities: A Study of Diversity Among Men and Women* (New York: Simon and Schuster, 1978), pp. 308, 9; see also Bell, Weinberg and Hammersmith, *Sexual Preference* (Bloomington: Indiana University Press, 1981).

[11] Shidlo, A., 1994, "Internalized Homophobia: Conceptual and Empirical Issues", in Greene, B., Herek G., *Lesbian and Gay Psychology*, Thousand Oaks: CA: Sage, pp. 176-205.

[12] McWhirter, D. and Mattison, A., *The Male Couple: How Relationships Develop*, Prentice-Hall, 1984.

[13] Ibid.

[14] Fergusson, D.M.; Horwood, L.J.; Beautrais, A.L., 1999: Is sexual orientation related to mental health problems and suicidality in young people? *Arch. Gen. Psychiatry* 56, pp. 876-880.; Herrell, R.; Goldberg, J.; True, W.R.; Ramakrishnan, V.; Lyons, M.; Eisen, S.; Tsuang, M.T., 1999: Sexual Orientation and Suicidality: a co-twin control study in adult men. *Arch. Gen. Psychiatry* 56, pp 867- 874.; Sandfort, T.G.M.; de Graaf, R.; Bijl, R.V.; Schnabel, 2001: Same-sex sexual behavior and psychiatric disorders. *Arch. Gen. Psychiatry* 58, pp. 85-91.; Bailey, J.M. (1999): Commentary: Homosexuality and mental illness. *Arch. Gen. Psychiatry* 56, pp. 876-880.

[15] Herrell, R.; Goldberg, J.; True, W.R.; Ramakrishnan, V.; Lyons, M.; Eisen, S.; Tsuang, M.T., 1999: Sexual Orientation and Suicidality: a co-twin control study in adult men. *Arch. Gen. Psychiatry* 56, pp 867-874.

[16] Richard Howe, "Homosexuality in America: Exposing the Myths", *The American Family Association*, 1994, p. 14.

[17] *Whatever Will Be Will Be (Que Sera, Sera)*, by Livingston, Jay; Evans, Ray, ©1956 St. Angelo Music

[18] *Mansion Builder*, by Annie Herring; ©1978 Latter Rain Music

Author bio:

You may not recognize Dennis Jernigan's name or face, but there is a good chance you know some of his music. Songs like, *We Will Worship the Lamb of Glory, Thank You, Great is the Lord Almighty, Who Can Satisfy My Soul* (or *There is a Fountain*), *I Belong to Jesus, Nobody Fills My Heart Like Jesus*, and *You Are My All in All* have been sung widely in churches since the early 1990s. Dennis has recorded and performed hundreds of songs, but does not consider himself a song writer, but rather a song "receiver" Both he and Christian singer Annie Herring of the group *Second Chapter of Acts* (visit: www.annieherring.com) often hear God singing over them and simply remember and write down the words and music. They have scriptural support for this. ***The LORD your God is with you, he is mighty to save. He will take great delight in you, he will quiet you with his love, he will rejoice over you with singing*** (Zephaniah 3:17, NIV).

Dennis has given a great deal of his life to setting the spiritually captives free. Having walked out of a homosexual identity and by the grace of God became a new creation; he is convinced that with God NOTHING is impossible. Through the sharing of his story and the sharing of the stories behind the songs, Dennis Jernigan has watched literally thousands walk out of all manner of spiritual bondage and has watched thousands more desperate, wounded people find healing through intimacy with Jesus Christ. Indeed, Jesus is STILL setting the captives free. He currently serves on the Board of Directors for the world's largest ministry to those who struggle with unwanted same-sex attraction, Exodus International. Each year Exodus related ministries receive over 350,000 calls for help from strugglers as well as family members and friends of those who struggle.

Dennis Jernigan has been married to Melinda for 26 years. Together they have nine children and some of them are already into ministry. His priority is to first be a loving husband and father. It is for that reason he has not made long music tours. The life message of Dennis Jernigan can be summed up in one word: Freedom. And he wants everyone else to know about the possibility of freedom in their own lives. Check out his series of testimony videos at http://www.youtube.com/watch?v=Z0oQuL8Gog4 and you'll see what we mean. Every song, every blog, every video, and every book he has written are all interlaced with the theme and hope and possibility of freedom. Jernigan says, "If you lead people to freedom, they WILL worship"

To find Jernigan's vast library of recorded music, books he has written, and various ministry tools, go to www.dennisjernigan.com. Dennis also has an online ministry called *DJinsiders* which serves to minister to those who desire to walk in freedom as Kingdom Seekers and Overcomers. There is also a vast array of resource material on his website. You may email Dennis directly: mail@dennisjernigan.com

Creation or Evolution? Consider the Evidence Before Deciding

E. Norbert Smith, Ph.D.

Synopsis: Central to the book are over 1,500 Bible verses related to God as Creator and Sustainer. Evidence for and against evolution are given and details of modern religious persecution with over 3,000 professors being denied tenure or fired for doubting evolution. A must read in these increasingly anti-Christian times.

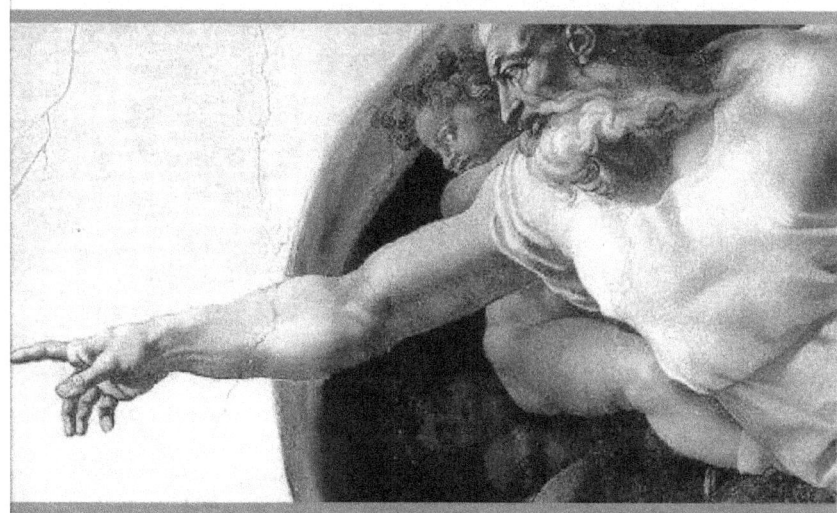

www.ingramcontent.com/pod-product-compliance
Lightning Source LLC
Chambersburg PA
CBHW081105170526
45165CB00008B/2328